Engineering for Sustainable Development

Engineering for Sustainable Development

Theory and Practice

Wahidul K. Biswas and Michele John
Sustainable Engineering Group, School of Civil and Mechanical Engineering
Curtin University
Perth, Australia

This edition first published 2023

Registered Office
John Wiley & Sons, Inc., 111 River Street, Hoboken, NJ 07030, USA

Editorial Office
111 River Street, Hoboken, NJ 07030, USA

For details of our global editorial offices, customer services, and more information about Wiley products visit us at www.wiley.com.

Wiley also publishes its books in a variety of electronic formats and by print-on-demand. Some content that appears in standard print versions of this book may not be available in other formats.

Library of Congress Cataloging-in-Publication Data applied for:

Hardback: 9781119720980

Cover Design: Wiley
Cover Image: © Pavel Chagochkin/Shutterstock

Set in 9.5/12.5pt STIXTwoText by Straive, Chennai, India
Printed and bound by CPI Group (UK) Ltd, Croydon, CR0 4YY

C9781119720980_220922

Dedicated to my late father
Professor Mozibur Rahman Biswas, PhD (Texas A&M University)
31/10/1939 – 06/07/2020

Though you never got to see this,
You're in every page.

Contents

Preface

Engineering for sustainable development brings both hope and challenges for the twenty-first century.

The latest IPCC (2021) Report states that global surface temperatures will continue to increase under all emission scenarios until at least the mid-century. It is clear that global warming of 1.5–2 °C will be exceeded during the twenty-first century unless significant reductions in carbon dioxide and other greenhouse gas emissions are made in the next few decades.

What does this mean for the engineering profession? It means that the technologies, material choices, and engineering designs will all face increasing pressures to improve energy efficiency, to move towards only renewable energy, reduce embodied energy and material intensity in engineering products and services, and consider reducing end-of-life waste management issues through improved design, resource recovery, and remanufacturing. It also will mean that the engineering profession will need to seriously consider the stewardship and corporate social responsibilities of living in a world destined for 9 billion people by mid-century and all the attendant pressures of a warming climate, increased population density in our major cities, and the constant trade-offs between increased economic growth and the conservation of global resources and the natural environment.

All these challenges are grounded in engineering decision-making. As highlighted by the Natural Edge Project (2002–2006), 70% of modern sustainability challenges fundamentally involve engineering decision-making.

Engineering covers a wide variety of industry applications, manufacturing sectors, consumer good production, and housing and energy systems. Engineers were central to the seventeenth century Industrial Revolution which catalysed the intensive production systems previous century have benefited from in terms of improved health, higher living standards and phenomenal global economic growth and technological development. We have the engineering profession to thank for this.

On the other side of the same coin, we have seen two centuries of significant carbon emissions resulting in global warming from fossil fuel consumption, we have large-scale clearing of vast tractions of global vegetation in the race for increased agricultural production and urban development and in doing so we have simply relied on the benefits of the attendant economic growth to justify our voracious need for expansion, market domination, and profitability. The yardsticks for the success of such achievements were largely measured through increased profits, shareholder return, and/or technological development.

However the tide is now turning. Climate pressures and a long winded global policy discussion on carbon mitigation, increasing environmental and marine degradation, extended periods of drought and water shortages and global resource issues such as critical metal availability are now raising questions that the engineering profession have not really had to previously seriously consider.

Our world is changing and will provide many new challenges for the twenty-first century engineer.

This book is about that very challenge. As we engineer a future of sustainable development, what is the role of the engineering in helping to mitigate, adapt, and add resilience to the climate, resource, and stewardship challenges that governments, community, and the younger generation will expect from the profession in coming decades? Zero carbon production, 100% renewable energy, circular economy use of product waste and resource recovery, mandatory sustainability assessments that perhaps blockchain instruments will audit on our behalf and increasing pressures on any associated negative impact on the community and the environment. Welcome to Engineering for Sustainable Development. The challenge is ours. The responsibility is ours.

The authors thank Dr Gordon Ingram of Chemical Engineering Department of Curtin University for his careful review of some chapters of this book.

July 2022 *Wahidul K. Biswas and Michele John*

Part I

Challenges in Sustainable Engineering

1

Sustainability Challenges

1.1 Introduction

Sustainability is the goal or endpoint of a process known as (ecologically) sustainable development. Sustainable development consists of a large number of pathways to reach this endpoint that sees a balance between the provision of ecosystem services, and human access to natural resources to meet the basic needs of life. Engineering sustainability challenges are focused on managing this challenge and coming up with innovative technological solutions to help sustain the earth, given the fact that the earth's existing resources will be inadequate in meeting the demands of future estimated population growth. The latest data from the global footprint network suggests that the humanity used an equivalent of 1.7 earths in 2016 (Vandermaesen et al. 2019), while the United Nations predicted that the global population will increase from 7.7 billion in 2019 to 11.2 billion by the end of this century (United Nations 2020). At the rate at which we consume the earth resources, future generations will require approximately $(1.7 \times 11.2/7.7)$ or 2.4 planets to provide equivalent resources by the end of this century. However, we only have one planet.

Worldwide human population growth has been supported by the industrial revolution and the invention of steam engine in the eighteenth century and mass production. This industrial revolution gave birth to our modern civilization and systematically improved living standards resulting in a population explosion from 0.5 to 7.7 billion only over 253 years (1776–2019) (Cilluffo and Ruiz 2019). The exploitation of minerals, fuels, biomass, and rocks for transport, agriculture, building, and manufacturing increased rapidly during this time to deliver the goods and services necessary to support the growth of modern civilization. Technologies have advanced over these years significantly to exploit rare-earth materials and scarce resources to meet the growing demand of an increasing population and to run the modern economy. The scarcity of important materials that are limited resources is only now being understood (Whittingham 2011).

Engineering for Sustainable Development: Theory and Practice, First Edition.
Wahidul K. Biswas and Michele John.

Humanity currently thus uses resources 1.75 times faster than they can be regenerated by nature or provided by our planet (GFN 2019). Apart from population growth, factors which are causing the rapid decline of the earth's resources are our increased dependences on non-renewable resources, energy and material intensive technologies, and uncontrolled production and consumption. Global demand for materials has increased 10-fold since the beginning of the twentieth century and is set to double again by 2030, compared with 2010 (European Commission 2020). Resource producers have been increasingly able to deploy a range of technological options in their operations, even mining and drilling in places that were once inaccessible, increasing the efficiency of extraction techniques, switching to predictive maintenance, and using sophisticated modelling tools to identify, extract, and manage resources. The major emphasis has been on economic growth to meet the demands of a growing population, technological progress based on throughput-increasing (or resource exploitation) without consideration of the bio-physical limits of our non-renewable resources (e.g. coal, gas, ore, rocks). These resources require hundreds of thousands of years to form below the earth, and it raises questions as to what will happen to future generations when all finite non-renewable resources are exhausted due to uncontrolled production and consumption. In addition to the exponential growth of resource use, technology that is used for converting earth resources to products (e.g. construction, automobiles, electronic items) and services (e.g. electricity, internet, transportation, communication system, water supply) to meet our growing demands have resulted in emissions of global warming gases (mainly CO_2). The consequence of global warming includes flooding, increased bushfire, and the destruction of ecosystems. By 2050, between 70% and 80% of all people are expected to live in urban areas (United Nation 2018), which are resource intensive and artificial environments made by man, to further improve living standards. The engineering challenge is to minimise land use and conserve resources whilst meeting the demands of the world population through energy efficient buildings, water conservation, compact cities, and efficient transportation systems in our built environment.

Population control, rapid technological innovation, and behavioural change are also required to enhance resource efficiency. It is now crucial for the present generation to change their behaviour and mindsets, which will enable them to sustain adequate resources for future generations (inter-generational equity). According to the Brundtland report (1987), 'Sustainable development is development that meets the needs of the present without compromising the ability of future generations to meet their own needs'. The widely used Brundtland's definition on sustainable development, was published in '*Our Common Future*' in 1987.

While population density of developed nations is far less than that of developing nations, overconsumption by the former has already exceeded their bio-capacity resulting in their need to source resources from developing nations. The UN Development Program reports that the richest 20% of the world's population consume 86% of the world's resources while the poorest 80% consume just 14% (UN 1999). This highlights the intra-generational social equity aspects of sustainability and the increased gap between rich and poor people. The rapid progress in technology has fuelled this social inequality. According to David Grusky, Director of Stanford's Center on Poverty and Inequality, 'One of the largest and most prominent debates in social sciences is the role of technology in inequality' (Rotman 2014). The biggest social inequity is that the technology-driven economy greatly favours a small group of people by amplifying their inherent skills and wealth. Human capital being continuously replaced with man-made capital (e.g. self-service cash register, food processors) has increased unemployment. Increased unemployment on the other hand increased social problems, such as poverty, crime, corruption, and domestic violence.

Secondly, technologies have not only enabled wealthy nations to control world resources but have also increased the overconsumption (luxurious pollutions), which is responsible for further environmental degradation. Poverty in a poor nation that causes environmental degradation is known as the pollution for the survival. For example, many children in developing nations are sent outside to collect low grade fuels like leaves and twigs as their parents cannot afford to purchase high quality fuel like gas or wood. Therefore, their children do not go to school spending the whole day gathering fuels to meet the daily cooking energy demand. The collection of low grade fuels not only affects the children's education but also causes ecological imbalance by depriving soil from nutrient rich organic matter.

Thirdly, sea level rise (SLR) due to global warming will affect a large portion of land of developing nations in densely populated countries in the Asia Pacific region.

Planned obsolescence of business strategy in recent times have made technologies obsolete, unfashionable or no longer usable before their natural end of life (EoL), which has created unsustainable consumption. For at least half a century, the mainstream fashion industry has purposely produced goods of inferior quality to increase sales to gain short term financial benefits. In essence, it means that a company is deliberately designing and manufacturing products with a shorter life span, by making them non-functional or unfashionable earlier than necessary and increasing the waste sink if these items are not designed for disassembly or reuse or remanufacturing.

Addressing inter- and intra-generational social inequities requires a reduction in the investment in unnecessary luxury items, controlled economic growth, sustainable behaviour and life style changes, and to design technology/products for repurposing and dematerialisation (e.g. accessing materials online reduced to need of hard copies, virtual conferences reduce travelling). A paradigm shift is urgently necessary to switch from resource intensive technologies that are currently being used (e.g. power plant, car, infrastructure) to more resource saving technologies (e.g. replacing a new engine with a remanufactured engine, super light car with reduced fuel consumption reduces long run costs and emissions). Secondly, it is important to encourage the technological race to enhance both inter- and intra-generational social equity. More dependence on technology means we need more energy and material resources to produce, operate, and maintain them in an increasingly resources scarce world. We need to achieve a balance between technology and human capital for enhancing intra-generational social equity while maintaining economic growth. In a nut shell, social equity means 'equal opportunity of access to basic needs' for all people on earth.

Innovative technological design for converting EoL product to new product will reduce land, energy, and the material consumption associated with virgin material consumption. We need planning and management of sourcing, procurement, conversion, and logistics activities involved during pre-manufacturing, manufacturing, use, and post-use stages in the product life cycle. For example, Renault is a remanufacturing company which requires more labour for remanufacturing gearboxes than making new ones, but there is still a net profit because no capital expenses are required for machinery, and no cutting and machining of the products, resulting in no waste and a better materials yield (Ellen MacArthur Foundation 2014), enhancing both inter- and intra-generational equity by creating jobs and by importantly conserving virgin resources for future generations.

1.2 Weak Sustainability vs Strong Sustainability

There are many different approaches to achieve sustainability, which may result in either 'weak' or 'strong' sustainability outcomes. The core ideas that are widely discussed in the literature such as engineering innovation should not only consider conservation of resources for future generations but also the technology needs to be designed to enhance social well-being (Table 1.1). Secondly, achieving social, economic, and environmental performance in a product or delivering a service to a particular sector could result in unplanned adverse consequences. The weak definition of sustainability illustrated in the interlocking diagram (Figure 1.1a) allows the trade-off between sectors, and does not take into account carrying capacity or the resource limitations of earth. The first priority for living within the world's carrying capacity is to achieve a strong definition of sustainability, as represented by the nested egg diagram (Figure 1.2b). Societal demand should consider inter- and intra-generational social equity issues within ecological limits or finite resources levels. Engineering design and innovation

Table 1.1 Approaches for addressing sustainability

Core idea	Comments and related concepts
Meeting needs of present without compromising needs of future	• Can be thought of as over-riding • Includes 'intergenerational equity', 'intergenerational discounting', recognition of ecological limits
Harmonising/integrating social, economic and environmental objectives	• Can tend towards 'weak' sustainability (assumes assets in one sector can be traded off against others) • Includes 'triple bottom line'
Living within the world's carrying capacity	• Tends to be associated with 'strong' sustainability (ecological assets cannot be traded off beyond a certain point) • Includes 'ecological footprint'

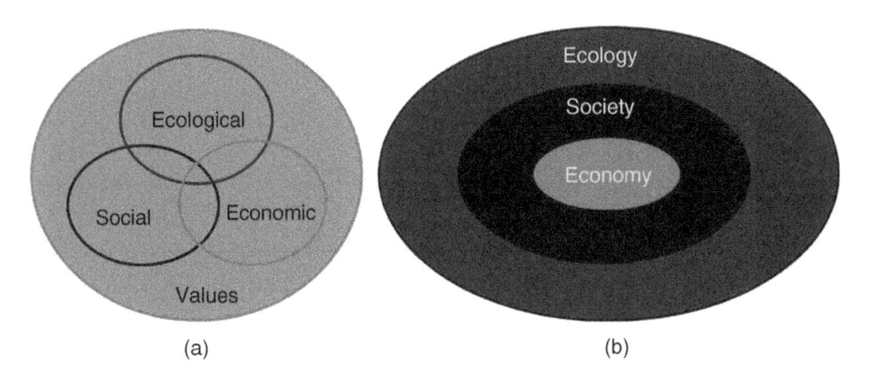

Figure 1.1 Weak (a) and strong (b) sustainability. Source: Modified from Lim et al. (2015)

must make sure to address social needs while using resources within the earth's carrying capacity and not exploit resources beyond the earth's carrying capacity.

The difference between 'weak' and 'strong' definitions of sustainability as shown in Table 1.2 highlights the factors that need to be taken into account to achieve strong sustainability. As it appears in Table 1.2, weak sustainability focuses on economic growth, pollution control rather than pollution prevention, ignores the earth resources limits and the risks associated with technological development, and ignores the fact that exponential growth is taking place in a world. On the hand, 'strong sustainability' focuses on pollution prevention instead of pollution control, considers the ecological or bio-physical limits of earths resources, and takes into consideration of consequences that may be resulted from human activities and technological development.

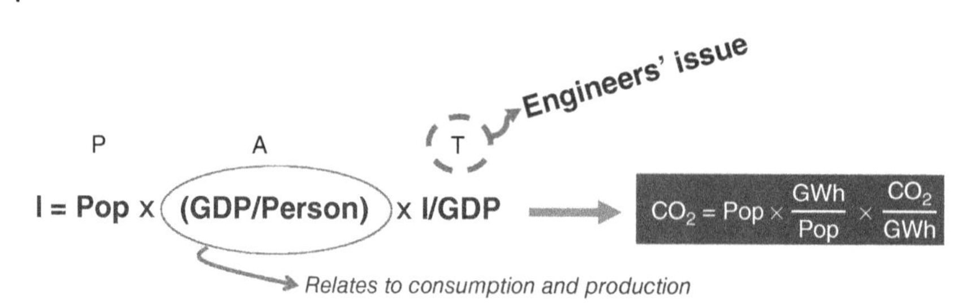

Figure 1.2 Engineers' challenge to combat climate change

Table 1.2 Difference between weak and strong sustainability

'Weak' sustainability	'Strong' sustainability
'Brown' agenda – pollution focus	'Green' agenda – focus on resilience of ecosystems
Environmental focus	Ecological focus
Degradation of one group of assets can be compensated by improvement in another	Not a balancing act, but an integrating act
Interlocking circles	Nested egg diagram
Evolutionary change required	Radical change required
Starts with economic imperatives	Starts with ecological imperatives
'Weak' sustainability	'Strong' sustainability
Can be accommodated within the traditional economic paradigm	Challenges the traditional economic paradigm
Downplays risk and uncertainty, although consistent with the precautionary principle	Highlights risk and uncertainty Identifies the need for better modelling of systems, while recognising full understanding unlikely
Favours 'pressure-state-response' model (linking cause and effect) for developing indicators	Argues that 'pressure-state-response' model oversimplifies dynamics of complex ecological (or social) systems

Source: Adapted from Diesendorf (2001)

1.3 Utility vs Throughput

Current engineering practice mainly meets the human needs of the current generation by providing utility services including energy, infrastructure, electronics, transportation, water, food processing, etc. At the same time, they also need to consider 'throughput' in their engineering design process. The throughput is the flow of raw materials and energy from the global ecosystem's sources of low entropy (mines, wells, fisheries, croplands) through the economy, and back to the global ecosystem's sinks for high entropy wastes (atmosphere, oceans, dumps). Engineers need to make sure that more utility is provided per unit of throughput (e.g. more MWh of electricity per unit of throughput consisting of mining, processing, transportation, and the combustion of fossil fuel; building of a 400 m^2 house with reduced level of quarrying, crashing, construction activities and the use of virgin materials).

Utility ignores the bio-physical limit of earth resources. This utility should be non-declining for future generations as the future should be at least as better off as the present in terms of its utility or happiness. On the other hand, physical throughput is also to be non-declining (Daly 2002) meaning that the capacity of the ecosystem to sustain the flows of food, fuel, minerals, water are not to be exhausted. The use of energy and material efficiency and all Rs (recycle, reuse, reduce, remanufacturing, redesign, recovery) can increase utility per unit of throughput, thus leaving adequate natural resources for future generations.

Natural capital (i.e. clean air, water and non-contaminated or non-toxic soil) is to be maintained and it is the capacity of the ecosystem to yield both a flow of natural resources and a flux of natural services (i.e. grains for food, water for drinking, irrigation). Maintaining natural capital constant is often referred to as 'strong sustainability' in distinction to 'weak sustainability' in which some natural resources are lost due to manmade capital (i.e. converting forest to an industrial area). Ecological limits are rapidly converting economic growth into uneconomic growth – i.e. throughput growth that increases costs by more than it increases benefits, thus making most of us as well as future generation poorer (Daly 2002). For example, climate change resulting from the GHG emissions from human activities to run the modern economy have already caused uneconomic consequences such as bushfire, SLR, drought, and the extinction of species. Uncontrolled growth without considering engineering innovation for resource conservation cannot possibly increase everyone's relative income as few people tend to control most of world resources. Use of increasingly finite, yet unowned ecosystem services (e.g. soil, water, forest) by few imposes opportunity costs on future generations. For example, rapid conversion of forest to cropland in Western Australia increased the mixing of rainwater with salty water in deep aquifers due to the fact that the roots of crops are not long enough like deep rooted perennial species to reach deeper aquifers. Consequently, saline water table rise and affect the cultivation of crops. This is an opportunity cost for future generations as they will not be able to use this salt degraded land for crop production. More hard evidence of over exploitation is in the mining industry where the ore grade in some mining operations has begun to decline resulting in increased energy consumption and costs, particularly with the extraction of small amount of minerals from a large amount of ore. This is not only making the mineral processing expensive but it is also causing a high level of land degradation, water pollution, and the loss of biodiversity.

1.4 Relative Scarcity vs Absolute Scarcity

Cleveland and Stern (1998) defined resource scarcity as the extent to which human well-being is affected by the quality and availability of natural resource stocks. On this basis, relative scarcity can be considered to be the extent to which the quality and availability of a particular resource type impacts upon human well-being relative to other resource types. Therefore, if the availability of a particular resource declines relative to other resources, then, its positive contribution to human well-being comparatively declines and, as such, its relative scarcity increases and people may switch to other resources. For example, if gas is scarce or runs out, then we can go for oil and then when oil runs out, we can go for coal. This actually gradually increases the price of resources due to the increased demand with the rapid decline of these energy reservoirs on earth.

Relative scarcity does not take into consideration the bio-physical limitations of non-renewable finite resources. Ecological economists consider the difference between relative and absolute scarcity to be a critical. Absolute scarcity means that there is a bio-physical limit of the resource. The consideration of absolute scarcity in the engineering design process should increase the use of renewable resources to address the scarcity associated with non-renewable resources.

1.5 Global/International Sustainability Agenda

The global sustainability agenda is a list of tasks or actions, which vary from country to country, based on their respective socio-economic and environmental sustainability goal. Sustainability is a global issue and it requires the participation of all nations across the globe to develop collaborative and collective action plans to manage earth resources in a socially equitable manner for both current and future generations. Pollution and GHG emissions and increased human refugee movement across country boundaries highlight the importance of conducting global sustainability management. Some popular slogans that established sustainable development as a global issue include:

> Think globally and act locally – The conservation of finite resources locally could help enhance the security of indispensable resources like food, minerals, fuel globally. Also, the reduction of GHG emissions locally using renewable energy sources could help reduce global warming impacts.

> Light tomorrow with today – Our combined effort and actions today or the use of environmental friendly and resource efficient technologies that can conserve resources and leave a better world for our future generations.

Some political events in recent decades have brought sustainable development firmly into the public arena, and established it as an accepted goal for international policy or decision makers. World Sustainable Development Summits brought together people from all walks of life, including nobel laureates, political leaders, decision makers from bilateral and multilateral institutions, business leaders, the diplomatic corps, engineers, scientists and researchers, media personnel, members of civil society and even celebrities; on a common platform to deliberate on issues related to environmental, social, and economic agendas for achieving sustainable futures.

The first Earth Summit on sustainable development in Rio de Janeiro in 1992 agreed on a global plan of action, known as Agenda 21, designed to deliver a more sustainable pattern of development. Agenda 21 focused on preparing the world for life in the twenty-first century. Signed by 178 national governments, Agenda 21 provides a comprehensive plan of action to attain sustainable development at local, national, and global levels (UN 1992). The issues that were discussed in Agenda 21 report covered both inter- and intra-generational equity issues, including uneven development, gender equity, poverty alleviation, and unsustainable consumption. Secondly, it focused on the impact of population growth and the need for developed nations to extend cooperation to developing nations to build local capacity building in addressing sustainability challenges. Finally, it discussed the need for research and innovation to achieve social, economic, and environmental objectives of sustainability. Engineers can potentially address sustainability challenges by innovating socially equitable, accessible, resource efficient, and affordable technologies, utilising indigenous or local natural and human resources.

The target of next World Summit on Sustainable Development, in Johannesburg in 2002 was to eradicate poverty for developing nations and to attain sustainable consumption and production for developed nations (UN 2002). Engineers can design affordable technologies using indigenous resources to create income-generating activities for poor people to meet the basic needs of life, while reducing environmental problems, like deforestation associated with the use of fuelwood and indoor air pollution from cooking with poor-quality fuels. On the other hand, engineers can design products for disassembly so that the EoL product can be given a new life and consequently emissions, wastes generation and land use, and energy and material consumption associated with upstream activities (i.e. mining to material production) can be avoided and conserve resources for the future generations. For example, a remanufactured compressor (i.e. EoL compressor turned into a usable compressor) costs one-third of the cost of a new compressor while offering the same durability or service life (Biswas and Rosano 2011).

The United Nations Conference on Sustainable Development – or Rio+20 – took place in 2012, 20 years after the first earth summit in Rio de Janeiro. World leaders decided to develop a set of sustainable development goals (SDGs) built upon the millennium development goals. The purpose of these goals was to promote sustainable development in an organised, integrated, and global way. Nations agreed on exploring different measures of wealth other than gross domestic product (GDP) that also considers environmental and social factors. This was a clear attempt to achieve ecologically sustainable development taking into account the bio-physical limits of earth resources. Standard neoclassical economics sees a tight coupling between GDP and welfare and a loose coupling between GDP and throughput or the flow of energy and materials from the global ecosystem (Daly and Farley 2004). Whereas ecological economics sees a tight coupling between GDP and throughput, with a loose coupling between GDP and welfare beyond basic sufficiency. Engineers play a pivotal role in maximising the use of renewable energy by integrating it on a mandatory basis into their engineering design. Reducing the use of fossil fuel could significantly contribute to the decrease of throughputs or resources. Renewable energy technologies are not completely environmental friendly due to the fact that its production process requires the consumption of non-renewable resources and emission intensive processes. Engineers face difficulty in recycling solar panels because of the fact that the materials they are made from are hard to recycle as they are constructed from many different parts and combine together to make one complex product. Without the recycling of photovoltaic (PV) cells, some rare earth elements (REEs) in PV like gallium and indium are being depleted from the environment over time (Energy Central 2018).

In 2015, in New York, the 193-Member United Nations General Assembly adopted the 2030 Agenda for Sustainable Development. This program is divided into 17 SDGs and 169 targets. All of these goals and targets are integrated with the social, environmental, and economic dimensions of sustainable development. Table 1.3 illustrates using practical examples of how engineering strategies can develop to achieve and support these SDGs.

1.6 Engineering Sustainability

Engineers are key members of the community often responsible for unsustainable technological development, but they also have power to reverse this problem by taking account of sustainability issues in their engineering design process. Engineers also play a pivotal role in implementing sustainable development agendas. A group of engineers after the Earth Summit in 1992 identified that about 70%

Table 1.3 Example of successful engineering solutions to achieve SDGs

SDG	Directly related
GOAL 1: No poverty	The application of Information and Communication Technology (ICT) has enabled the grassroots producers in developing nations to access market and price information and create employment opportunities to eradicate poverty (World Bank 2013)
	The application of a human operated pedal pump for pumping water from ponds for treatment using low-cost hollow fibre membranes and granular activated carbon columns not only overcomes limitations in existing water technologies to provide affordable water supply in arsenic-contaminated villages in developing nations but also creates local employment by involving local people in the water business (Biswas and Leslie 2007)
	Engineers Without Borders (EWB) Australia, is involved in humanitarian engineering as they are working with communities in Cambodia, Vietnam and Timor Leste to design appropriate technologies using locally available indigenous resources to create change in four key thematic areas, including water, sanitation and hygiene, shelter, energy and education to alleviate poverty (EWB 2018)
GOAL 2: Zero Hunger and GOAL 15: Life on land	In the Sahara Forest Project, engineers produced electricity from solar power more efficiently to operate energy- and water-efficient saltwater-cooled greenhouses for producing high value crops in the desert, to desalinate seawater using solar radiation to produce freshwater for irrigating crops, safely manage brine which was produced as a by-product in seawater desalination to harvest useful compounds from the resulting salt to grow salt tolerant biomass for energy purposes without competing with food cultivation, and also to revegetate desert lands (Sahara Forest Project 2020)
	Telecommunication and digitalisation in agriculture has improved farmers' knowledge of the efficient use of chemicals and water to increase productivity (Light 2009)

(Continued)

Table 1.3 (Continued)

SDG	Directly related
GOAL 3: Good health and well-being	Using 3D printing technologies, a range of medical products were manufactured to address the shortages of personal protective equipment (PPE) during the COVID19 pandemic (Richardson 2020)
	United Kingdom and Norway are considering the use of clean energy powered electric vehicles as a replacement of diesel and petrol cars to reduce air pollution (Coren 2018)
GOAL 4: Quality education	ICT has been used to develop a professional development program for teachers/trainers to assist them to integrate project-based learning into six schools in Chile, India, and Turkey to enhance students learning outcomes and apply theory to practice (Light 2009)
	Pacific islands are mountainous and Appropriate Technology for the Community and Environment (APACE) took the advantage of the use of these slopes to generate electricity through hydropower using local indigenous designed resources (e.g. wooden dam and a penstock supported by trees). The micro-hydro projects have turbines of 10–12 kW capacity and generate 5 kW or less to provide lighting in order to allow children to study at night (Bryce and Bryce 1998)
GOAL 5: Gender Equality GOAL 6: Clean Water and Sanitation	Rural women have been found to walk long distances to collect water, as there is usually only one or two deep tube water wells in villages in some developing parts of the world. SkyJuice came up with engineering solutions suitable for supplying safe drinking water for humanitarian programs in remote locations in developing nations including emergency and disaster relief using both surface and ground water supplies to reduce this drudgery on rural women. They introduced ultrafiltration membranes providing low cost physical filtration including disinfection to remove bacteria, protozoa, and pathogens greater than 0.04 μm. This filtration technology is known as 'SkyHydrants' which are lightweight and easy to transport, require no power to operate and can be setup and run by non-technical persons. It offered safe and clean water for less than $1 per person per year (SkyJuice Foundation 2020)

(Continued)

Table 1.3 (Continued)

SDG	Directly related
GOAL 7: Affordable and Clean Energy	There is now more than one solar panel installed per person in Australia due to significant reductions in the cost of solar electricity with the increase in the maturity of the market and manufacturing facilities of this system. Systems cost in 2020 is over 50% less to install than in 2012 (Sykes 2020) Since the energy intensity of renewable energy is less than fossil fuel based electricity and also as these energy sources are intermittent, both demand and supply side management is crucially important to further improve the sustainability of these systems. Smart grid technology has been used in integrating demand side management into renewable power system operation, making it possible to monitor and integrate diverse energy sources into power systems while facilitating and simplifying their interconnection. It has cost efficiently integrated the behaviour and actions of all users connected to it – generators, consumers, and those that do both – in order to ensure economically efficient power systems with low losses and high levels of quality and security of supply and safety (Australia Trade and Investment Commission 2017)
GOAL 8: Decent Work and Economic Growth GOAL 10: Reduced Inequality	An integrated ecological, economic, and social model was developed to assist sustainable rural development in villages in Bangladesh to create income-generating activities using renewable energy technologies (RETs) for male landless and marginal farmers and for women from such households, while reducing environmental problems, like deforestation and indoor air pollution from cooking with poor-quality fuels. With the assistance of an External Agency composed of NGO, business, government and university representatives, such groups of villagers form Village Organizations, comprising cooperatives or other forms of business, borrow money from a bank or large NGO, and purchase a RET based on biogas, solar or wind, depending upon location. By selling energy to wealthier members of the village, the Village Organizations repay their loans, thus gaining direct ownership and control over the technology and its applications (Biswas et al. 2001)

(Continued)

Table 1.3 (Continued)

SDG	Directly related
GOAL 9: Industry, Innovation and Infrastructure	Industrial symbiosis is one way to achieve industrial sustainability to reduce disposal of wastes or emissions to water and atmosphere through the realisation of regional resource synergies. The CSBP chemical works which is a fertilizer company in Western Australia supplies its gypsum by-product for residue area amelioration at an alumina refinery. Built in 1999, a cogeneration facility (40 MW), owned by Verve Energy, provides superheated steam and electricity for process needs at the nearby Tiwest pigment plant. These resource synergies are beneficial for both companies (van Beers et al. 2007)
GOAL 11: Sustainable Cities and Communities	Ground Water Recycling is the process by which secondary treated wastewater undergoes advanced treatment to produce recycled water which meets Australian guidelines for drinking water prior to being recharged to an aquifer for later use as a drinking water source (Simms et al. 2017). This way people do not have any psychological problem in drinking wastewater.
GOAL 12: Responsible Consumption and Production	The introduction of an energy efficient Building Management System (BMS) in a new engineering building at of Curtin University reduced energy consumption significantly. The specific energy consumption of the usage stage of the university is 0.92 GJ/m^2/year which is 18% higher than the specific energy consumption of the new engineering building (i.e. 0.74 GJ/m^2/year). GHG emissions from this engineering building are 63% lower than the university building 'usage stage' average (i.e. 0.16 ton CO_2 e-) (Biswas 2014)
	The replacement of conventional concrete with mixes using construction and demolition wastes and industrial by-products (Fly Ash and micro-silica) offers the same compressive strength with economic savings (i.e. 7–18%) and environmental savings (i.e. 14–31%) (Shaikh et al. 2019)

(Continued)

Table 1.3 (Continued)

SDG	Directly related
GOAL 13: Climate Action GOAL 14: Life Below Water	Remanufacturing a compressor reduced CO_2 emissions by almost 1.5 ton/unit, equivalent to taking a small car off the road (Biswas and Rosano 2011). Engineers built a seawall to separate land and water areas, and this wall was primarily designed to prevent erosion and other damage due to wave action. However, rich and vibrant habitats of marine species have been replaced with seawalls and degraded by plastic pollution. In recent years, Volvo developed seawall tiles after its research found that one rubbish truck of plastic enters the world's oceans every minute (Yalcinkaya 2019). More than half of Sydney' s shoreline is made of artificial seawalls. The Seawall consists of 50 hexagonal tiles with small corners and recesses that are designed to imitate the root structure of native mangrove trees – a popular habitat for marine wildlife. Each tile is made from marine-grade concrete that has been reinforced with recycled plastic fibres
GOAL 16: Peace and Justice Strong Institutions	Today, the engineering profession seems to have preserved the sense that technology is almost by necessity a force for good. Engineers focused on the technical and managerial sides of technology – how to design algorithms; how to build machines – but not so much on the context of the technologies deployment or its unintended consequences. Engineers are typically are not very interested in politics and social dynamics. They do not make weapons for a specific war or algorithms for a specific surveillance activity. As a result, engineers who build these devices usually operate totally removed from the consequences of their actions (El-Zein 2013)
GOAL 17: Partnerships to achieve the Goal	Woodside, an oil exploration company, engaged their engineers with the local Aboriginal indigenous communities in the gas exploration site of their Pluto project to minimise impacts to Aboriginal cultural heritage, access to land and rights to land (Woodside Energy Ltd. 1999). They talked with the aboriginal communities and traditional owners about project plans, and involved them in monitoring non-ground disturbing works and conducted heritage surveys. The project design had to be changed to avoid or minimise impacts to heritage based on work programs clearance reports

of the issues listed in Agenda 21 concerned engineering design, and at least 10% of these issues had major engineering applications (Smith et al. 2007). The skills required for sustainable engineering represent a meta-disciplinary endeavour, combining information and insights across multiple disciplines and perspectives with the common goal of achieving a desired balance among economic, environmental, and societal objectives (Mihelcic et al. 2003). This is an urgent task for current and future engineers to innovate resource efficient and environmentally friendly technologies to assist future generations to live within earth's carrying capacity. As we approach the future, more resources will become scarce and so it will become increasingly challenging or difficult tasks for engineers to address sustainability issues such as global warming, deforestation, social inequity, and resource scarcity. Engineers need to improve their thinking process and change their mind set now as to what needs to be done in order to deliver products and services to society within a resource and carbon constrained economy.

This book deals with the contribution of engineering to the development and implementation of sustainable solutions and is based on the popular ideas of Boyle (2004): The engineering context of sustainability involves the design and management of sustainable technology, research into environmental and social impacts and limitations, living within global limitations, and management of resources from cradle to cradle. This book discusses life cycle assessment tools to enable engineers to think out of the box so that they are able to find or select processes, chemicals and energy used during the life cycle from cradle to cradle of products or services which can be included in their engineering design to avoid negative social, economic, and environmental consequences. Incorporating life cycle concepts into their engineering design will enhance recycling, remanufacturing, and recovering of EoL products/components, therefore contributing to the reduction in the size of landfill as well as the reduced use of land for upstream activities, including mining, processing, and manufacturing. The concept of industrial ecology enables engineers to achieve a zero waste solution by exchanging wastes and by-products between neighbouring industries in an industrial park, reduce ecological footprint or land for human consumption, and to enhance carrying capacity. Engineers will be able to understand how the use of cleaner production strategies can achieve environmental solutions in an economically feasible manner. Finally, they will be able to assess the social, economic, and environmental implications of an engineering innovation.

The book content is broadly organised in three clusters:

- *The Engineering Sustainability Challenges*: An introduction to the sustainable development agenda and debate, covering key sustainability issues (local, national, and global) and key government and corporate response strategies, engineering's impacts on community, society, and culture.

- *Sustainability Assessment Tools*: The 'nuts and bolts' of efficient resource utilisation in 'industrial' and service operations, identification of social, economic, and environmental and triple bottom line (TBL) 'hotspots', cause diagnosis, and sustainability implications/consequences.
- *Sustainable Engineering Solutions*: An exploration of the role of technology (engineering) in achieving sustainable development or TBL objectives, covering both sustainability-driven innovations as well as sustainability-applications of emerging technologies.

1.7 IPAT

In a series of papers during 1970–1974, Paul Ehrlich and John Holdren proposed the IPAT equation to estimate the overall impact of our economic activities on the environment (Holdren 1993):

$$I = PAT$$

where

I = the impact (total impact of mankind on the planet),
P = the population (total population size),
A = affluence (number of products or services consumed per person), and
T = technology (impact per unit consumed, often called technology efficiency).

This equation has also a significant bearing on engineering innovation, as it helps discern the level of technological improvement required to reduce the environmental impacts associated with increased population levels and economic growth.

As shown in Figure 1.2, affluence is represented in terms of GDP per capita, which is the total value of goods produced and services provided in dollar terms in a specific country during one year. In order to develop GDP growth, engineers are required to develop technologies to explore for resources and then convert them to products and services for society. When these technologies use energy and material resources to produce products and then deliver services, emissions and wastes are created together with increased environmental impacts. The use of renewable energy technologies and energy efficiency could help engineers to achieve a low carbon economy.

For example, consider a situation where the human population is increased by 200% and the level of affluence by 500% against current levels. If the environmental impacts of this future scenario are required to be only 50% of current levels, the IPAT equation can work out the level of technological improvement (what value of technological factor) that is required.

The equation for the current scenario is

$$I_{\text{current}} = P_{\text{current}} \times A_{\text{current}} \times T_{\text{current}} \tag{1.1}$$

The ingredients for developing future scenario are as follows:

$$\begin{aligned} P_{\text{future}} &= 2P_{\text{current}} \\ A_{\text{future}} &= 5A_{\text{current}} \\ I_{\text{future}} &= 0.5I_{\text{current}} \\ T_{\text{future}} &= X\,T_{\text{current}} \end{aligned} \tag{1.2}$$

where 'X' is a technological factor.

The equation for the future scenario is

$$\begin{aligned} 0.5I_{\text{current}} &= 2P_{\text{currant}} \times 5A_{\text{currant}} \times X\,T_{\text{currant}} \\ 0.5I_{\text{current}} &= 10XP_{\text{currant}} \times A_{\text{currant}} \times T_{\text{currant}} \end{aligned} \tag{1.3}$$

$$X = 0.5/(2 \times 5) = 1/20 \text{ (Using Eq. (1.1))}$$

$$T_{\text{future}} = 1/20T_{\text{current}} \text{ (Using Eq. (1.2))}$$

Therefore, future technology must be 20 times more efficient than current technologies. It is a challenge for future engineers to come up with the innovations required to achieve significant technological development; however, this is the level of innovation required to reduce the associated environmental impacts associated with GDP growth.

1.8 Environmental Kuznets Curves

Environmental Kuznets curve (EKC) represents the environmental consequences of economic activities. It is the inverted U curve of the Kuznets' curve of income vs social equity, where the social equity (i.e. income gap between people in the same generation) decreases with income up to a certain point and then social equity increases with further increase in income which then creates more business and employment generation activities. In the case of EKC, the situation is completely reversed as the environmental emissions increase with income up to a certain point due to increases in the production of products and growth of service sectors, but after this point, pollutions levels decrease with the increase in

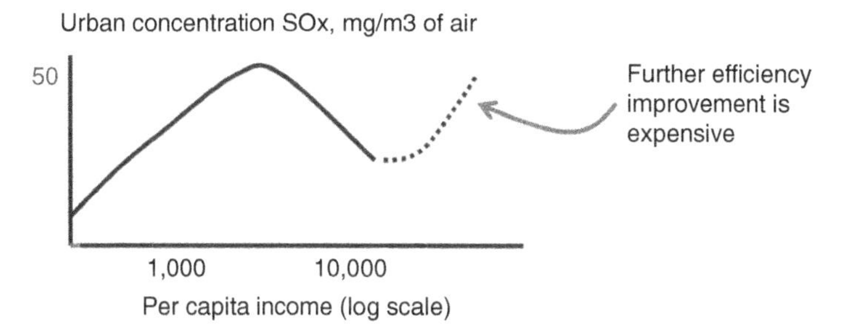

Figure 1.3 Environmental Kuznets curve

income as people can afford to manage and control pollution levels. However, the inverted U relationship may not hold for the long term. As Figure 1.3 shows, an increase of pollution occurs again with the increase in income. This is because of the gradual conversion of natural resources to man-made capital often causing the irreversible and irreparable damage to earth, when even a large amount of investment in technology cannot control emissions and waste levels. At this point, the level of pollutants in the air and soil contamination have far exceeded the critical limits. Super technology or engineering innovation will reach its maximum limit at this point.

1.9 Impact of Engineering Innovation on Earth's Carrying Capacity

Dematerialisation through ICT, renewable energy technologies, waste management, 7Rs (reuse, recycling, reduce, remanufacturing/repurpose, retrofit/refurbish, recovery, redesign), energy efficiency, green chemistry principles can increase the carrying capacity of the earth and conserve adequate material resources to sustain earth's population. The carrying capacity is in fact the maximum population (i.e. red dotted line in Figure 1.4) that a region can sustain indefinitely, given the food, habitat, water, and other necessities available in the environment. The over consumption of earth resources associated with population growth and the overuse of resource intensive technologies are reducing earth's carrying capacity significantly (Meadows et al. 2004). We are currently consuming 50% more resources from the planet than it can replenish. What happens if we continue to consume more than the earth's carrying capacity over the long term? We will not be able to meet the minimum basic needs of life even by increasing expenditures on technological development by the required order of magnitude. This could lead to a likely paradigm shift from overshoot and oscillation to overshoot and collapse (Meadows et al. 2004). Overshoot and oscillation are

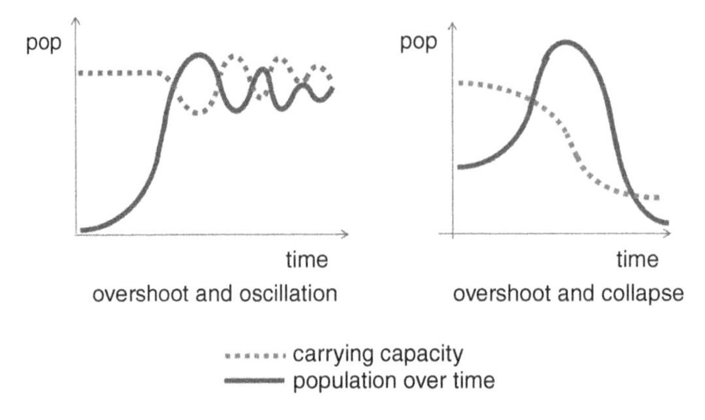

Figure 1.4 Carrying capacity scenarios

natural phenomenon when a region's population and carrying capacity balance out each other. The excessive exploitation of resources will cause irreversible and irreparable damage to earth resources resulting in scarcity of ecosystem services such as water, food, and energy. This is known as the overshoot and collapse scenario, representing a situation when even the investment of a large amount of capital in technological development cannot restore the ecosystem resulting in permanent damage to the ecosystem and reducing its carrying capacity.

The man-made artificial ecosystem is unrealistic to consider once the earth's carrying capacity is collapsed. Biosphere 2 was developed in 1991 in the Arizona desert to demonstrate the viability of a man-made self-sustaining closed ecological systems to support and maintain human life in Outer Space. At the time it cost US $150 million dollars to construct this system to meet the demand of only eight people, which meant that it would have cost US $121 Billion Millions (6.5 billion people × US$150 million/8 people) to meet the demand of the global population of 6.5 billion people at that time. However, the value of earth resources is US $5 Billion Millions (Messenger 2020). This raised a question as to whether it is possible to meet the demand of a growing population artificially like Biosphere 2 given also the fact that nearly half the world population live on less than $5.50 a Day (The World Bank 2018).

1.10 Engineering Challenges in Reducing Ecological Footprint

A community's ecological footprint is the total resource area of land in a given ecosystem required to support the community's need for food, water, wood, energy, and waste processing capacity. On the other hand, the ecological footprint will decrease due to increases in carrying capacity. Using innovative engineering design (e.g. dematerialisation, video conferences), resource efficient technologies (e.g. remanufacturing, recovery), and renewable energy resources, the amount of land used during the upstream processes or activities including mining and processing of virgin mineral and energy resources, and manufacturing activities and associated fossil consumption during transportation can be avoided or reduced.

System thinking approaches, thus, need to be considered in all engineering design. A significant level of importance needs to be given to the design stage so that the products after their EoL can be remanufactured, reused, refurbished, and recycled. These strategies will avoid EoL products going to landfill, which not only reduces the size of the landfill or 'sink' but also slows the extraction of virgin materials or non-renewable resources (e.g. iron, coal, rocks) from the earth. The reduction of the 'sink' is therefore a reduction of the 'source', the ecological footprint is reduced during upstream and downstream activities of a product or service life cycle and this gives birth to the circular economy by increasing the use and circulation of materials in society and the economy. A circular economy is one that exchanges the typical cycle of make, use, and dispose in favour of as much re-use and recycling as possible.

As we have seen in the IPAT equation, both population levels and behavioural change along with technological improvement are desperately needed to reduce either ecological footprint or increase carrying capacity. The first two variables need to be controlled at a reasonable level, otherwise, they will increase the level of challenge for future engineers to achieve sustainability. Policymakers, industries, and communities should engage engineers in the consultation of any decision-making process concerning resource exploitation or any sort of development projects requiring significant resources. The global hectares available per person in developed countries is three to six times higher than the global average (1.8 ha available per person). On the other hand, India's footprint is half of the global average (Raj et al. 2012). Sustainable consumption and production is necessary in developed countries as these nations are not only exploiting their own resources but also from other nations to maintain their level of affluence, thereby affecting earth's carrying capacity causing natural resources degradation and associated environmental impacts. A large-scale gold mine owned by a New Zealand company created a 200 ft deep pit in the old village of Didipio in the Philippines, where there was previously a mountain and fields where small farmers planted rice and grew fruit but is now an industrial site, and traditional wooden houses have now been abandoned (The Guardian 2015). The Filipino mine, guarded by high fences and bitterly contested by the indigenous Bugkalot people who fear pollution, mining spills, and ill-health, is just one of a number of major new gold and copper mines opened in the last few years to meet soaring world demand for minerals used in electronic devices such as smartphones and laptops. The Philippines that was well endowed with minerals, including copper, gold, nickel, and silver had to be heavily dependent on developed countries for investment and exploration activities to sustain her economic growth without taking into account the inter- and intra-generation social equity aspects of sustainability such as land degradation, loss of biodiversity, water pollution, health problems, and resource scarcity for current and future generations (Ingelson et al. 2009).

1.11 Sustainability Implications of Engineering Design

Environment, economics, and society are the three pillars of sustainability. These three objectives need to be met in order for any product or service or technology to be considered sustainable.

The engineering profession in fact evolved to serve mankind. Engineering design is critical in considering the associated side effects of both production and consumption on the three sustainability pillars.

In the case of civil and construction engineering, climate change and energy resource scarcity impacts have resulted from the use of energy intensive materials such as Ordinary Portland Cement, bitumen, aggregates, and steel. Secondly, deforestation and the loss of biodiversity have resulted from the conversion of pristine land to quarries for sourcing raw materials for construction materials production, such as limestone for cement production and rocks for aggregates production. Biologists think up to 50% of species will be facing extinction by 2050 (The Guardian 2017). The increased use of impervious pavements due to rapid urbanisation have reduced the recharge of underground aquifers with rainwater, which instead flows down to nearby lakes or ocean with toxic wastes generated from urban activities. Another observed environmental impact from civil and construction engineering is the urban heat island (UHI) effect, which is due to the replacement of natural land cover with dense concentrations of pavement, buildings, and other surfaces that absorb and retain heat for a city to experience much warmer temperatures. This effect increases energy costs and the associated GHG emissions (e.g. for air conditioning), air pollution levels, and heat-related illness, mortality, and increased energy consumption costs from air cooling, all with TBL implications.

In the case of mechanical engineering, the combustion of petroleum products in the internal combustion engines produces volatile organic compounds (VOCs), carbon monoxide (CO), nitrogen oxides (NOx), particulate matter (PM10), sulfur oxides (SOx), methane (CH4), nitrous oxides (N_2O), and carbon dioxide (CO_2), all resulting in global warming, acidification (soil degradation), eco-toxicity (urban air pollution), eutrophication (water pollution), and photo-chemical smogs (loss of visibility) (Dutton 2017). Secondly, in manufacturing operations, production, use, and disposal activities predominantly cause global warming effects, eutrophication, and eco-toxicity. Cutting fluids, which are produced from petroleum products, involve an energy intensive process resulting the GHG emissions. In addition, cutting fluids become contaminated with heavy metals, oil from the machine tools and waste food during machining operations. The used cutting fluids contain fungi and living organisms and cause health effects such as skin disorders, respiratory diseases, and skin cancer through skin contact with contaminated materials, spray, or mist and through inhalation from breathing as mist or via aerosol. After use, cutting fluids are typically disposed to landfill/soil, contaminates both ground and surface water, and this is known as 'eco-toxicity'.

While the accelerating growth of communication and information technologies (CITs) worldwide has brought advantages in life such as education, communication, banking, entertainment, or navigation, through the widely increasing availability of internet access, the hardware components used have caused exponential growth of end-of-life waste levels with minimum potential for reuse, recycle, and remanufacturing. The increasing shortage of landfill space and expanding regulation of waste disposal together have further increased the costs of EoL computers and e-waste disposal significantly in past few years. The planned obsolescence of many electronic products in recent times have made many technologies obsolete, unfashionable, or no longer usable. The electronic industries are gaining enormous amounts of short-term benefits from the planned obsolescence without considering that these electronic devices are made of rare earth elements (REEs), in rare supply (Bradshaw et al. 2010). The low concentration of these REES in the earth's crust makes economic exploitation difficult, while some of them are already considered as critical elements (i.e. neodymium, dysprosium, indium, tellurium, and gallium) as their supply is believed to be most at risk because of the exponential growth in demand of electronic products. The known and accessible reserves of these elements will be depleted quickly because of the current lack of appropriate methods of recycling, reducing, or reusing REES (de Boer and Lammertsma 2013). Due to the high cost of recycling in an environmentally sound way in industrialised countries, much of the E-waste is sent to poorer countries, even though this practice is banned by the Basel Convention and the European directive on Waste for Electrical and Electronic Equipment (WEEE). E-waste frequently contains valuable as well as potentially toxic materials. It is evident that the highest recovery value comes from metals with the lowest concentration. A small percentage of gold is found in mobile electronics, and yet it contributes to the highest portion of the total recoverable value of e-waste. These materials require special process recovery treatment when the devices reach their EoL in order to avoid environmental contamination and accumulation of hazardous substances in the human body. Hazardous materials contained in cell phones include brominated flame retardants, arsenic, antimony, beryllium, cadmium, copper, lead, mercury, nickel, and zinc. They have a long life-span and they can accumulate in animal tissues, increasing their amount in the body over time and thus leading to contamination through the food chain. In humans, they can lead to cancer as well as reproductive, neurological, and development disorders. Thus, it is a challenging task for the future engineers to decrease the dependency on REEs, for example, by identifying possible replacements or increasing their efficient use, increased recovery represents another possibility.

While chemical engineering has brought a number of important products to the market to improve health, provide greater safety, and enhance the quality of life, in doing so it has also released toxic and hazardous substances into the environment (or its substances have been released during use) that have led to negative impacts on society and the environment (e.g. CFCs [chlorofluorocarbons], PCBs [polychlorinated biphenyls], PBBs [polybrominated biphenyls]). Over the entire life of a chemical product ('cradle to grave'), there is significant potential for detrimental impact.

First, as a major user of raw materials, both for energy consumption (14% of world energy use) (IEA 2020) and as feedstock, the chemical industry have significantly increased the use of non-renewable resources. In fact, these materials are, in general, based on hydrocarbons, combustion of these sources have led to emissions of – global warming gas (carbon dioxide, CO_2), VOCs, and nitrogen oxides (NO_x) which contribute to the formation of tropospheric ozone or 'smog' affecting visibility losses and human health like respiratory problems. Certain pollutants released by the chemical industry in its production processes can lead to a direct or indirect impact on man and/or the environment (e.g. hydrogen sulphide separated during the crude oil refining process causes eco-toxicity, respiratory problems). Exposure to certain hazardous substances – such as PCBs, DDT (dichlorodiphenyltrichloroethane), PBBs, heavy metals, endocrine disrupting substances – can lead to a direct toxicological effect on human or the surrounding environment due to short- or long-term exposure. Other substances – such as VOCs, NO_x, and SO_x – are gases for concern after they react with other substances. For instance, VOCs and NO_x promote the formation of smog and SO_x is responsible for the formation of acid rain, which acidifies soil and water, thus affecting crop production and the marine life. Next, the processing of the raw materials and feedstocks can result in the release of hazardous pollutants to the environment (e.g. propylene) from emission stacks, discharge pipes, flanges, waste ponds, storage tanks, and other equipment. Of all the sectors of the chemical industry, the basic chemical sector is generally the largest emitter (by volume) of such pollutants because these bulk chemicals are usually produced in high volumes at large plants. For example, the health costs associated with fine particle pollution (particulate matter or PM2.5) from coal sources (coal mines and coal fired power stations) in Singleton in the Hunter Valley region of Australia is AUD 47 million per annum (Climate and Health Alliance 2015). Thirdly, workers can be exposed to pollutants in a gaseous or liquid form, for example by inhaling a pollutant emitted from leaks in equipment or splashing the substance on their skin or in the eyes. Larger accidents involving chemicals can also occur due to equipment failures. Major spills can result in inadvertent releases to workers, the surrounding neighbourhood or perhaps even communities and the environment at some distance from the plant. The official immediate death toll of 3928 that resulted from the gas leak in 1984 at the Union Carbide India Limited (UCIL) pesticide plant in Bhopal, Madhya Pradesh, India, is much lower than the many lives in future generations that may have been affected by this leak.

The final user of a chemical product can be another chemical company, other industries, or consumers. Depending on the product, and how it is used, there also can be exposure during the end use phase. For instance, chemicals such as plasticisers and stabilisers found in plastics can leak during consumer use. Similarly, leakage of brake fluids from automobiles and disposal of these substances (generally classified as hazardous waste) can impact on the environment. The use of some consumer products can have global impacts, as is the case with refrigerants containing CFCs that have led to a depletion of the ozone layer. Just weeks after the deal to purchase Monsanto was completed in 2018, a jury in a California state court awarded $289 million to a school groundskeeper, after concluding that glyphosate caused his cancer. This is because Monsanto had failed to warn consumers of the risk (The New York Times 2020). There was another similar case in Australia, where Asbestos was used as a popular building product from the 1940s to 1987 due to its resistance to high temperatures and fire and effective insulation properties, but was later found to increase the risk of developing cancers of the lung, ovary, and larynx as well as mesothelioma (Cancer Council 2020).

1.12 Engineering Catastrophes

A good number of engineering catastrophes that have taken place over the last 70 years have given birth to environmental laws or acts, ISO (International Standard Organization) guidelines and innovation in clean technologies. The first example started with an oil slick that caught fire on the Cuyahoga River just southeast of downtown Cleveland, Ohio, USA, in 1969 after decades of industrial waste became a rallying point for the environmentalists in their fight for cleaner water (Rotman 2017). Whilst Cleveland city saw water pollution as a necessary consequence of industrial competition, this catastrophe ultimately led to a federal action to introduce the Clean Water Act in the United States in 1972.

In the middle of the twentieth century, when the residents of London were heavily dependent on poor quality coal for domestic heating and electricity generation in metropolitan areas, large amount of pollutants were produced from this combustion process and remained in the atmosphere for about four days, where 370 tons of sulphur dioxide haze turned into 800 tons of sulphuric acid and resulted in the deaths of 4000 people (The Washington Post 2016). This catastrophe then led to the development of the Clean Air Act of 1956 in London.

An explosion of a nuclear power plant in Chernobyl, Ukraine, USSR in 1986 occurred due to poor coordination and safety precautions, design faults, and the inherent instability of the plant. As a result, 2 million people had to live on radioactively contaminated land, 270,000 people still need protective measures, 1.6 million lives with enhanced radioactive monitoring, and 54,600 thousand km^2 of land remain contaminated, thus increasing the scarcity of land and affected food production in Russia (World Nuclear Association 2020). About the same time, Union Carbide India Ltd. in Bhopal exploded as rainwater accidently entered into a corroded pipeline of the methyl isocyanate (MIC) storage tank, which resulted in an exothermic reaction which caused the explosion at the plant. MIC gas that was released to atmosphere left 3800 dead and 11,000 people disabled.

An environmental outcry occurred in 1998, when Shell was planning to sink the Brent Spar platform, which had been used as a loading buoy and storage tank for crude oil in the Atlantic Ocean. Greenpeace opposed this action and many people boycotted Shell products as a result of this plan. The company lost millions of dollars in sales due to this campaign (Reuters Events 2010) and the platform was not sunk.

DDT (Dichloro diphenyl trichloroethane) – A chemical which was invented to kill mosquitos and to prevent malaria transmission was also found to make its way into the food chain which threatened bird and fish populations and eventually caused illness in children. Rachel Carson, an American citizen scientist, gathered this evidence on the side effects of the use of chemicals. Carson was able to present these consequences to the general public which spawned a revolution which gave birth to the Environmental Protection Agency, in the United States (Carson 1962).

Love Canal is a neighbourhood in Niagara Falls, New York, which became the dump site for municipal refuse for the city of Niagara Falls in 1920 and after 20 years, the canal was purchased by a chemical company to dump roughly 19,800 tons of chemical by-products from the manufacturing of daily essential chemicals (e.g. dyes, perfumes, and solvents). Later on, Love Canal was sold to a local school district in 1953. Since then, it has attracted national attention for the public health problems that it caused over the next three decades due to the former dumping of toxic waste in the canal.

All these environmental catastrophes show that there is a critical need for selecting non-toxic and non-hazardous chemicals and for the use of clean energy sources and efficient technologies to prevent negative social, economic, and environmental consequences. Industry needs to move away from older preventive approaches, including waste minimisation, pollution prevention and toxics use reduction, and focus on engineering design strategies that eliminate these problems. As shown in Figure 1.4, the newer preventive approaches explicitly follow industrial ecology principles by targeting the reduction of environmental impacts along the upstream processes of a product's life cycle, by focusing on product design (in case of design for environment) or on new approaches for value adding activities (in case of eco-efficiency or doing more with less) by taking account of disassemblability, recyclability, and remanufacturability into the engineering design process. The cleaner production concept involves preventative approaches as it considers technological efficiency, innovative product design (modular, disassemblable, multi-functional), use of renewable and non-toxic materials, recycling and good housekeeping strategies to increase productivity with less costs and environmental impacts (known as Eco-efficiency). According to the World Business Council for Sustainable Development (WBCSD), eco-efficiency is achieved by the delivery of competitively priced goods and services that satisfy human needs and bring quality of life, while progressively reducing ecological impacts and resource intensity throughout the life cycle, to a level at least in line with the earth's carrying capacity (WBCSD 2006).

Prevention is better than cure in achieving eco-efficiency for two reasons (Figure 1.5). Firstly, the use of energy efficient, non-toxic processes automatically reduces the amount of emissions and wastes, thus avoiding the need for of pollution control and waste management technologies. Secondly, the efficiency improvement saves operational costs and maximises the company's profit. Cleaner production strategies are business driven rather regulation driven. The

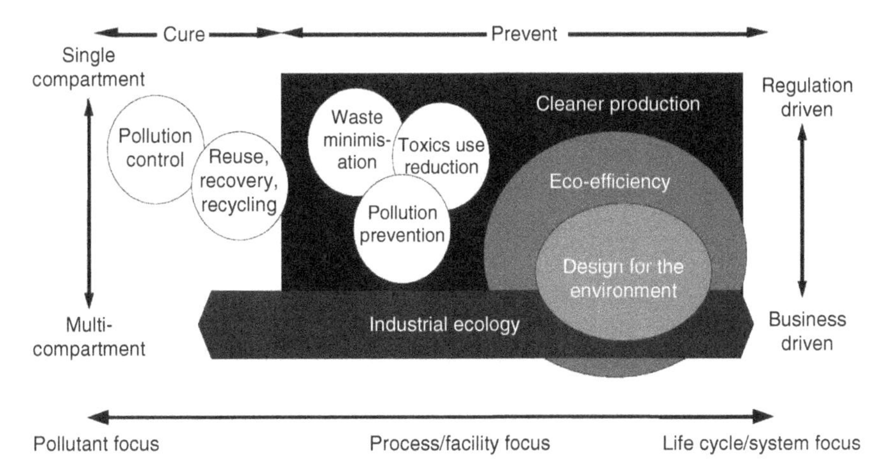

Figure 1.5 Achieving eco-efficiency using preventative approaches Source: Modified from van Berkel (2000)

life cycle approach along with preventative strategies reduce both upstream and downstream impacts or the amount of 'sinks' and 'sources' to follow industrial ecology principles.

1.13 Existential Risks from Engineering Activities in the Twenty-First Century

1.13.1 Artificial Intelligence (AI)

Whilst the industrial revolution brought far reaching changes to firms and employment over the last 250 years, it also resulted in a significant amount of environmental degradation and resource scarcity. On the other hand, digital technology in the twenty-first century has started to slowly replace these industrial activities, which is undoubtedly causing dematerialisation (e.g. use of virtual library, digitised video, use of on-line diagnostic systems, development of new sensor elements), enabling the effective control of all important decisions for people to use resources efficiently. Artificial intelligence (AI) is transforming these digital strategies, which are used for many good causes, including making better medical diagnoses, finding new ways to cure cancer, and making transportation safer. There are self-driving vehicles, nurse-robots taking care of the elderly, and 'Google Search' which may know better than us what we are looking for! AI also performs new tasks like facial recognition, language translation, and internet engine searches.

At the core, AI is about building machines that can think and act intelligently and includes tools such as Google's search algorithms or the machines that make self-driving cars possible. Self-driving delivery vehicles like the Chinese Modai seemed the perfect solution to COVID as it was used for delivering online orders for grocery shopping and reduced interpersonal contact to slow disease spread. In addition, there are a number of areas where AI can be applied to obtain environmental benefits, such as the use of robots in recycling, environmental monitoring and remediation, and clean-up after nuclear and chemical accidents (Lin 2012). The use of robotic platforms for the remote decommissioning of offshore underwater structures was found to prevent people from being exposed to dangerous elements (streams, pressure, temperature, pollution, etc.). This dismantling process not only enables operators offshore to remotely perform the dismantling operation but also helps restore the marine habitat (Cavallo et al. 2004).

The downside of digital technology is that it is likely to continue the decrease of employment in agriculture and manufacturing while contributing to strong increases in services, in particular in computers, the internet and the mobile phone markets. The digital revolution has also resulted in a decrease of employment in large industrial firms.

The expected changes being brought by AI technologies will be just as, or even more significant as those of the Industrial revolution. Firstly, AI technologies will be faster than humans in performing non-repetitive mental tasks. Secondly AI will develop their own programs without humans to develop this program for them. There is no doubt, that AI technologies coupled with the exponential growth of the Internet will affect how firms operate, how they sell their products/services as well as how they are managed, further influencing employment patterns. The speed of the recently developed quantum computer expected revolutionise AI by solving problems millions of times faster than conventional computers (UNSW 2018). The social pillar of the TBL will actually be impacted by the future growth of AI.

The growth of AI could also increase inequalities as fewer people will hold well-paying jobs and the majority will depend on part time work or limited employment opportunities at lower incomes. The same will happen as in the industrial revolution. Aristocrats did not work at all, devoting all their time to leisurely activities, hobbies, holidays and travel, and things they were interested in by controlling world resources through sophisticated technologies and investment (Makridakis 2017). According to E. F. Schumacher – For the rich countries, they say, the most important task now is 'education for leisure' and, for the poor countries, the 'transfer of technology' (Schumacher 1973).

The impact of the use of AI will probably be more pronounced in developing countries than in advanced countries for two reasons. First, as unskilled and semiskilled labour will be replaced by computers and robots, there will be no reason for a firm to move their production to developing nations to exploit their cheap supply of labour as they can achieve the same or cheaper costs utilising AI technologies, thus increasing the trend towards 'reshoring' back to advanced countries (Ford 2016). Secondly, developing countries will be at a disadvantage by not being able to invest in expensive AI technologies, particularly since such technologies will reduce the demand for human labour further increasing unemployment The Ready-Made Garments (RMG) industry is the main source of manufacturing employment in Bangladesh. However, according to the International Labour Organisation (ILO) around 60% (5.38 million) of garment workers in Bangladesh will become unemployed by 2030 and be replaced by robots due to automation in the RMG sector (FW 2020).

Some notable individuals such as legendary physicist Stephen Hawking and Tesla and SpaceX leader and innovator Elon Musk suggested that AI could potentially be very dangerous; Microsoft co-founder Bill Gates also believes there is reason to be cautious, but that the good can outweigh the bad if managed properly. Since recent developments have made super-intelligent machines possible much sooner than initially thought, the time is now to determine what dangers AI poses as well as what could be the opportunities for sustainability benefits. One of the key challenges for engineers is to address the environmental impact of the entire AI and robot production cycle. This would include mining for rare-earth elements and other raw materials, the energy needed to produce and power the machines, and the waste generated during production and at the end of product life cycle. Robotics is likely to add to growing concerns about the increasing volumes of e-waste and the pressure on rare-earth elements generated by the electronics industry (Alonso et al. 2012).

1.13.2 Green Technologies

To support the transition to a low-carbon economy, governments, businesses, and consumers around the world are investing considerable amounts of money in clean technologies, including electric vehicles (EVs), solar panels, and wind turbines. Cobalt and lithium are central to the development and deployment of these technologies – largely due to their use in lithium-ion batteries – and as such, the demand for both minerals has increased substantially. In the case of EV's, it is important to have efficient energy storage systems to increase the vehicle driving range, while the need for an efficient storage system is important for renewable energy technologies due to their intermittent nature in electricity generation. The supply of rare earth minerals, however, is not projected to meet demand, as shortfalls are expected to occur in the coming decades (Church and Wuennenberg 2019). It is an important task for engineers to incorporate recycling into metal and mineral supply chains – and especially those of cobalt and lithium in order to switch to a responsible, sustainable, and stable transition to a low-carbon economy.

The recycling of cobalt and lithium however could offer a solution to overcome these supply chain challenges by extracting metals and minerals from products and infrastructure no longer in use. Moreover, increased mineral recycling could contribute to the commitments set out by the Paris Agreement and the 2030 Agenda for Sustainable Development. However, many actors do not think lithium recovery is currently economically viable. In Japan, for example, estimates suggest that if all used mobile phones were collected and recycled, it would reduce Japanese annual consumption of gold, silver, and palladium by only 2–3% (Mishima et al. 2016). This indicates that a high level of engineering innovation is still required to improve the recovery rates of both the cobalt and the lithium from the lithium-ion batteries through the use of mechanical and hydrometallurgical methods.

If not managed properly, lithium and cobalt mining, like other forms of mining, can also cause environmental and health problems related to the waste generated. The quantity of lithium that is contained in 1 ton of ore in Western Australia's Pilgangoora region is only 12.6 kg (1.26%), meaning that not only a large residue area is required to store lithium by-product but also a large amount of land will be required to obtain small amounts of mineral, causing deforestation, loss of bio-diversity, and land degradation (Pilbara Minerals Ltd. 2018). Beyond waste, mining can also place considerable demands on the local resource base; lithium extraction, for example, can use up to 500,000 gal of water per ton of lithium extracted, which can potentially affect surrounding vegetation, including pasture and grain industries. In addition, the water pollution due to toxic chemical leakage (i.e. hydrochloric acid) from lithium mining in Tibet polluted local water sources (Katwala 2018). Due to the finite, non-renewable nature of these minerals, continued consumption at current (and expected) demand levels could create considerable waste, exacerbate problems associated with natural resource use, and jeopardise global supply chains and production patterns.

The prices of cobalt and lithium are projected to increase in line with growing EV battery demand. This has a dichotomous effect. If prices for raw materials increase, this makes recycled, secondary minerals more price competitive and provides economic grounds for investing in research and development (R&D) for EoL innovation. On the other hand, these resources are not uniformly distributed across the globe and so an increase in raw material prices and the possibility of persistent political fragility, conflict, or violence along the supply chains of lithium and cobalt extraction, should encourage manufacturers of electronic equipment to substitute the affected minerals for more commonly available resources.

However, if manufacturers begin designing lithium-ion batteries with low-value materials capable of replacing cobalt, these products may not be economically worthwhile to recycle, designating them instead for permanent waste disposal and harming the profitability of recycling operations. There are also significant barriers relating to the current low recovery rates and the expense of lithium and cobalt recycling. Also, a permanent assembly method, like spot welding, presents a significant challenge to mineral recovery processes. Insufficient labelling on products and their components at the primary production stage places an additional burden on collection services and secondary producers. There is also an ongoing perception, for example, that recycled materials may not be of the same quality as virgin minerals and metals.

A lack of consumer awareness regarding what to do with electronics at the end of their first-life presents major barriers to the increased recycling of lithium and cobalt. In order to address these engineering challenges, it is important to first understand Lithium-ion battery material flow and enhance the reuse, remanufacture, or recycling process. This process is essential in designing better technologies for resource recovery and efficiency. Manufacturers should be encouraged to consider eventual recyclability in the initial design of their products.

In addition, effective battery collection services require a stable supply of batteries and other lithium and cobalt-intensive products before they can be successfully recycled. This recycling process may be hindered if the batteries being supplied are not of similar composition or chemistry. To help develop this important element in the supply chain, some actors – collection services, battery testers, recyclers, manufactures should partner to establish a single point of contact for battery end users. This collaboration could reduce the complexity in recycling supply chain and can help to provide specialised processes where the industry may be underdeveloped or market signals slow in suggesting the potential market for resource recycling and recovery.

1.14 The Way Forward

Engineering students need to be introduced to sustainability assessment tools so that they are able to assess the sustainability implications of products, services, and engineering design using region specific TBL indicators. Sustainability case studies across a wide spectrum of engineering disciplines, including civil, mechanical, electrical, and chemical, have been provided in this chapter. These indicators are commonly measured by using a life cycle assessment approach. LCA is a diagnostic tool allowing engineers to improve sustainability performance by improving processes, inputs or technologies affecting or contributing to environmental, social, and economic impacts. As a doctor cannot prescribe medicine without diagnosing the patient. Likewise, engineers cannot provide sustainable engineering solutions without doing an LCA analysis of products or services. The theories and examples of environmental life cycle assessment, economic and social life cycle assessment, discussed in subsequent chapters help improve the sustainability performance of engineering products and services. Life cycle sustainability assessment is also discussed as this combines the TBL objectives of sustainability in determining the overall performance of products or services and assists in the comparison between alternative options to determine the most sustainable engineering strategies or product/service.

Sustainable engineering strategies can only be applied once engineers are able to identify TBL 'hotspots' or problematic areas using an LCA analysis. Cleaner Production strategies were discussed that enabled engineers to choose preventative approaches to improve both environmental and economic outputs. Secondly, engineering discipline specific case studies of environmental management system (EMS) are discussed to help industries to achieve environmental targets using engineering strategies. Thirdly, design for the environment is discussed and how it assists engineers in designing technologies or products for disassembling, recovery, and remanufacturing of EoL products so that both landfill size and pressure on virgin resources can be reduced. The principles of Green Engineering covering both green chemistry and engineering were discussed to reduce risks, costs, energy, materials, and wastes. The chapter on Sustainable Energy discussed three pathways to achieve sustainable energy solutions, including the deployment

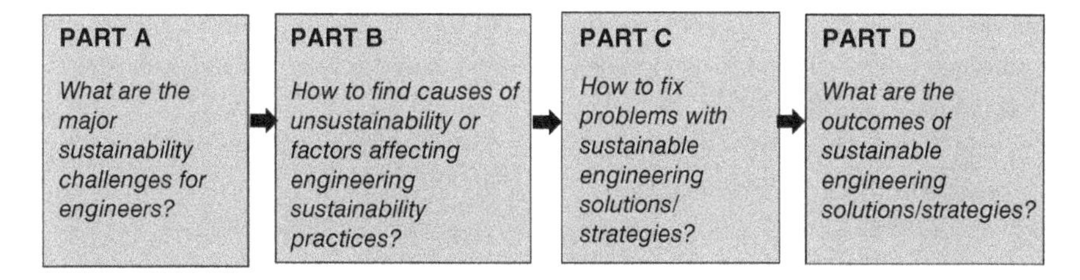

Figure 1.6 Structure of the book for addressing engineering sustainability challenges

of renewable energy, energy efficiency and clean conventional technology. Then, the principles and synergies of solid, gaseous, and by-product Industrial Symbiosis and their TBL implications were discussed. Finally, there is a chapter on engineering technology and innovation, particularly focusing on the environmental, social, and economic implications of recent or cutting edge technologies or concepts.

This flow chart (Figure 1.6) shows the link between consecutive parts of this proposed book.

References

Alonso, E., Sherman, A.M., Wallington, J.T. et al. (2012). Evaluating rare earth element availability: a case with revolutionary demand from clean technologies. *Environmental Science & Technology* 46 (6): 3406–3414. https://doi.org/10.1021/es203518d.

Australia Trade and Investment Commission (2017). Microgrids, smart grids and energy storage solutions. Australian Government, Sydney NSW 2001.

van Beers, D., Corder, G., Bossilkov, A., and van Berkel, R. (2007). Industrial Symbiosis in the Australian minerals industry the cases of Kwinana and Gladstone. *Journal of Industrial Ecology* 11 (1): 35–72.

van Berkel, R. (2000). Cleaner production in Australia: revolutionary strategy or incremental Tool? *Australian Journal of Environmental Management* 7 (3): 132–146. https://doi.org/10.1080/14486563.2000.10648495.

Biswas, W.K. (2014). Carbon footprint and embodied energy consumption assessment of building construction works in Western Australia. *International Journal of Sustainable Built Environment* 3: 179–186.

Biswas, W.K. and Leslie, G. (2007). Technologies for safe water supply in arsenic affected villages of Bangladesh utilizing a pedal pump. *AIP Conference Proceedings* 941: 218–227.

Biswas, W. and Rosano, M. (2011). A life cycle greenhouse gas assessment of remanufactured refrigeration and air conditioning compressors. *International Journal of Sustainable Manufacturing* 2 (2/3): 222–236.

Biswas, W.K., Bryce, P., and Diesendorf, M. (2001). Model for empowering rural poor through renewable energy technologies in Bangladesh. *Environmental Science and Policy* 4 (6): 333–344.

de Boer, M.A. and Lammertsma, K. (2013). Scarcity of rare earth elements. *ChemSusChem* 6 (11): 2045–2055.

Boyle, C. (2004). Considerations on educating engineers in sustainability. *International Journal of Sustainability in Higher Education* 5 (2): 147–155.

Bradshaw, A.A., Reuter, B., and Hamacher, T. (2010). The potential scarcity of rare elements for the Energiewende. *Green* 392: 93–111.

Brundtland, G.H. (1987). *Our common future.* World Commission on Environment and Development. Oxford University Press.

Bryce, P. and Bryce, D. (1998). Small-scale village electrification; an NGO perspective. Presented at *Village Power 98 Scaling Up Electricity Access for Sustainable Rural Development*, Washington, D.C., October 6–8.

Cancer Council (2020). Asbestos. www.cancer.org.au/cancer-information/causes-and-prevention/workplace-cancer/asbestos (accessed 5 July 2020).

Carson, R. (1962). *Silent Spring*. Houghton Mifflin Company.

Cavallo, E., Michelini, R.C., and Molfino, R.M. (2004). A remote operated robotic platform for underwater decommissioning tasks. 35th International Symposium on Robotics ISR 2004, Paris, 23–26 March.

Church, C. and Wuennenberg, L. (2019). Sustainability and second Life: the case for cobalt and lithium recycling. The International Institute for Sustainable Development, Canada.

Cilluffo, A. and Ruiz, N.G. (2019). World's population is projected to nearly stop growing by the end of the century. Pew Research Center, USA.

Cleveland, C.J. and Stern, D.I. (1998). Indicators of natural resource scarcity: review, synthesis, and application to US agriculture. In: *Theory and Implementation of Economic Models for Sustainable Development*, Economy & Environment, vol. 15 (ed. J.C.J.M. van den Bergh and M.W. Hofkes). Dordrecht: Springer.

Climate and Health Alliance (2015). Coal and health in the Hunter: Lessons from one valley for the world. https://d3n8a8pro7vhmx.cloudfront.net/caha/legacy_url/53/Climate-and-Health-Alliance_Report_Layout_PRINTv2.pdf?1439938112 (accessed 10 July 2020).

Coren, M. (2018). Nine countries say they'll ban internal combustion engines. So far, it's just words. Quartz. https://qz.com/1341155/nine-countries-say-they-will-ban-internal-combustion-engines-none-have-a-law-to-do-so (accessed 25 June 2020).

Daly, H. (2002). Sustainable development: definitions, principles, policies. Invited Address, World Bank, April 30, 2002, Washington, DC.

Daly, H.E. and Farley, J. (2004). *Ecological Economics – Principles and Applications.* Island Press.

Diesendorf, M. (2001). Models of sustainability and sustainable development. *International Journal of Agricultural Resources, Governance and Ecology* 1: 109–123. https://doi.org/10.1504/ijarge.2001.000007.

Dutton, J.A. (2017). Products of combustion. https://www.e-education.psu.edu/egee102/node/1951 (accessed 6 August 2020).

Ellen MacArthur Foundation (2014). Towards circular economy. https://www.open.edu/openlearn/ocw/pluginfile.php/1142963/mod_resource/content/1/Reading%209_%20EMF%20circular%20economy.pdf (accessed 30 June 2020).

El-Zein, A. (2013). As engineers, we must consider the ethical implications of our work. *The Guardian*. 13 December.

Energy Central (2018). The recycling challenge of solar power. https://energycentral.com/c/um/recycling-challenge-solar-power (accessed 30 August 2020).

Engineers Without Borders Australia (2018). Water. Sanitation. Hygiene. https://ewb.org.au/project/wash (accessed 15 May 2020).

European Commission (2020). Global demand for resources. https://knowledge4policy.ec.europa.eu/foresight/topic/aggravating-resource-scarcity/global-demand-resources-materials_en (accessed 10 March 2020).

Ford, M. (2016). *Rise of Robots: Technology and the Threat of a Jobless Future*. New York: Basic Books.

FW (Fashioning Forward) (2020). Training, education to prepare Bangladesh workers for RMG automation. Fashion world, 30 January 2020.

GFN (The Global Footprint Network) (2019). Earth Overshoot Day 2019 is July 29th, the earliest ever. https://www.footprintnetwork.org/2019/06/26/press-release-june-2019-earth-overshoot-day (accessed 2 March 2020).

Holdren, J.P. (1993). A brief history of "IPAT" (Impact = Population × Affluence × Technology). The Millennium Alliance for Humanity and the Biosphere. Stanford University. https://www.mdpi.com/2313-4321/1/1/122/htm (accessed 14 March 2020).

Ingelson, A., Holden, W., and Bravante, M. (2009). Philippine environmental impact assessment, mining and genuine development. *Law, Environment and Development Journal* 5 (1): 1. http://www.lead-journal.org/content/09001.pdf.

International Energy Agency (2020). Chemicals. https://www.iea.org/fuels-and-technologies/chemicals (accessed 15 August 2020).

Katwala, A. (2018). The spiralling environmental cost of our lithium battery addiction. www.wired.co.uk/article/lithium-batteries-environment-impact (accessed 25 July 2020).

Light, D. (2009). The role of ICT in enhancing education in developing countries: findings from an evaluation of the Intel teach essentials course in India, Turkey, and Chile. *Journal of Education for International Development* 4 (2): 1–15.

Lim, C., Biswas, W., and Samyudia, Y. (2015). Review of existing sustainability assessment methods for Malaysian palm oil production. *Procedia CIRP* 26: 13–18.

Lin, P. (2012). Introduction to robot ethics. In: *Robot Ethics: The Ethical and Social Implications of Robotics* (ed. P. Lin, K. Abney and G.A. Bekey), 3–16. London: MIT Press.

Makridakis, S. (2017). The forthcoming artificial intelligence (AI) revolution: its impact on society and firms. Working Paper January 2017.

Meadows, D.H., Randers, J., and Meadows, D.L. (2004). *The Limits to Growth: The 30-Year Update*. White River Junction, Vt: Chelsea Green Pub. Co.

Messenger, S. (2020). New Formula Values Earth at $5,000,000,000,000,000. Treehugger and Dotdash. https://www.treehugger.com/new-formula-values-earth-at-4858691 (accessed 12 March 2020).

Mihelcic, J.R., Crittenden, J.C., Small, M.J. et al. (2003). Sustainability science and engineering: the emergence of a new metadiscipline. *Environmental Science & Technology* 37: 5314–5324.

Mishima, K., Rosano, M., Mishima, N., and Nishimura, H. (2016). End-of-life strategies for used mobile phones using material flow Modelling. *Recycling* 1 (1): 122–135. https://doi.org/10.3390/recycling1010122.

Pilbara Minerals Ltd. (2018). Pilgangoora – the world's leading lithium development project. www.asx.com.au/seminars/ceo-connect/documents/ken-brinsden-pilbara.pdf (accessed 20 August 2020).

Raj, S., Goel, S., Sharma, M., and Singh, A. (2012). Ecological footprint score in university students of an Indian city. *Journal of Environmental and Occupational Science* 1 (1): 23–26.

Reuters Events (2010). Brent spar: battle that launched modern activism. https://www.reutersevents.com/sustainability/business-strategy/brent-spar-battle-launched-modern-activism

Richardson, R. (2020). *3D Printing of Medical Equipment Can Help in the Pandemic – But Is Only a Stopgap*. Philadelphia, Washington, DC: The Pew Charitable Trusts.

Rotman, D. (2014). Technology and inequality. MIT Technology Review.

Rotman, D. (2017). Cuyahoga river fire. https://clevelandhistorical.org/items/show/63

Sahara Forest Project (2020). Technologies, Oslo, Norway.

Schumacher, E.F. (1973). *Small Is Beautiful: Economics as if People Mattered*. New York: Harper & Row. Chicago.

Shaikh, F.U., Nath, P., Hosan, A. et al. (2019). Sustainability assessment of recycled aggregates concrete mixes containing industrial by-products. *Materials Today Sustainability online* 1–1.

Simms, A., Hamilton, S., and Biswas, W.K. (2017). Carbon footprint assessment of Western Australian groundwater recycling scheme. *Environmental Management* 59: 557–570.

SkyJuice Foundation (2020). Sustainable water solutions for humanitarian & disaster relief. NSW, Australia.

Sykes, J. (2020). Is Solar still worth it in Australia in 2020? Solar choice.

Smith, M., Hargroves, K., Paten, C., and Palousis, N. (2007). *Engineering Sustainable Solutions Program: Critical Literacies Portfolio - Principles and Practices in Sustainable Development for the Engineering and Built Environment Professions.* The Natural Edge Project (TNEP), Australia.

The Guardian (2015). How developing countries are paying a high price for the global mineral boom. https://www.theguardian.com/global-development/2015/aug/15/developing-countries-high-price-global-mineral-boom

The Guardian (2017). Biologists think 50% of species will be facing extinction by the end of the century. https://www.theguardian.com/environment/2017/feb/25/half-all-species-extinct-end-century-vatican-conference

The New York Times (2020). Roundup maker to pay $10 billion to settle cancer suits. https://www.nytimes.com/2020/06/24/business/roundup-settlement-lawsuits.html

The Washington Post (2016). Studying Chinese pollution unraveled the mystery of London's lethal Great Smog.

United Nation (1992). Agenda 21. The United Nations Conference on Environment and Development (UNCED), Rio de Janeiro, Brazil.

United Nation (1999). Changing consumption and production patterns in developed and developing countries discussed in commission on sustainable development. Environmental Issues and Sustainable Development. https://www.un.org/press/en/1999/19990423.ENDEV509.html

United Nation (2002). World Summit on Sustainable Development (WSSD), Earth Summit Johannesburg.

United Nation (2018) 68% of the world population projected to live in urban areas by 2050, says UN. Department of Economic and Social Affairs. UN.

United Nations (2020). Populations. Peace, dignity and equality on a healthy planet. https://www.un.org/en/sections/issues-depth/population

UNSW (2018). What's ahead for Professor Michelle Simmons, Australian of the Year? https://www.inside.unsw.edu.au/news-unsw/whats-ahead-professor-michelle-simmons-australian-the-year

Vandermaesen, T., Humphries, R., Wackernagel, M. et al. (2019) EU Overshoot Day Living beyond limits, World Wide Fund For Nature (formerly World Wildlife Yalcinkaaya 2019). Volvo creates Living Seawall to combat pollution and promote biodiversity.

WBCSD (2006). Eco-efficiency learning module. http://wbcsdservers.org/wbcsdpublications/cd_files/datas/capacity_building/education/pdf/EfficiencyLearningModule.pdf

Whittingham, M.S. (2011). Materials challenges facing electrical energy storage. Published online by Cambridge University Press: 31 January.

Woodside Energy Ltd (1999). Conservation agreement. Commonwealth of Australia.

World Bank (2013). Harnessing the transformative power of technology to end poverty. Discussions paper. 5th October.

World Bank (2018). Nearly half the world lives on less than $5.50 a day. Press Release No: 2019/044/DEC-GPV. Volvo NSW.

World Nuclear Association (2020). Chernobyl accident 1986. https://www.world-nuclear.org/information-library/safety-and-security/safety-of-plants/chernobyl-accident.aspx

Yalcinkaya, G. (2019). Volvo creates Living Seawall to combat pollution and promote biodiversity. Autocad Design Pro.

Part II

Sustainability Assessment Tools

2

Quantifying Sustainability – Triple Bottom Line Assessment

2.1 Introduction

In addressing sustainability challenges, engineers are required to focus on what is described as the 'Triple Bottom Line' of sustainable development, including environmental equality, social justice, and economic prosperity. The design of any new technology or engineering innovation should consider the minimisation of environmental impacts during the life cycle from cradle to cradle, cost-competitiveness, business development potential, and other socio-economic issues such as affordability, accessibility, acceptability, employment creation, and community empowerment. The triple bottom line (TBL) is typically performed using a sustainability accounting framework that differs from traditional reporting frameworks as it includes ecological (or environmental) and social measures that require a systematic procedure to assign appropriate means of measurement (Slaper and Hall 2011).

TBL enables engineers to report how they are responding to sustainability in the development, operational, and management phases of engineering projects.

Applying a TBL lens enables multiple community objectives to be fulfilled and this adds value to the project. Engineers will be able to monitor and evaluate their actions, thereby improving technological design to address sustainability challenges. They will also be able to compare the sustainability performance of their design, innovation, and services with the conventional designs. Apart from considering TBL in the design and operational stages of engineering projects, this knowledge will assist engineers in the preparation of sustainability reports for their organisation and the development of sustainable project proposals.

Engineering for Sustainable Development: Theory and Practice, First Edition.
Wahidul K. Biswas and Michele John.

2.2 Triple Bottom Line

2.2.1 The Economic Bottom Line

Economic variables ought to be variables that deal with the flow of money. In the case of engineering projects, it mainly deals with capital and operational costs of technology, replacement costs of the main components, and the benefits associated with the production and use of technologies, products, and services. Important parameters that are often calculated in any engineering design project for assessing the economic viability or feasibility of engineering projects are life cycle cost, benefit cost ratio, internal rate of return, and the payback period. This information is useful for any investor prior to the financing of any engineering projects. Investors are concerned with the environmental implications of the project due to the increasing emergence of environmental laws and regulations. Engineers need to find solutions to reduce environmental costs to enhance the long-term financial feasibility of the project. Some other related indicators are personal income, cost of underemployment, establishment cost, establishment sizes, investment capacity, and limit. These indicators can help engineers to make sure that engineering projects are able to create activities for more employment and wage increases. Life cycle cost (LCC) analysis provides a significantly better assessment of the long-term cost-effectiveness of a project than alternative economic methods that focus only on first costs or on operating-related costs in the short run. A light emitting diode may cost twice as much as the compact florescent lamp, but the amount of electricity that can be saved by the former will be significantly higher than the incremental cost. Similarly, the cost of a front loader washing machine is about twice as much as the cost of the top loader, but reduces the water consumption by about 50%.

2.2.2 Environmental Bottom Line

The Environmental Bottom Line is the outcome of assessing the degradation of natural resources and environmental impacts made during the life cycle of products or services. The environmental performance is based on air and water quality, energy consumption, generation of solid and toxic wastes, natural resource depletion, and the loss of biodiversity associated with typical engineering activities, including mining, processing and manufacturing. The long-range trends available for each of the environmental variables will help engineers to identify the impacts resulting from a project or engineering design. For example, the use stage of buildings or cars accounts for a significant proportion of environmental impacts compared to the initial manufacturing or construction stages. Engineers are able to understand the trade-off between the short-term profits from 'dirty' technologies and the long-term benefits of a clean environment. Therefore, environmental life cycle assessment is usually conducted to estimate the environmental impacts of products or services. Some examples of environmental impacts that are estimated during the life cycle of a product or service are the global warming potential, eutrophication, water depletion, photochemical smog, ozone layer depletion, land use changes, ecological diversity, human toxicity, terrestrial eco-toxicity, freshwater eco-toxicity and marine eco-toxicity, ionising radiation, abiotic depletion, ozone depletion, acidification, and respiratory inorganics. The variation of inputs in the form of chemicals and energy used during the product life cycle results in a variation in emissions and a variation in emissions means a variation in impacts. These impacts vary from product to product, service to service, region to region due to variations in inputs, socio-economic processes, energy mix, resource availability, and climatic condition. For example, the energy mix in Australia (i.e. 39% oil, 30% coal, 25% natural gas and 6% electricity) (Australian Government 2019) is different to the energy mix in Denmark (37% oil, 15% natural gas, 13% coal, 2% waste, non-renewable and 33% renewable), which results in a variation in environmental emissions associated with the production of the same product in two different countries. The domestic energy use pattern in Thailand will be different to the energy use pattern in Siberia due to variations in temperature in these two different climatic zones.

2.2.3 The Social Bottom Line

Social variables refer to social dimensions of a community or region that involve the measurement of social indicators, including education, access to social resources, affordability, accessibility, acceptability, empowerment, poverty alleviation, health and well-being, quality of life, and social capital (e.g. good will, business reputation, interpersonal relationship). Further examples include unemployment rate, female labour force participation rate, median household income, relative poverty, percentage of population with a post-secondary degree or certificate, average commute time, violent crimes per capita, and health-adjusted life expectancy. The social indicators could vary from product to product or service to service due to the variation in stakeholders, nature of work, community, and the type activities in the surrounding areas. Different social indicators of a product will be applicable to different stakeholders directly or indirectly involved in the product supply chain. Stakeholder selection is a critical issue. Stakeholders should have their interests in and expectations from the project and have ability to influence the project either positively or negatively (e.g. producers, legislative body). Also, social licence needs take into account how their livelihoods could be impacted by the project either positively or negatively (e.g. workers, surrounding community, businesses) (IUCN 2016).

2.3 Characteristics of Indicators

TBL indicators should be able to represent the characteristics of products or services. The indicators that will be developed for assessing the TBL performance of a product or service should be able to confirm if an engineering solution or design is socially, economically, and environmentally performing well. Indicators can be considered as key parameters in a system that should be monitored, measured, and analysed to assess the sustainability status and performance. TBL indicators are able to identify areas (e.g. chemicals, processes, management, policy, price, workforce, consumer), where the links between the economy, environment and society are weak. They allow engineers to find where the problem areas are and help find technical solutions or to redesign the product/service (e.g. renewable energy, green chemicals, recycling, energy efficiency) to come up with a sustainable engineering solution.

An indicator is specifically developed for a particular product or service and it should be able to show changes or progress of an engineering action, innovation or design towards achieving a specific outcome. It should be able to be consistently measurable over time, in the same way by different observers. It should be precise and operationally defined in clear terms. It should be timely by providing a measurement at time intervals relevant and appropriate in terms of programme goals and activities.

Many indicators in common use are not well-defined in clear terms, or at least include terminology that could be improved to add greater understanding. For instance, this can be achieved by providing clear units of indicators (e.g. MWh of electricity used/ton of iron ore production/year) (UN Women 2010). The more defined an indicator, the less room there will be for later confusion or complications. Ideal indicators may not be practical, the feasibility of using certain indicators can be constrained by the availability of data and financial and human resources (e.g. number of fish killed per year due to discharge of cooling water to a nearby river).

The cost of collecting appropriate data for ideal indicators is often prohibitive. These indicators might require collecting data to calculate an unknown denominator, or national data to compare with local area data, or tracking lifetime statistics for an affected and/or control population. Human resources and technical skills may also be a constraint in determining ideal indicators involving high tech laboratory equipment and highly paid scientists. When quantitative indicators of success cannot be identified, qualitative methods offer a valuable alternative. It is sometimes difficult or not possible to measure the quantitative values of 'benefits' or 'risks', but it is almost always possible to gather qualitative data, such as information on the perspectives of workers and community, who are directly or indirectly affected by the production of a product or service. In many cases, qualitative indicators provide more relevant information with respect to the success and effectiveness of the intervention.

There are two types of indicators:

- Tangible indicators, which can be measured and stated in term of numbers (e.g. kg of CO_2 per kWh of electricity generation). Any tangible indicator must have a numerator and a denominator (e.g. Giga litre of water consumed per ton of gold production).
- Intangible indicators are not measurable, but they can be felt (i.e. level of satisfaction, agree or disagree, fair or unfair, etc.) and they have influence on the system. Qualitative feedback (i.e. level of satisfaction) that are used for calculating intangible indicators are typically converted to numerical values using a Likert scale.

2.4 How Do You Develop an Indicator?

The following examples in Box 2.1 show how engineers can develop indicators by reviewing case studies/circumstances of the use of products/services.

Box 2.1 Development of TBL indicators

Social indicator

Step 1: Review the case study: You have information on social aspects, which says that 360 jobs were created over a period of 10 years by developing the market for biofuels.
Step 2: Development of relevant indicators (e.g. Job creation).
Step 3: Development of a unit for each indicator (e.g. Number of jobs/year).
Step 4: Calculate the value of an indicator by using the information. In this case, the indicator value is 36 jobs created/year.

Environmental indicator

Step 1: Review the case study (for example, you have an information on the environmental impact, which says that about 100 tons of CO_2 e- greenhouse gases were emitted over a period of 10 years to produce 10 tons of metal A).
Step 2: Development of relevant indicators (e.g. carbon footprint).
Step 3: Development of a unit for each indicator (e.g. tons of CO_2/year/ton).
Step 4: Calculate the value of an indicator by using the information matrix. In this case, the indicator value is 1 ton CO_2 e-/year/ton.

(Continued)

Box 2.1 (Continued)

Economic indicator

Step 1: Review the case study (for example, you have an information on the economic impact, which says that the capital cost was $100K and the yearly profit is $50K/year due to the generation of electricity from wind mills.

Step 2: Development of relevant indicators (e.g. payback period).

Step 3: Development of unit for each indicator (e.g. year).

Step 4: Calculate the value of an indicator by using the information matrix. In this case, the indicator value is 2 years.

2.5 Selection of Indicators

As mentioned before, the TBL indicators vary with products as the materials, energy, costs, management, stakeholders, skills, emissions, and wastes associated with the production of products vary. Also, the indicators for the same product will vary with location and country of production as the socio-economic situation, resource utilisation pattern, energy mix, and availability of resources (i.e. transportation) vary with location. Indicators need to be selected that are relevant to the product or service. The consideration of a large number of indicators for assessing a product will not only complicate the assessment process but it will make the assessment lengthier and more expensive. Therefore, a considerable amount of effort should be made in selecting the most appropriate indicators that also characterise sustainability well.

2.6 Participatory Approaches in Indicator Development

A participatory approach involves a consensus survey that can be considered face to face, online, or via postal survey to engage with stakeholders and area experts in the selection of indicators for improving the creditability of the sustainability assessment. Stakeholders who are directly and indirectly involved in the development, use, and research of the product are given an opportunity to give their opinion in the selection and development of the indicators. Since these stakeholders will ultimately be directly or indirectly affected by the development or use of the product or service, it is important to get their opinion regarding the relevance and importance of the indicators. Stakeholder engagement is an important means for developing sustainability indicators in terms of social learning ethical and moral perspectives (Mathur et al. 2008). A structured method for developing indicators is described below.

2.7 Description of Steps for Indicator Development

2.7.1 Step 1: Preliminary Selection of Indicators

Prior to engaging stakeholders in the consultation process, a thorough literature review of international journals, government documents, and reports published by recognised organisations (e.g. World Bank, Dell Computers) and a large number of industry sustainability reports need to be carried out to preliminarily select the social, economic, and environmental indicators. The preliminary selection of an indicator focused questionnaire can provide a structured and organised guideline for the survey.

2.7.2 Step 2: Questionnaire Design and Development

A questionnaire for the survey should be designed based on the preliminarily selected indicators to collect feedback and opinions from the stakeholders and area experts on the relevance and importance of these indicators. Firstly, the respondents need to be asked as to whether the indicator is relevant for the sustainability assessment of a product or service. If the response is 'yes', the respondent needs to tell how important is the indicator is (e.g. very important, important, moderately important, and somewhat important). If the response is 'no', then the respondent needs to provide a brief reasoning for the irrelevance. At the end of the questionnaire, when the respondent has already provided feedback on the preliminarily selected indicators, there is a provision for respondents to suggest additional indicators. This is an optional task for the respondents. If the respondents provide additional indicator(s), then they need to provide reasoning and understanding the level of importance of the indicator(s) suggested.

2.7.3 Step 3: Online Survey Development

Considering their diverse backgrounds, the participants may be contacted by their preferred survey mode (i.e. hard copy, face to face conversation, or online survey). In some instances, people in the remote areas may have limited access to the internet and so an online survey may not be convenient for them. In other cases, an online questionnaire, using Google forms, can be an appropriate mode for the survey.

2.7.4 Step 4: Participant Selection

It is also important to identify the right participants in the indicator selection process. In an 'expert consensus' survey, participants or stakeholders should have significant experience in the field, some influence on decision-making or have authority for judgement in the production process, or have a sound research track record in the area (Linstone and Turoff 1975).

Criteria for the selection of the right participants are as follows (Lim and Biswas 2018):

- Individual who is directly/indirectly involved with the production
- Individual who is affected by the production
- Individual who is the beneficiary of this production
- Individual who incurs losses from the production
- Individual who has authority to influence, control, and terminate the production
- Individual with subject matter expertise

Additional feedback from the participants, including suggestions for new indicators can also be compiled for evaluation and consideration.

2.7.5 Step 5: Final Selection of Indicators and Calculation of Their Weights

The final selection of indicators is based on following criteria (Lim and Biswas 2018; Janjua et al. 2020):

- the indicators receiving votes as 'relevant' by more than 50% of the participants; and
- the additional indicators provided by the participants are compared with existing indicators to see if they are overlapping. The researcher or sustainability assessment practitioner also need to make sure that these suggested indicators are relevant for the assessment of the product or service.

These indicators are also known as performance measures. The weight of each performance measure (PM) was calculated using the following equations.

The total point value of each PM was calculated using Eq. (2.1).

$$W_j = n_{j1} * 1 + n_{j2} * 2 + n_{j3} * 3 + \ldots + n_{j5} * M \quad (2.1)$$

where

j = 1, 2, 3, …, M, is the indicator,
n_{j1} = number of 'least responses' for indicator of j,
n_{j2} = number of 'somewhat important' responses for indicator of j,
n_{j3} = number of 'moderately important' responses for indicator of j,
n_{j4} = number of 'important' responses for indicator of j,
n_{j5} = number of 'very important' responses for indicator of j.

Total weight for M number of PMs has been calculated as follows:

$$W_{\text{Total}} = \sum_{j=1}^{M} W_j \quad (2.2)$$

The weight (W'_j) for each PM has been calculated as follows:

$$W'_j = \frac{W_j}{W_{\text{Total}}} \quad (2.3)$$

Box 2.2 gives an example for calculating the weights of PMs.

Box 2.2 Example for calculating the weight of PMs (W'_j)

Suppose you have asked experts to rank five PMs in below Table from 1 to 5, in which the most important impact is marked as 5 and the least important as 1. After conducting this survey, you have obtained the data that is presented in below Table.

Ranks provided by the experts in the survey

Indicators	Expert 1	Expert 2	Expert 3	Expert 4	Expert 5
PM 1	5	4	3	4	4
PM 2	3	5	4	3	5
PM 3	1	3	1	5	1
PM 4	2	1	2	2	3
PM 5	4	2	5	1	2

Calculate the weight of each PM

Below Table shows how W_j, W_{Total}, and W'_j have been calculated using Eqs. (2.1–2.3), respectively.

Calculation of W_j, W_{Total}, and W'_j

	Expert 1	Expert 2	Expert 3	Expert 4	Expert 5	W_j	W'_j
PM 1	5	4	3	4	4	20	0.27
PM 2	3	5	4	3	5	20	0.27
PM 3	1	3	1	5	1	11	0.15
PM 4	2	1	2	2	3	10	0.13
PM 5	4	2	5	1	2	14	0.19
					W_{Total}	75	1.00

Once the PMs have been assigned weights, their values are determined for the product or service. Some PMs are quantitative (e.g. global warming impact [kg of CO_2/ton of iron ore production], net benefit [$/kWh of electricity production]), and they are determined from the amount of chemicals and energy used and cost involved in different processes. Other qualitative PMs depend on stakeholders judgements such as level of the satisfaction or agreement.

In the case of qualitative PMs, they are typically ranked from 1 to 5 or 1 to 7 where level 5 or 7 is considered as a threshold value/expected performance. The gap would be the difference between the rank of a PM and 5 or 7. The highest value can go much bigger (i.e. >7), but it depends on what level of preciseness is required for assessing the sustainability performance. A threshold value is defined as the targeted sustainability performance. The assessment framework would measure how close the existing performance of a product is to the threshold values for a particular PM to present sustainability. For example, about 750 kg of CO_2 equivalent greenhouse gas (GHG) is emitted due to the production of 1 MWh of electricity from a particular fuel. The lowest possible value is 250 kg CO_2 and the highest possible value is 1250 kg CO_2. On a 1–5 Likert scale, the position of this particular fuel will be 3 (i.e. 250, 500, 750, 1000, and 1250 kg CO_2 will correspond to points 5, 4, 3, 2, and 1 on a Likert scale).

Gap G, is the difference between the position value and the threshold value of each PM. The gap is multiplied by the corresponding weight of the PM to determine the score of each PM. The highest gap for an PM identifies the hotspot requiring the most improvement.

In the case of qualitative PMs, the response in terms of level of satisfaction or agreement are compared with the highest value on the scale. If the level of satisfaction by a respondent on a 1–7 Likert scale is 5 and the highest value is 7, then the gap is '−2' (or 5–7) for a particular PM. The average response of all participants completing the questionnaire for that particular PM is then multiplied by the corresponding weight determined by the experts/participants to determine the score of PMs. The largest gaps identify the most prominent impacts. Box 2.3 gives an example for calculating the gaps for PMs.

Box 2.3 Example of calculating the scores of PMs

Once you have ascertained the weights of these PMs in Box 2.2, you then conduct the survey of the people who are involved in the product supply chain as well as affected by the production, including management, workers, local community, government officials, and researchers. Using a scale from 1 to 7, where 1 means totally disagree and 7 means totally agree, you have asked respondents how they would rate these impact categories. The responses that you had received from 5 respondents (R) are presented in the below table.

Responses provided by the respondents in the product supply chain

	R1	R2	R3	R4	R5
PM 1	3	2	4	3	5
PM 2	6	5	5	7	5
PM 3	3	2	4	5	1
PM 4	6	4	5	4	5
PM 5	3	2	4	3	5

Calculate the score for each PM

Following table shows the average gaps of PMs resulting from the response of respondents 1, 2, 3, 4, and 5. The average gap is then multiplied by the corresponding weights in Box 2.2 to determine the score of PMs.

Calculation of score of PMs

	R1	R2	R3	R4	R5	Average gap	Score
PM 1	−4	−5	−3	−4	−2	−3.6	−1.0
PM 2	−1	−2	−2	0	−2	−1.4	−0.4
PM 3	−4	−5	−3	−2	−6	−4	−0.6
PM 4	−1	−3	−2	−3	−2	−2.2	−0.3
PM 5	−4	−5	−3	−4	−2	−3.6	−0.7

2.8 Sustainability Assessment Framework

In a sustainability assessment framework, the overall sustainability assessment is segregated into three sustainability objectives i.e. environment, economy, and social. Each TBL objective consists of a number of headline performance indicators (HPIs), which form the highest aggregation level for the performance measurement against sustainability objectives. Each HPI is the aggregation of key performance indicators (KPIs), which further describe the key impact areas of each HPI for the product or service. The performance measures (PMs), which are the lowest level of aggregation, are established to provide quantitative values that could either directly or indirectly affect both KPIs and HPIs. The smallest units or PMs are calculated using the procedure discussed in Section 2.7.5 (i.e. Box 2.3).

The following formulae are used to calculate the KPI, HPI, and overall sustainability performance score (Lim and Biswas 2018).

- The performance of a KPI = the average of ranking value of PMs related to the KPI.

$$\text{Performance of KPI}\, j = \frac{\sum_{i=1}^{i=I} \text{PM}_{ij}}{\sum_{i=1}^{i=I} W'_{ij}} \tag{2.4}$$

  where PM_{ij} refers to ith PM for KPI_j, I = total number of PMs of KPI;
- Performance of the HPI = average performance of KPIs under this HPI.

$$\text{Performance of HPI}_k = \frac{\sum_{j=1}^{j=J} \text{KPI}\, jk}{J} \tag{2.5}$$

  where KPI jk refers to jth KPI for HPI_k, and $j = 1,2,3, \ldots, J$; J = total number of KPI of HPI_k.
- Performance of each TBL sustainability objective = the average of ranking values of HPIs under this objective.

$$\text{Performance of TBL objective} = \frac{\sum_{i=1}^{i=K} \text{HPI}_k}{K} \tag{2.6}$$

  where HPI k refers to k type of HPI for each TBL sustainability objective, $k = 1, 2, \ldots, K$; K = total number of HPI.
- Overall sustainability performance = the average performance of TBL objectives

$$\text{Overall sustainability performance} = \frac{\sum \text{TBL objectives}}{3} \tag{2.7}$$

where the TBL objectives refer to the ranking values for environment, economic, and social sustainability objectives.

Tables 2.1 and 2.2 show a hypothetical example of the TBL sustainability assessment for biofuel production.

This involves two steps: an expert survey (Table 2.1) and a stakeholder survey (Table 2.2).

2.8.1 Expert Survey

As can be seen in Table 2.1, the expert responses in terms of the importance of the PMs are converted to weights using Eqs. (2.1–2.3). The scale which is used in this expert survey is a 1–4 point scale, where, 1, 2, 3, and 4 represent somewhat important, moderately important, important, and very important PM references, respectively.

Table 2.1 Scores provided by experts to calculate W_j, W_{Total}, and W'_j

KPI	PM	Expert 1	Expert 2	Expert 3	Expert 4	Expert 5	W_j	W'_j
Human rights	Free from the employment of child labour	2	1	3	4	4	14	0.061
	Equal opportunities	1	2	3	4	2	12	0.052
Working conditions	Fair salary	3	2	3	1	2	11	0.048
	Decent working hours	4	1	3	3	1	12	0.052
Cultural heritage	Land acquisition	1	2	3	4	2	12	0.052
	Community engagement	3	4	3	4	2	16	0.070
Governance	Fair competition	2	2	3	4	1	12	0.052
	Free from corruption	1	2	3	4	2	12	0.052
Productivity efficiency	Plantation yield	2	2	3	1	2	10	0.044
	Mill production efficiency	1	2	1	4	2	10	0.044
Consistent profitability	Actual growth rate	4	2	3	3	3	15	0.066
Biodiversity	Plantation practice	1	2	2	4	2	11	0.048
	Land use	3	2	3	4	3	15	0.066
	Species loss	2	2	3	3	2	12	0.052
Climate change	GHG emission	4	2	3	4	4	17	0.074
Air pollution	NO_x emission intensity from palm oil mill	3	4	3	4	2	16	0.070
Eco-toxicity	BOD of water	1	2	1	1	2	7	0.031
	Soil nitrate level	3	3	3	4	2	15	0.066
						W_{Total}	229	1.00

Table 2.2 Calculated values of scores for PMs, KPIs, HPIs, and sustainability gap

PM	Stakeholders	Average gap	Weight	Score	KPI	Score	HPI	Score	TBL	Score	Sustainability gap
Free from the employment of child labour	Workers	−2	0.061	−0.12	Human right'	−2.46	Social equity	−2.51	Social	−2.26	−2.33
Equal opportunity	Workers	−3	0.052	−0.16							
Fair salary	Workers	−1	0.048	−0.05	Working condition	−2.57					
Decent working hours	Workers	−4	0.052	−0.21							
Land acquisition	Community	−0.9	0.052	−0.05	Cultural heritage	−0.67	Social regulation	−2.01			
Community engagement	Community	−0.5	0.070	−0.03							
Fair competition	Value chain actor	−5	0.052	−0.26	Governance	−3.35					
Free from corruption	Value chain actor	−1.7	0.052	−0.09							
Plantation yield	Producer	−3.5	0.044	−0.15	Productivity efficiency	−3.00	Business continuity and resiliency	−2.45	Economic	−2.45	

Mill production efficiency	Producer	−2.5	0.044	−0.11						
Actual growth rate	Producer	−1.9	0.066	−0.12	Consistent profitability	−1.90				
Plantation practice	Regulatory body	−4.3	0.048	−0.21	Biodiversity	−1.94	Natural capital conservation	−1.94	Environmental	−2.29
Land use	Regulatory body	−0.8	0.066	−0.05						
Species loss	Regulatory body	−1.2	0.052	−0.06						
GHG emission	Producer	−3.75	0.074	−0.28	Climate change	−3.75	Air pollution	−2.38		
NO_x emission	Producer	−1	0.070	−0.07	Air quality	−1.00				
BOD of water	Producer	−4.25	0.031	−0.13	Water and soil quality	−2.55	Eco-toxicity	−2.55		
Soil nitrate	Producer	−1.75	0.066	−0.11						

For the first PM in Table 2.1 (i.e. free from the employment of child labour), 2 experts considered this PM as very important (4), 1 expert considered this as important (i.e. 3), 1 expert considered this as moderately important (i.e. 2), and the last expert considered this as somewhat important (i.e. 1). Then using Eq. (2.1), the total point value of this PM is calculated as follows:

$$W_1 = 1 \times 1 + 1 \times 2 + 1 \times 3 + 2 \times 4 = 14$$

Likewise, the total points for the 18 PMs are then calculated. The total points for these 18 PMs are added to find the total weight (W_{Total}) using Eq. (2.2). The value of the total weight is 229 (see Table 2.1).

Finally, the weights for the individual PMs are calculated using Eq. (2.3). For example, the weight of the first PM has been calculated as follows:

$$W'_1 = \frac{14}{229} = 0.061$$

It should be noted that the summation of the weights of 18 PMs should be 1.

In some cases, the weights are determined on the basis of the number of PMs considered. Since the number of PMs in this example is 18, the number that is ranked first is 18, while 1 is the lowest number. The experts ranked the PMs accordingly. The same procedure that was demonstrated in Box 2.2 can be followed to calculate the weights.

2.8.2 Stakeholders Survey

The next step is to involve stakeholders in the supply chain to obtain data in order to determine the sustainability performance Gaps of the PMs (Box 2.3). In this particular example, the gap was calculated by comparing the values of PMs with the threshold value on a 1–5 point Likert scale, which means that the maximum point is 5.

For example, the gap of the 15th PM (i.e. GHG emissions) has been calculated as follows:

$$\text{Gap}_{15} = 1.25 - 5 = -3.75$$

Suppose the GHG emissions from the production of biofuel in this case study is 50 kg CO_2/GJ but the best (or threshold value) and worst values are 20 kg CO_2/GJ and 60 kg CO_2/GJ, respectively. In this case, the PM value on a 1–5 point Likert scale can be calculated as follows:

The interval between points on the 1–5 point scale has been calculated as follows:

$$\text{Interval} = \frac{60 - 20}{5} = 8$$

The following scale has been developed on the basis of this interval.

	Best					Worst
Scale	5	4	3	2	1	0
kg CO_2/GJ	20	28	36	44	52	60

The point of 50 kg CO_2/GJ on this scale has been calculated as follows:

$$= 1 + \frac{52 - 50}{8} = 1.25$$

Once the Gaps of these PMs are determined using the procedure in Box 2.3, the value of PMs were calculated by multiplying the gap with the corresponding weights. The score for the15th PM has been calculated as follows:

$$(-3.75) \times \text{corresponding weight (in Table 2.1)} = -3.75 \times 0.074 = -0.28$$

Once the PMs for KPI are calculated, the value of KPI can be determined. For example, human rights are one of the KPIs in Table 2.2. The score of this KPI has been calculated using Eq. (2.4):

$$\text{KPI}_1 = \frac{-0.12 - 0.16}{0.061 + 0.052} = -2.46$$

Then, Eqs. (2.5)–(2.7) are used to calculate the values for the HPI, TBL objectives, and the overall sustainability score.

Using Eq. (2.5), the HPI for social equity has been calculated as follows:

$$\text{HPI}_1 = \frac{-2.46 - 2.57}{2} = -2.51$$

Using Eq. (2.6), the social objective of the TBL has been calculated as follows:

$$\text{Social objective} = \frac{-2.51 - 2.01}{2} = -2.26$$

Finally, the sustainability score has been calculated using Eq. (2.7)

$$\text{TBL} = \frac{-2.26 - 2.45 - 2.29}{3} = -2.33$$

This means that biofuel is 2.33 points away from the threshold value (i.e. 5) to reach the required level of sustainability. If we investigate further, more emphasis needs to be given to the economic objectives as the productivity efficiency has been found to be the hotspot with the larger gap (i.e. −3.00).

It should be noted that in real-life assessments, the number of respondents as well as experts are far more than the numbers considered in these hypothetical examples. The number of experts considered in some recent studies ranged from 30 to 40, but the number of respondents depends on the type of stakeholders, size of the supply chain, the nature of the production and the location of the products.

2.9 TBL Assessment for Bench Marking Purposes

The comparison of TBL indicators can be used for benchmarking purposes by calculating the ratio of values of the indicators for the product and the sector-average. Table 2.3 shows the TBL performance of 10 economic, social, and environmental indicators for soft drinks produced by a company in Norfolk island together with a typical Australian soft drink producer (Lenzen 2008). This case study shows how sustainable the soft drink company in Norfolk island is in comparison with an Australian soft drink producer. The value of the indicators in the Table 2.3 are used to develop a spider diagram to more clearly convey an overview of the TBL performance of the soft drink company in Norfolk island on 10 economic, social, and environmental indicators in one visual representation. The ratios of indicator values of these two companies were used to develop this spider diagram. While calculating these ratios, the following rules were considered.

- The Cascade soft drink is benchmarked or compared against an average Australian soft drink producer.

Table 2.3 Comparison of TBL intensities for cascade soft drinks and the average Australian soft drink producer

Indicators	Unit	Cascade soft drinks	Total sector intensity	Ratio (R)	$Log_{10}R$
Environmental bottom line					
Material flow	g/$	178	1126	0.16	−0.80
Energy consumption	MJ/$	2.27	2.58	0.88	−0.06
Greenhouse gas emission	g CO_2 -e/$	197	209	0.94	−0.03
Water use	l/$	9.67	38.5	0.25	−0.60
Land disturbance	m^2/$	0.06	0.14	0.43	−0.37
Social bottom line					
Family income	¢/$	56.1	21.7	0.39	−0.41
Employment	emp-min/$	6.7	0.59	0.09	−1.06
Economic bottom line					
Government revenue	¢/$	1.16	2.62	2.26	0.35
Gross operating surplus	¢/$	6.36	21	3.30	0.52
Total intermediate uses	¢/$	15.6	89.9	5.76	0.76

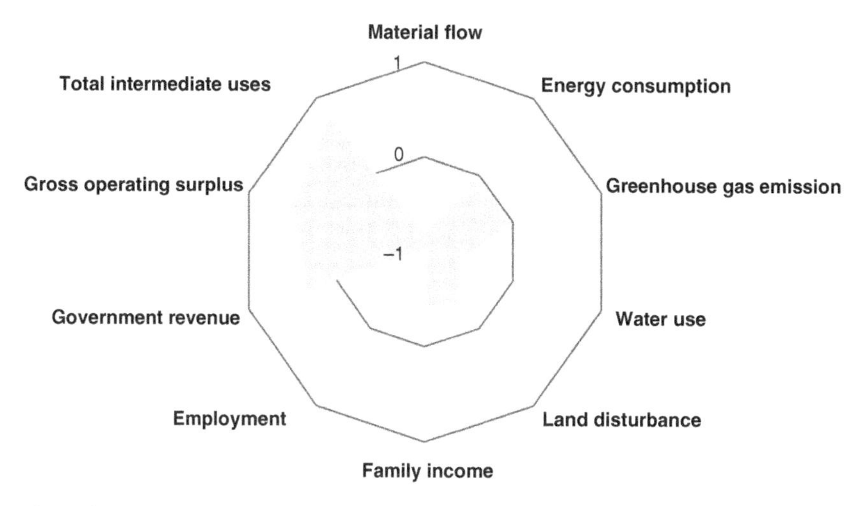

Figure 2.1 Spider diagram

- The ratios divide Cascade drink indicator values by those for an average Australian producer for negative indicators listed in Table 2.3 (less is good, e.g. GHG emissions), so that better performance leads to lower ratios.
- For positive indicators (more is good, e.g. income), these ratios have been inversed, so that once again better performance leads to lower ratios.
- In Figure 2.1, it is clear that 'dents are goods, spikes are bad'.

For this particular analysis of Cascade soft drinks, there is an overall positive TBL outcome except for three indicators due to island-specific circumstances. Within the inner polygon, a soft drink company offers better performance than an average Australian producer. On this polygon, the ratio of indicator values for these two companies are 1 (i.e. $\text{Log}_{10}\ 1 = 0$) meaning that both companies attain the same level of sustainability performance. When the ratios are <1, the Cascade soft drink performs better than the average Australian soft drink company. The centre locates 10-times-better performance than the average Australian soft drink producer, while the outer rim represents 10-times-worse performance.

2.10 Conclusions

This chapter provided a sustainability assessment framework. These frameworks will enable the calculation of an overall sustainability score and will identify the performance measures or KPIs requiring further improvement. Both qualitative and quantitative indicators and tangible and intangible indicators were discussed. The indicators vary by location, product to product, and service to service and so a procedure for selecting indicators was discussed. In addition, different indicators have different weights or levels of importance. This chapter discusses how the weights of the indicators can be developed through a consensus survey process. The indicators that were discussed in this chapter are to be calculated using a life cycle assessment approach. Chapters 3 and 4 discuss in more detail environmental life cycle assessment, social life cycle assessment, and life cycle costing for calculating environmental, social, and economic indicators, using similar methodologies to those broadly outlined in the chapter.

References

Department of Energy and Environment (2019). Australian energy update 2019. www.energy.gov.au/sites/default/files/australian_energy_statistics_2019_energy_update_report_september.pdf (accessed 5 December 2020).

IUCN (2016). Social impact assessment (SIA). https://www.iucn.org/sites/dev/files/iucn_esms_sia_guidance_note.pdf (accessed 10 December 2020).

Janjua, S., Sarker, P., and Biswas, W. (2020). Development of triple bottom line indicators for life cycle sustainability assessment of residential buildings. *Journal of Environmental Management* 264: 110476.

Lenzen, M. (2008). Sustainable island businesses: a case study of Norfolk Island. *Journal of Cleaner Production* 16: 2018–2035.

Lim, C.I. and Biswas, W.K. (2018). Development of triple bottom line indicators for sustainability assessment framework of Malaysian palm oil industry. *Clean Technologies and Environmental Policy* 20: 539–560.

Linstone, H.A. and Turoff, M. (1975). *The Delphi method: Techniques and Applications*. Reading, MA: Addison-Weshley.

Mathur, V.N., Price, A.D.F., and Austin, S. (2008). Conceptualizing stakeholder engagement in the context of sustainability and its assessment. *Construction Management & Economics* 26 (6): 601–609.

Slaper, T. and Hall, T.R.J. (2011). The Triple Bottom Line: What Is It and How Does It Work? Indiana University Kelley School of Business, Indiana Business Research Center, USA.

UN Women (2010). Indicators. https://www.endvawnow.org/en/articles/336-indicators.html (accessed 25 November 2020).

3

Life Cycle Assessment for TBL Assessment – I

3.1 Life Cycle Thinking

Life cycle thinking (LCT) is a concept that emphasises the need to understand the environmental, social, and economic impacts of a product or service over its entire life cycle. This encourages the examination of the entire product value chain – involving designers, planners, manufacturers, engineers, consumers, and recyclers – to consider from a holistic perspective the life cycle of products/services, and more specifically, to understand the inputs (including resources such as energy and water) and outputs (emissions to the environment) that result from the transformation of resources into a product, from a product to service, and from service/use to end of life disposal.

LCT holistically assesses the sustainability of products and services and can be used by decision makers in the private and public sector in the development of sustainable products, in green procurement (e.g. organic fertiliser, bio-fuels, recycled composites, solvent-free paints) and in the provision of sustainable services (e.g. solar electricity, 3D printing) (Klöpffer 2003). This can enhance the environmental supply chain by avoiding the burden shifting that can occur when we attempt to reduce environmental impacts that then result in increasing other environmental impacts (European Union 2010). This involves the minimisation of impacts at one stage of a product life cycle, or in a particular geographic region or impact category, that then helps to avoid increasing impacts in other production stage(s).

The most well-known application of LCT is environmental life cycle assessment (ELCA), normally referred to as life cycle assessment (LCA) analysis, which was developed by industrial ecologists to evaluate the environmental impacts of products and services across all phases of their life cycle (McConville 2006; UNEP 2012).

Engineering for Sustainable Development: Theory and Practice, First Edition.
Wahidul K. Biswas and Michele John.

3.2 Life Cycle Assessment

The life cycle is the consecutive and interlinked stages of a product system, from raw material acquisition or generation from natural resources right through to final disposal. LCA is also known as a sustainability assessment tool as it can help estimate environmental, economic, and social indicators using the information on materials and energy used and the emissions and wastes generated during the life cycle of a product or a service. It is the compilation and evaluation of the inputs (e.g. chemicals, electricity, costs, jobs) and outputs (e.g. emissions and wastes) and their potential impacts (e.g. global warming impact, acid rain, net benefits, social equity) for a product system throughout its life cycle. These impacts could also be termed as environmental, social, and economic indicators. For example, global warming potential and job losses can be considered either as impacts or indicators. LCA encompasses three assessment tools, including ELCA for estimating environmental impacts or indicators, social life cycle assessment (SLCA) for estimating social impacts or indicators, and life cycle cost (LCC) for estimating economic indicators.

The LCA can help identify the hotspot in production, processing during the life cycle of a product that contributes the most significant portion of the total impacts. Once this hotspot(s) have been identified, improvement strategies can be incorporated into the product life cycle to improve the sustainability performance. Secondly, it helps us to compare the sustainability performance of different versions of the same product. LCA is also known as a decision-making tool. One example is the comparison between fast food and dine-in restaurant meals, where the former consumes six times less energy, seven times less water, and generates five times less waste than the latter during the use or customer service stage (Viere et al. 2020). When students compare the overall life cycle results that take into account the mining, processing, farming, packaging, disposal of pre-restaurant waste, use and disposal stages of food production, they discover that the dine-in restaurant performs better than the fast food restaurant. LCA can be an eye opener for young engineering students in fully understanding sustainable production and consumption in engineering decision-making. Dow Corporation were unsure which of the two different technologies that they had developed for making a foam-core film was better from an environmental perspective (Helling 2015). Therefore, they had to do a LCA to determine the environmental impacts of three options, including a conventional polyethylene film and two options for a foamed-core film. The LCA results showed that option 2 performed 10–15% better than option 1 and 30% better than the current option. This LCA confirmed that both ways of making foam-core film would likely offer environmental advantages over replacing the current method, and that option 2 was the best, giving a clear decision to the

Research and Development (R&D) division that both these new options would be worth pursuing provided they are economically feasible.

LCA can help enhance engineers' critical thinking ability and enable them to challenge conventional thinking. For example, in a workshop case study, students initially believe that solar photovoltaic energy and biodiesel energy production are carbon neutral as they are considered a renewable energy. However this perception changes when LCA results show that the manufacturing of solar PV panels and biodiesel crop production systems produce emissions that also need to be managed and minimised through cleaner production and eco-efficiency strategies. In another example, in a comparison between corn based plastic (PLA) cups and polystyrene (PS) cups, students discover that the latter is more environmentally friendly than the former. Students are unable to comprehend this initially as they consider corn a renewable resource. Corn-based plastic has been found to be environmentally friendlier than polystyrene on a weight basis, but not on a volume basis. The student's perception changes when they realise that the function of the cup is to hold liquid, so space/volume matters. Since the PS cup is lighter than PLA cup, it requires less material to provide the same utility volume and therefore, the PS cup turns out to be the more environmentally friendly option. Whilst this comparative LCA outcome can change dependent on the choice of transportation and method of environmental impact estimation (Van der Harst and Potting 2013), students are able to understand that something 'natural' is not always the most environmentally friendly option, which in turn suggests a need to investigate life cycle methodologies to more accurately support critical thinking in engineering decision-making.

A product can be environmentally sustainable, but it has to be also socially and economically sustainable to become a sustainable product. The treatment of hotspots of one bottom line can either positively or negatively affect other bottom lines. For example, incorporation of mitigating strategies to improve the environmental performance can increase the cost or have negative social (e.g. affordability) and economic (e.g. there will be net loss instead of net benefit) implications. Therefore mitigation strategies are chosen in a way that environmental sustainability objectives can be attained without compromising the social and economic objectives. Therefore, LCA helps decision makers, policymakers, or inventors to choose the right option, pathway, technology or inputs to achieve sustainable solutions.

3.3 Environmental Life Cycle Assessment

This chapter discusses ELCA, its application, guidelines, the types of ELCA, uncertainty analysis, and environmental product declarations.

3.3.1 Application of ELCA

LCA is a decision-making tool (cradle to grave or cradle to cradle) to help identify processes or inputs or the stage causing the most emissions or waste during the life cycle of products or services. Engineers or researchers use LCA to find the right strategy to improve the overall sustainability performance. In medical science, body fluid tests, non-invasive scans – such as X-ray examinations, magnetic resonance imaging (MRI), ultrasound, and computer tomography (CT) are used to find the cause of diseases to help doctors to prescribe the right medicine. ELCA plays the role of a diagnostic tool in an engineering project to find the drawbacks or root causes for unsustainable technological performance, enabling engineers to innovate, implement, and bring improvements to technological design.

LCA can help improve the environmental performance of products to reduce their environmental impacts. For example, the cradle to cradle LCA was successfully used to estimate the benefits that occurred at the end of life stage of prefabricated buildings due to the reusability of materials, which reduces the space required for landfill as well as the need for additional resource requirements. Building design for disassembling and recovery could offer an 80% net energy saving and a net saving of 50% in material consumption during the life cycle of the prefabricated building (Aye et al. 2012). If this building is not prefabricated, the embodied energy of the building would have been lost to landfill and so LCA enabled the quantification of these benefits for facilitating the sustainability decision-making process.

LCA can also help in new design and material selection processes. The LCA of jars conducted for NIVEA Face Care products helped the management team switch from glass to plastic jars from an environmental perspective. This LCA took into account all stages of the product life cycle including raw materials, manufacturing processes, transport, product application, recycling, and disposal. The results of the study indicated that using jars made of polyethylene terephthalate (PET) and polypropylene (PP) instead of glass reduced the packaging's greenhouse gas potential by up to 16% and 28%, respectively (McDougall et al. 2001).

LCA can help improve production processes. An LCA of the Southern Seawater Desalinisation Plant (SSDP), in Western Australia identified that the reverse osmosis process would cause the most significant greenhouse emissions as a result of the use of electricity generated from fossil fuels. Once this grid electricity had been identified as a hotspot, the LCA then showed that about 90% of the total greenhouse gas emissions from the national grid-powered desalination plant can be avoided by switching to wind energy (Biswas 2009). This assisted the Western Australia Water Corporation to purchase energy from a wind farm and solar farm in the Mid-West to help offset the electricity needs of this desalination plant.

LCA can help influence policymaking and it can assist in the development of environmental policies. In 2008, when the Australian government was first planning to introduce energy saving compact fluorescent lamps by phasing out the sale of inefficient incandescent lamps, the public raised concerns regarding the disposal of mercury containing CFL at the end of their life. The author was one of the respondents who had to answer which would be the most sustainable option to follow. The author was not able to answer this question as he did not carry out an LCA of these two competing technologies. However, a few months later, the LCA conducted by one of his research students confirmed that the amount of mercury emissions avoided due to the reduction in electricity consumption due to use of CFL are actually significantly lower than the amount of mercury in the CFL.

LCA can help in environmental labelling and taxation. Typically, LCA is the method that calculates the impacts which are used in the development of environmental product declarations (EPDs). The completed EPD therefore serves as an environmental label which has the overarching goal of stimulating the potential for market-driven continuous environmental improvement. EPDs, as discussed later in this chapter, objectively and transparently communicate information on the environmental impacts of products according to their life cycle assessment (i.e. based on LCA) and allow purchasers and/or users to make a fair comparison of the environmental performance of products within a life cycle and encourages improvements in environmental performance.

LCA can also assist stakeholders in the supply chain in niche market targeting as this tool involves the calculation of environmental impacts resulting from the production of inputs used during the life cycle of the products. Therefore, inputs need to be sourced from places or industries that are producing them in an environmentally friendly way. For example, the replacement of diesel with biodiesel does not always result in the reduction in greenhouse gas emission mainly due to the emissions of GHG from the production of chemicals and their applications to soils for biofuel feedstock production. The greenhouse emissions from canola production in Western Australia are significantly lower than other regions across the globe due to lack of presence of denitrifying bacteria in WA's soil. As a result, German delegates were in consultation with their Australian counterparts to discuss the possibility of sourcing canola seeds from WA as feedstock for biodiesel production in Germany. WA's canola in the biodiesel supply chain produces half the greenhouse gas emissions of canola produced in Europe. Europe had tightened its renewable energy greenhouse gas savings target to be 50% of the emissions from mineral diesel (ABC 2017). LCA can, therefore, enhance the supply chain management by exerting up- and down-stream pressure on environmental performance in the supply chain. Some companies have actively collected supply chain environmental performance information for use in their business decision-making. Puma conducted a detailed LCA of the environmental impacts of its operations and supply chain. Only 6% of the impacts came from Puma's offices, warehouses, stores, and logistics (Balkau et al. 2015). The rest came from its supply chain, with more than half from the production of raw materials for manufacture. The findings were used by Puma to review where its raw materials came from, and what alternative sources of materials could be found.

LCA provides useful information to customers and facilitates public debate. Few studies in the United States have systematically compared the life cycle GHG emissions associated with food production locally against long-distance distribution, 'food-miles', due to the significant recent public debate and media attention on the environmental impacts of the food supply chain. GHG emissions associated with food production were found to be dominated by the production phase (83%), while transportation as a whole represented only 11% of life cycle GHG emissions, and final delivery from producer to retailers accounts only 4% (Weber and Matthews 2008). Therefore, growing local is not going to make any significant difference to GHG mitigation. Instead, a change in dietary habits can be a more effective means of reducing an average household's food-related carbon footprint than buying local due to the fact that red meat is around 150% more carbon-intensive than items providing protein such as chicken or fish (Weber and Matthews 2008). Interestingly, shifting less than 1 day per week's worth of calories from red meat and dairy products to chicken, fish, eggs, or a vegetable-based diet could achieve more GHG reduction than buying all locally grown food.

3.3.2 ISO 14040-44 for Life Cycle Assessment

There are four steps in the ELCA framework as per ISO 14040-44 (2006), including goal and scope, life cycle inventory (LCI), life cycle impact assessment (LCIA), and interpretation (Figure 3.1).

3.3.2.1 Step 1: Goal and Scope Definition

The goal examines decision-making strategies for the stakeholders across the environmental supply chain of a product or service. It can be either estimation or reduction of environmental impacts taking place during the life of a product (e.g. building, car, computer) or delivering services (e.g. electricity, water, transport). The scope defines the depth and breadth of the study to address the goal. It depends on the availability of data, locally available emission factors, interest by the industry, time and resources and cut-off criteria. This step allows LCA practitioners to discuss limitations on the LCA as it is indeed a challenging task to gather real-world data and to obtain a complete set of field or laboratory data. In some cases, data may not be accessible due to confidentiality constraints and also there could be absence of funding in experimental research to help

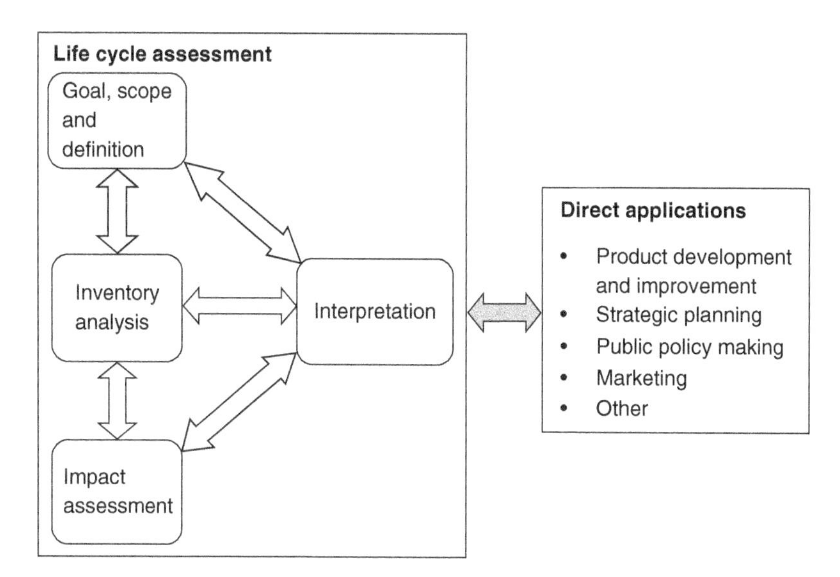

Figure 3.1 Steps of life cycle assessment. Source: Permission to reproduce extracts from British Standards is granted by BSI Standards Limited (BSI). No other use of this material is permitted. British Standards can be obtained in PDF or hard copy formats from the BSI online shop: https://shop.bsigroup.com/.

develop the complete database. Sometimes experiments are time and resource intensive. For example, it took one year to obtain one value of N_2O emission factor associated with the production of canola seeds and was an expensive study to complete. In instances, when the laboratory experiment is out of the scope of the research, the available data for similar process is obtained from either peer reviewed publications, or some other authentic sources are used. These data are known as surrogate data. It is not necessary to consider everything that has taken place and used during the life cycle of a product. A cutoff criterion is used to omit non-relevant life cycle stages, activity types, specific processes and inputs and outputs, and elementary flows from the system model. As a result, LCA cannot provide an absolute value and there is always a truncation error in any LCA result. However, the magnitude of this error varies with the quality and completeness of the data. LCA produces a relative value, which means that it is used for comparison purposes. Using the same system boundary and LCA framework, the environmental performance of different versions of the same products can be compared.

The scope in fact determines the functional unit of the LCA, which is important in conducting a mass balance to develop a LCI. The functional unit can be a residential house or 1 km of road or 1 MWh of electricity or the number of vehicle kms travelled by a vehicle or the production and combustion of 1 GJ of biodiesel.

The functional unit can vary for the same product or service depending on the information sought. The industry may be interested in knowing the environmental impacts of the products before leaving the factory gate or the overall life cycle performance of their products including end of life disposal to landfill.

The environmental impacts from the production of capital equipment, including factory buildings, warehouses, pipe infrastructure and machinery, are usually excluded from the system boundary of the LCA analysis due to their long life spans (Biswas 2009; Frischknecht et al. 2007). For example, a building life is say 50 years. If this building is not used, then there are no emissions from the building, but if machinery is installed inside the building to produce a product, then emissions will take place. It is another cause of the truncation error.

The goal of the LCA determines the system boundary or the number of stages considered during the life cycle. There is a range of system boundaries, including:

Cradle to Grave: This approach was followed when LCA was first established. It considers the full life cycle from raw material extraction to the end of life disposal stage. For example, the cradle to grave system boundary of a residential house includes extraction of clay, production of bricks, construction, use, demolition, and land fill of demolition waste. Consideration of transport between consecutive stages is a must across all system boundaries.

Cradle to Cradle: This boundary goes one step further than the previous approach. The full life cycle from raw material extraction to use and recycling and re-use at the end of life (EoL) stage (substantially recycled). For example, this boundary considers recycling, reconstruction, and re-use stages in addition to other life cycle stages of the residential building. Another example using the life cycle of auto parts includes bauxite ore mining, alumina processing, production of aluminium ingot, manufacturing of auto parts and the car, use and the recycling of the auto parts.

Cradle to Gate: This is a partial LCA from raw material extraction to the factory gate (i.e. before it is transported to the consumer or for further assembly). For example, the extraction of bauxite ore, the processing of alumina and then the production of the aluminium ingot.

Gate to Gate: This is also a partial LCA which only includes one value addition in the entire life. For example, assembling of the car, or powder coating of an auto part in a car manufacturing plant.

Streamed Line LCA: When the LCA does not consider all stages of the product life cycle, it is also known as streamlined LCA, i.e. LCA using a cradle to gate or a gate to gate system boundary is a streamlined LCA – a slimmed down version of a full LCA. It is important that the streamlined LCA has not streamlined out any core information that would be included in a full LCA. In the case of streamlined LCA, it will be more comprehensive if upstream activities are counted, it includes only selected environmental impacts and the inventory is linked to these impacts. Depending on the complexity of the object of the study, a full LCA can be a very complex, long, and difficult exercise. Streamlining LCA is a practice which has been widely adopted to make this type of assessment more manageable. Streamlined LCA can be achieved in a number of ways, including:

- Limiting the scope (e.g. eliminating life cycle stages which are considered not significant, or processes with negligible effect on the environment)
- Use of qualitative information
- Removal of upstream and/or downstream components
- Use of specific impact categories

Since a full LCA can be time- and resource-consuming, there is a need for developing simplified methods. Choosing a streamlined LCA-method involves a balance between the simplification of the method and the type of accuracy and results the user is looking for. The results of the full LCA and streamlined LCA usually identify the same output hotspot. Insufficient material or processes in the streamlined LCA databases were one of the reasons for the difference between the results of the streamlined-LCA and a full LCA.

The system boundary of LCA depends on industry's or business's or the researcher's need. For example, the Grain Research and Development Corporation (GRDC) of Australia were only interested in estimating the global warming impact of grain production, which only included upstream processes (i.e. pre-farm emissions from chemicals and energy production, and on-farm emissions from farm machinery and soil). This is a cradle to gate LCA, which does not consider downstream processes (e.g. bakery, supermarket, household waste etc.). The LCA that was conducted for GRDC is also known as a limited focused LCA as it only estimated one impact while there are other impacts such as acidification and eutrophication, which can potentially result from grain production (Frischknecht et al. 2007).

The whole life cycle can be broken down into upstream processes (from materials extraction to delivery at factory) and downstream processes (from factory despatch to disposal of product). The transportation processes are always included in both cases.

3.3.2.2 Step 2: Inventory Analysis

The inventory consists of the quantitative values of inputs (materials, energy) used and outputs (emissions and wastes) produced during the life cycle stages of a product. As stated above, a suitable functional unit needs to be defined in order to calculate the amount of inputs and outputs of all stages of the product life cycle to develop a LCI.

The steps for calculating the LCI are as follows:

- Quantification of all inputs/outputs in the process
- Summation of inputs/outputs by type/substance for each life cycle stage
- Summation of input/output by type/substance for all life cycle stages
- All transportation processes between consecutive stages

How do we do the mass balancing in order to develop an LCI?

An example has been provided in Box 3.1 to show how a LCI can be developed.

Box 3.1 Life cycle inventory development

Suppose in stage A, 10 kg of X and 4 kWh of Y are required to produce 1 kg of Q and 1.5 kg of Q is required to produce 1 kg of P (e.g. loss is 0.5 kg of Q). In stage B, 4 l of U and 3 MJ of V are required to produce 1 kg of P. The density of U is 0.8 kg/l and 1 kWh = 3.6 MJ.

If the quantity of P is 200 units, then what will be the material and energy inputs at both stages?

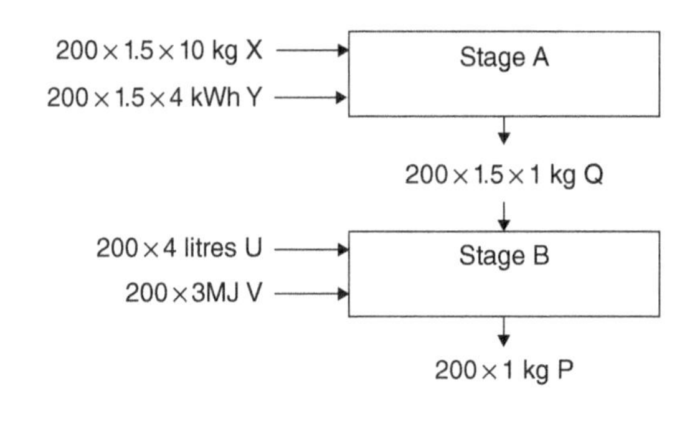

3.3.2.3 Step 3: Life Cycle Impact Assessment (LCIA)

Multiply the amounts of inputs (I) in the LCI by their respective emission factors (IF) to calculate the impacts, E (environmental impacts) = I (inputs) × EF (emission factor), associated with mining, processing, transportation, and production of these inputs. Make sure that the units of both inputs and the emission factors have the same denominators, otherwise they need to be converted using available conversion factors. For example, if you have an emission factor for transportation in terms of ton kilometres travelled (i.e. tkm), you need to convert the unit of the transportation input to tkm meaning that you need to multiply the tonnage of inputs by the amount of distance travelled (km) to carry this input from A to B.

There are different types of emissions resulting from the production and use of inputs that are available in the LCI of a product or a service. Different emissions cause different impacts, therefore, you need to classify these emissions in terms of the impact that it will have to help determine the type of environmental impacts that will occur during the life cycle of the product. For example CO_2 emissions cause global warming impact, NO_2 causes acidification, nitrification, and human toxicity impacts, and Hg causes eco-toxicity and human toxicity impacts. Once these emissions have been classified, they need to be characterised, meaning that within each impact category we need to convert contributing emissions into the equivalent amount of gas or pollutant representing the impact category. For example, global warming impacts are represented as kg CO_2 equivalent, which means that the greenhouse gases like N_2O, CH_4, and CO_2, if produced during the life cycle of a product or service, need into be converted to equivalent amounts of CO_2 (or CO_2 e-). The factors which are used for this conversion are known as characterisation factors. For example, N_2O is a greenhouse gas but that is 265 times powerful than CO_2, meaning that amount of N_2O emissions that are emitted during the life cycle of a product or a service need to be multiplied by 265.

These impacts are considered as 'midpoint' in the overall LCA and are responsible for causing effects known as 'endpoints' (Figure 3.2). For example, global warming is an impact, which results from GHG emissions, causes sea level rise and ultimately affects food production and the human health (Figure 3.2). Different emissions resulting from the use of inputs in the LCI cause different impacts and then different impacts cause different effects. The following section on eco-points

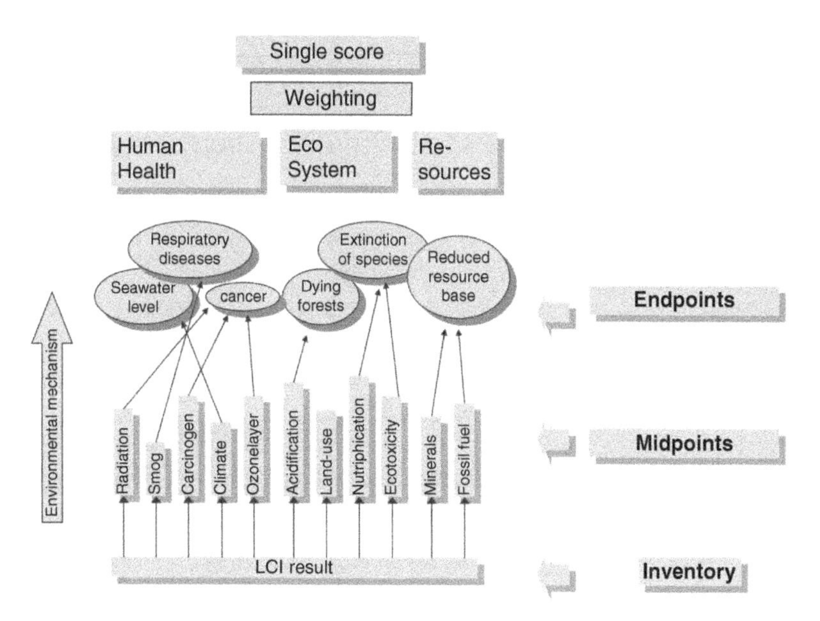

Figure 3.2 Relation between LCI, midpoints, and endpoints.

will show how different impacts are aggregated to obtain a single environmental score.

What is an Emission Factor? An emission factor is a representative value that attempts to relate the quantity of pollutants released to the atmosphere associated with the production of the input during its upstream processes (e.g. mining, processing, and manufacturing of inputs such as cement, caustic soda, urea, etc.). These factors are usually expressed as the weight of the pollutant divided by a unit weight or volume or distance or duration of the activity emitting the pollutant (e.g. kilograms of particulate emitted per ton of coal burned). For example, emission factors for ammonia, which is an indispensable input for urea fertiliser production, are 1 kg nitrogen dioxide per kg of ammonia, 0.1 kg carbon monoxide per kg ammonia (EEA 2019).

Box 3.2 shows how the inputs in Box 3.1 were converted to impacts using their emission factors.

Box 3.2 Conversion of LCI inputs to impacts

Emission factors of X, Y, U, and V

	kg CO_2	kg N_2O	kg CH_4	kg SO_2	kg NO_2
per kg of X	50	2	—	—	5
per kWh of Y	500	—		30	4
per kg of U			5	—	8
per kWh of V	450	—	—	5	—

Using this information, the impacts have been classified as follows:

Global warming potential (GWP) causing gases are kg CO_2, kg N_2O, and kg CH_4
Acidification potential causing gases are kg SO_2 and kg NO_2
Using the information in Box 3.1, the values of X, Y, U and V are calculated as 3,000 kg, 1,200 kg, 640 kg and 167 kWh, respectively. Since the emission factors for U and V are in kg and kWh, respectively, their values are converted from litre to kg and MJ to kWh by using conversion factors (1 kg of U = 0.8 litres of U; 1 kWh = 3.6 MJ). The total emissions associated with the use of X, Y, V and U were calculated. These emissions are multiplied by the corresponding equivalence factor to characterise these impacts.

$$\text{GWP}_{i\text{CO}_2\ \text{equivalent}} = I_i(\text{EF}_{i\text{CO}_2} + \text{EF}_{i\text{CH}_4} \times 28 + \text{EF}_{i\text{N}_2\text{O}} \times 265)$$

$$\text{AP}_{i\text{SO}_2\ \text{equivalent}} = I_i(\text{EF}_{i\text{SO}_2} + \text{EF}_{i\text{NO}_2} \times 1.4)$$

	GWP gases			Acidification			
I_i	kg CO_2	kg N_2O	kg CH_4	kg SO_2	kg NO_2	kg CO_2 e- GWP	kg SO_2 e- AP
X	150,000	6,000	—	—	15,000	1,740,000	21,000
Y	600,000	—	—	36,000	4,800	600,000	42,720
U	—	—	3,200	—	5,120	89,600	7,168
V	75,000	—	—	835	—	75,000	835
Total						2,504,600	71,721

Practice Example 1 – GHG emissions from bio-electricity production

The Shire of Riverwood uses canola based biodiesel to run its electricity generator. The company has appointed you as a consultant to determine the total GHG emissions associated with their electricity production from canola based biodiesel for the next 10 years. Per capita demand of electricity for the Shire is 2000 kWh/year. The number of residents in the Shire is 5000.

Electricity generation from their biodiesel generator has four stages:

1. The pre-farm stage considers GHG emissions from the production and transportation of fertiliser (50 kg/ha) and herbicide (3.5 kg/ha) for canola production.
2. The on-farm stage includes GHG emissions from machinery operation (40 l of diesel/ha) during seed, fertiliser, and herbicide application and harvesting of the canola.
3. The post-farm stage includes GHG emissions resulting from the conversion of canola seeds to biodiesel including transportation of canola seeds/biodiesel at 5 l diesel/ha.
4. The electricity generation stage includes emissions from the combustion of biodiesel to produce electricity.

Production factors

- The yield of canola seeds = 1000 kg/ha.
- 1 kg canola seeds are required to produce 0.3 l of canola oil. The density of canola oil is 0.88 kg/l. About 0.075 kWh of electricity is required to convert 1 kg of canola seeds to oil.

- 0.2 kWh electricity is used from national grid for converting 1 kg canola oil into 0.95 kg biodiesel.
- The calorific value of biodiesel = 40 MJ/kg.
- Efficiency of biodiesel generator = 40%.
- 1 kWh = 3.6 MJ.
- Heating value of diesel = 38 MJ/l.
- Density of biodiesel = 0.8747 kg/l.

Emission factors

- Biodiesel combustion in the plant – 0.1 kg CO_2 e-/l biodiesel
- Production and transportation of fertiliser – 40 kg CO_2 e-/kg fertiliser
- Production and transportation of herbicide – 23 kg CO_2 e-/kg herbicide
- Electricity from the national grid – 1.032 kg CO_2 e-/kWh
- Diesel production stage – 0.014 kg CO_2 e-/MJ
- Diesel combustion stage – 73.25 g CO_2 e-/MJ

Answer the following questions

i. Calculate the total GHG emissions in ton CO_2 e-.
ii. Using a bar chart, show which stage is creating the most GHG emissions.
iii. Draw a pie chart showing the breakdown of GHG emissions in terms of inputs.
iv. Identify the GHG production hotspot and suggest possible mitigation strategies.

Solution

Answer to question i Figure 3.3 shows the first two steps for calculating the inputs used in the life cycle stages for producing 100×10^6 kWh of electricity for this Shire from the canola based biofuel.

Step 3 – Development of LCI and the determination of impacts (Table 3.1).

The emissions values in the last column for pre-farm, on-farm, and post-farm stages were used in the bar chart below showing that the pre-farm stage contributed most of the emissions.

Answer to question ii The message received from this bar chart (i.e. Figure 3.4) is that the pre-farm stage accounted for the significant portion of the total GHG emissions.

Answer to question iii Further investigations have been made to find the inputs responsible for contributing to the significant emissions during the pre-farm stage (Figure 3.5).

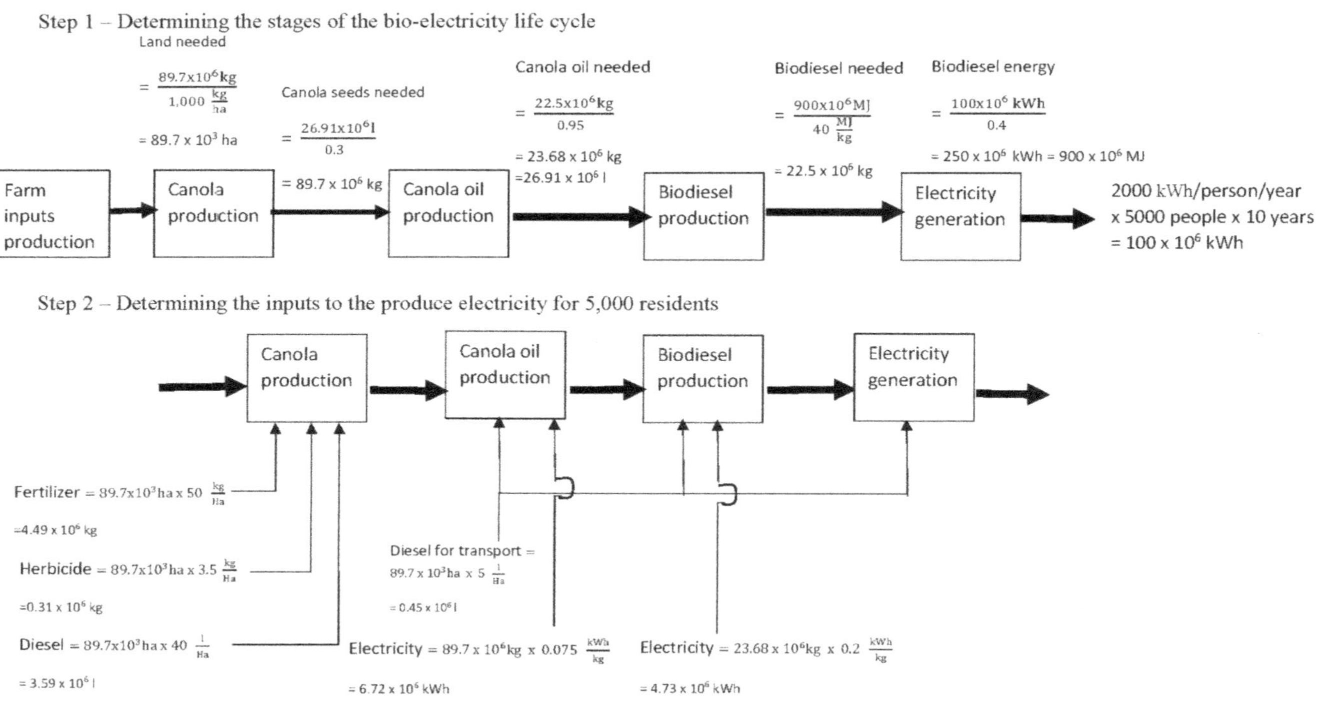

Figure 3.3 Developing an inventory using mass balance.

Table 3.1 Conversion of LCI data into GHG emissions.

Inputs	Amount	Units	Emission factors	Emissions (kg)
LCI			LCIA	
Pre-farm				
• Fertiliser	4,485,645.93	kg	40 kg/kg	179,425,837.32
• Herbicide	313,995.22	kg	23 kg/kg	7,221,889.95
			Subtotal	186,647,727.27
On-farm				
• Diesel for farm machinery	3,588,516.75 136,363,636.4[a]	l MJ	0.08725 kg/MJ[b]	11,897,727.27
			Subtotal	11,897,727.27
Post-farm				
• Diesel for transport	448,564.59 17,045,454.55[a]	l MJ	0.08725 kg/MJ[b]	1,487,215.909
• Electricity for converting seeds to oil	6,728,468.9	kWh	1.032 kg/kWh	6,943,779.904
• Electricity for converting oil to biodiesel	4,736,842.1	kWh	1.032 kg/kWh	4,888,421.053
• Biodiesel[c]	900×10^6 22.5×10^6 25,723,105.06[a]	MJ kg l	0.1 kg/l Subtotal	2,572,310.506 15,891,727.37
			Total	214,437,182 kg 214,437 tons

a) The value of diesel in terms of litres has been converted to MJ, because the emission factor for diesel has been given in terms of MJ. This is the same case with biodiesel where MJ is converted to kg and then to litres due to the fact that the denominator of biodiesel emission factor is in litres.
b) The emission factors of both production and combustion of diesel have been added.
c) The emissions associated with the production of biodiesel have already been calculated in this example. It is now considered as an input as it has been combusted in an electricity generator to produce electricity.

Answer to question iv This pie chart shows that the fertiliser is the hotspot, requiring further sustainability improvement. There can be a number of solutions, including the application of organic fertiliser and the introduction of a crop rotation system to reduce the amount of chemical fertiliser required to reduce the overall life cycle emissions.

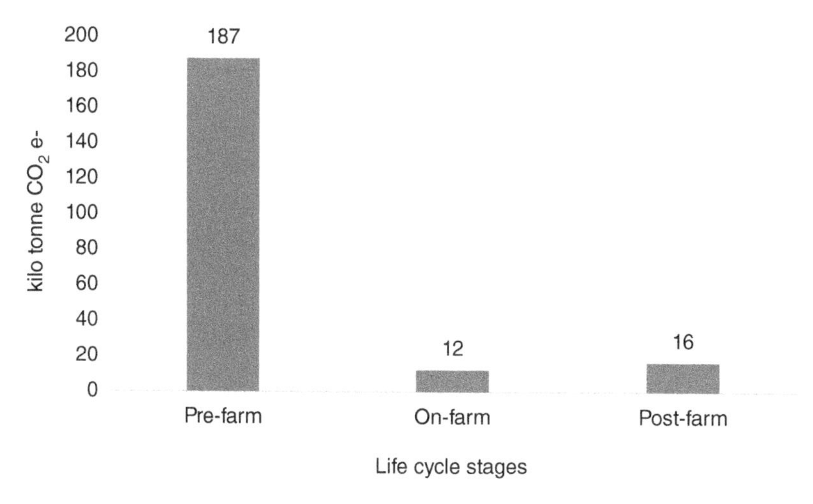

Figure 3.4 Breakdown of GHG emissions in terms of stages.

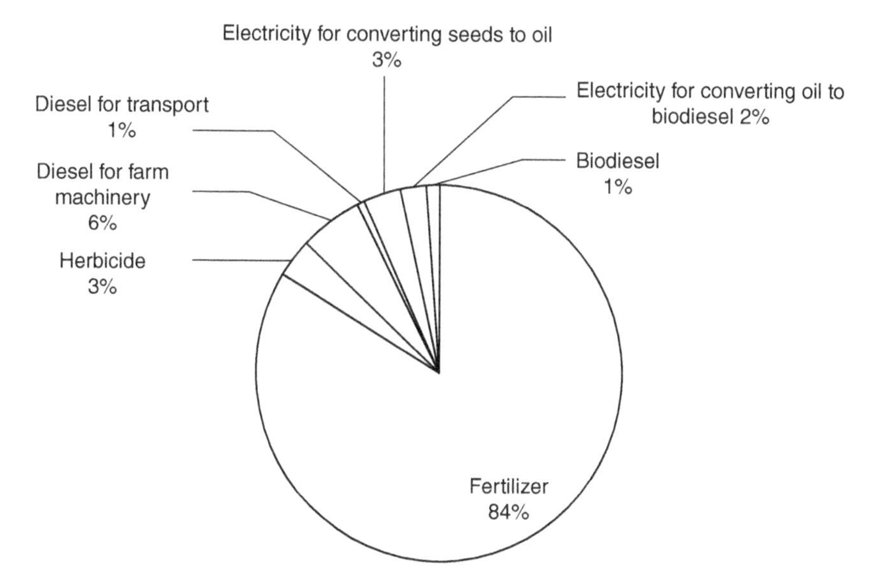

Figure 3.5 Pie chart showing the GHG hotspot.

Comment: This LCA is known as a single focused LCA as it dealt with one impact only. It is also known as an attributional LCA.

Practice Example 2: Environmental impacts of water supply

There is an increasing awareness of the effects of GWP and acidification on the environment and human health resulting from the combustion of fossil fuels.

The Green University is located along the sea coast and due to the salinity of ground water, it receives its water supply from bore wells (ground water) which are located at a distance of 40 km inland from the University's overhead water tank for subsequent distribution to the entire campus. The water from these bore wells is pumped up to University's water tanks using a series of diesel powered pumping and pressure boosting stations. The University is committed to improve the sustainability of all their operations and is considering various alternative options. The facility management department has submitted a proposal to set up a desalination plant within the campus that will convert seawater to potable water by using the latest technology and the water from the desalination plant will be pumped directly into the University's overhead water tank.

The University will base their decision on the environmental benefits associated with the substitution of ground water with desalinated water. You have been appointed as a consultant to compare the impacts associated with ground water versus desalinated water.

Your task is to compare both the options over a period of one year and provide an assessment on environmental impacts to the University.

I. The functional unit (FU) is the total annual water demand of the University and FU will be calculated using this equation, **FU = 400,000 m^3**.
II. Calculate the **GWP** and **acidification** impacts of the production of ground water and desalinated water for the above functional unit to determine the environmental implications of this substitution. In order to perform this task, complete the following activities.
    - Develop a LCI flow chart showing the quantitative values of all inputs for the **production** and **distribution** stages of groundwater supply and also show the calculations for the ground water option, which consists of production and distribution (i.e. amount of energy and material inputs required for extraction, treatment, and delivery)
    - Develop an LCI flow chart showing the inputs of the **production** and **distribution** stages of desalinated water and also show the calculations for the desalinated water option for production and distribution (i.e. amount of energy and material inputs required for processing, treatment, and delivery)
    - Calculate the emissions associated with all stages of ground water and desalinated water options.

The following data will be used to perform the above quantitative analysis

The energy and materials used in the production and maintenance of pumps, pipes, and desalination plant are not considered for this comparison.

A. Data for ground water
    - Diesel consumption for pumps = 2.9 MJ/m^3/km

- Chlorine dose for making ground water potable = 51.25 g/m^3 of ground water
- Heating value of diesel = 38 MJ/l

B. Data for desalinated water
- Distance from desalination plant to University's overhead water tank = 2 km
- Efficiency of desalination process = 70% (i.e. for 1unit of treated water 1/0.7 units of untreated water from the sea will be required)
- Electricity consumption for desalination = 8 kWh/m^3 of untreated water (including water intake, pre-treatment, reverse osmosis, post-treatment)
- Electricity consumption for brine pumping to sea = 3 kWh/m^3 of brine
- Electricity consumption for delivery into University's overhead tank = 0.6 kWh/m^3/km
- Chemicals required for desalination process = 79.8 g/m^3 of untreated water (sodium hypochlorite, antiscalant, lime, and liquid CO_2)

Solution

Box 3.3 shows the solution of the above problem.

Box 3.3 Estimation of environmental impacts with the production of desalinated water

		Emission factors				
		CO_2	CH_4	N_2O	SO_2	NO_2
Input	Unit	kg	kg	kg	g	g
Electricity	Per kWh	8.12E−01	2.94E−04	1.71E−04	2.25E+00	2.59E+00
Diesel – production and combustion	Per MJ	9.84E−02	8.68E−05	6.37E−06	2.00E−01	1.45E+00
Chlorine for ground water	Per g	1.10E−03	1.02E−06	1.44E−07	4.85E−03	3.48E−03
Chemicals for desalination	Per g	8.03E−04	6.27E−06	1.36E−08	2.60E−03	1.06E−03

Functional unit = 400,000 m^3

Diesel for pumping = 400,000 m^3 × 2.9 MJ/m^3/km × 40 km = 46.4 × 10^6 MJ

Chlorine required for ground water = 400 m^3 × 51.25 g/m^3 = 20.5 × 10^6 g

(Continued)

Box 3.3 (Continued)

Using the following equation, the CO_2 equivalent amount of GHG emissions from diesel and chlorine inputs (*I*) have been calculated.

$$GWP_I\ (kg\ CO_2\ e-) = I\,(1 \times kg\ CO_2 + 28 \times kg\ CH_4 + 265 \times kg\ N_2O)$$

	Diesel – Production and combustion I_{Diesel} **× Equivalence × EF**	**Chlorine for ground water** $I_{Chlorine}$ **× Equivalence × EF**
kg CO_2 e- CO_2	$46.4 \times 10^6 \times 1 \times 9.84E{-}02 =$ 4.57E+06	$20.5 \times 10^6 \times 1 \times 1.10E{-}03 =$ 22,550
kg CO_2 e- CH_4	$46.4 \times 10^6 \times 28 \times 8.68E{-}05 =$ 1.13E+05	$20.5 \times 10^6 \times 28 \times 1.02E{-}06 =$ 585.48
kg CO_2 e- N_2O	$46.4 \times 10^6 \times 265 \times 6.37E{-}06 =$ 7.83E+04	$20.5 \times 10^6 \times 265 \times 1.44E{-}07 =$ 782.28
	4,756.86 tons of CO_2 e-	23.92 tons of CO_2 e-
		Total = 4,780.77 tons of CO_2 e-
g SO_2 e- SO_2	$46.4 \times 10^6 \times 1 \times 2.00E{-}01 =$ 9.28E+06	$20.5 \times 10^6 \times 1 \times 4.85E{-}03 =$ 99,425
g SO_2 e- NO_2	$46.4 \times 10^6 \times 1.4 \times 1.45E{+}00 =$ 9.42E+07	$20.5 \times 10^6 \times 1.4 \times 3.48E{-}03 =$ 99,876
	103.47 tons of SO_2 e-	0.2 tons of SO_2 e-
		Total = 103.67 tons of SO_2 e-

Plant efficiency = 70%

Water requirement due to plant efficiency (untreated water) = 400,000 m^3/0.7 = 571,429 m^3

Amount of brine = 571,429−400,000 = 171,429 m^3

Electricity for desalination = 571,429 m^3 × 8 kWh/m^3 = 4,571,429 kWh

Electricity for brine pumping = 171,429 m^3 × 3 kWh/m^3 = 514,286 kWh

Electricity for pumping in to overhead water tank = 400,000 m^3 × 0.6 kWh/m^3/km × 2 km = 480,000 kWh

Total electricity consumption = 5.6×10^6 kWh

Chemicals for desalination = 571,429 m^3 × 79.8 g/m^3 = 45.6×10^6 g

Acidification impact (kg SO_2 e−) = 1 × kg SO_2 + 1.4 × kg NO_2

	Electricity $I_{Electricity}$ **× Equivalence × EF**	**Chemicals for desalination** **Chemicals × Equivalence × EF**
kg CO_2 e- CO_2	$5.6 \times 10^6 \times 1 \times 8.12E{-}01 =$ 4.52E+06	$45.6 \times 10^6 \times 1 \times 8.03E{-}04 =$ 36,616.8
kg CO_2 e- CH_4	$5.6 \times 10^6 \times 28 \times 2.94E{-}04 =$ 4.58E+04	$45.6 \times 10^6 \times 28 \times 6.27E{-}06 =$ 8005.536
kg CO_2 e- N_2O	$5.6 \times 10^6 \times 265 \times 1.71E{-}04 =$ 2.52E+05	$45.6 \times 10^6 \times 265 \times 1.36E{-}08 =$ 164.3424
	4817.4 tons of CO_2 e-	44.8 tons of CO_2 e-
		Total = 4862.2 tons of CO_2 e-
g SO_2 e- SO_2	$5.6 \times 10^6 \times 1 \times 2.25E{+}00 =$ 1.25E+07	$45.6 \times 10^6 \times 1 \times 2.60E{-}03 =$ 118,560
g SO_2 e- NO_2	$5.6 \times 10^6 \times 1.4 \times 2.59E{+}00 =$ 2.02E+07	$45.6 \times 10^6 \times 1.4 \times 1.06E{-}03 =$ 67,670.4
	32.7 tons of SO_2 e-	0.19 tons of SO_2 e-
		Total = 32.89 tons of SO_2 e-

From this analysis, it appears that the groundwater option is better than the desalination option from a global warming potential perspective, but not from an acid rain perspective. However, the selection of these options depends on the priority given to the impacts by different countries. Different countries give different priorities or weights to different impacts based on the type of emissions produced resulting from human or economic activities. Therefore, it is important to ascertain the country-specific weights for different environmental impacts so that they can be used to obtain a single environmental score to facilitate the decision-making process. The following section provides a procedure to address these issues.

Eco-points: This allows us to integrate all environmental impacts. Different environmental impacts have different units, which cannot be added. Therefore, all environmental impacts need to be converted to a single unit known as Eco-points. Then we can add all environmental impacts with the same unit to get a single score of environmental impact.

The steps for calculating eco-points for these impacts are as follows:

I. Environmental impacts are normalised using a corresponding normalisation factor (NF). This represents an impact as a percentage of the total of impact in a country per capita per annum.
II. The normalised values of all impact categories are then aggregated into a single indicator to take into account the relative importance of the different category results. The weighting factor (WF) shows the relative importance of each impact. WFs of impact categories are multiplied by corresponding normalised values to estimate the eco-points (Pt) of the impact.

The eco-points of a product or a service has been calculated using the following equation:

$$\text{Eco} - \text{point} = \sum_{n=1}^{N} \frac{\text{EI}_n}{\text{NF}_n} \times 100\% \times \text{WF}_n \tag{3.1}$$

where, I = impact categories, 1, 2, 3, …, N.

Practice Example 3: Calculation of eco-points

Box 3.4 Eco-point problem

Country C is one of the largest producers of product P. An industry B in this country produces 10 pieces of P a year where 1 piece = 1 kg.

In **Stage 1** of the life cycle of P, 2 kg of X and 2.5 kWh of Z are required to produce 1 kg of Q. In **Stage 2**, you need 2 kg of Y and 5 kWh of Z to convert 2 kg of Q to 1 kg P (i.e. loss is 1 kg of Q).

You need to determine the environmental impacts associated with the production of P by an industry B per year in terms of 'Eco-points'.

A systematic methodology to estimate the eco-point is as follows:

1. *Determine the functional unit.*
2. *Develop a life cycle inventory.*
3. The emission factors of X, Y, and Z are given in Table 3.2. *Calculate the amount of emissions associated with the production of product 10 Ps per year.*

Table 3.2 Emission factors of inputs.

	CO_2	CH_4	C_2H_2
X	5 kg of CO_2/kg of X	1 kg of CH_4/kg of X	
Y		0.5 kg of CH_4/kg of Y	0.01 kg of C_2H_2/kg of Y
Z	50 kg of CO_2/kWh of Z		

4. With the aid of Table 3.3, *classify the environmental emissions associated with the production of a P in terms of environmental impacts.*

Table 3.3 Gases of different impacts.

Impacts	GWI (global warming impacts)	Acid rain	Photochemical smog	Eutrophication
Gases	CO_2, N_2O, CH_4	SO_x, NO_x	CH_4, C_2H_2	PO_4, NO_x, NH_3

5. *Characterise these environmental impacts for Stages 1 and 2.* Note that 1 kg of CH_4 = 28 kg of CO_2 equivalent global warming impact and 1 kg of CH_4 = 0.006 kg of C_2H_2 equivalent photochemical smog.
6. *Calculate the total impacts for GWI and Photochemical smog.*
7. The per capita GHG emissions and photochemical smog of this country C are 10,000 kg CO_2 equivalent and 2 kg C_2H_2 equivalent, respectively. Country C gives five times more importance to photochemical smog than GHG emissions.
   Determine the overall environmental impacts in terms of 'Eco-points'.
8. *Identify the environmental hotspot in terms of inputs.*

Solution

Box 3.5 shows the steps to calculate the eco-points.

Box 3.5 Conversion of inputs into eco-points

1. Functional unit = 10 Ps
2. Life cycle inventory – A mass balance has been conducted to quantify the inputs in Stages 1 and 2 required to produce 10 Ps per annum

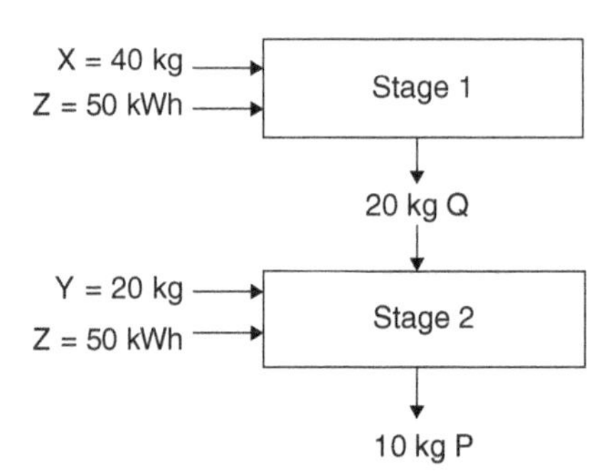

3. The amount of inputs in the LCI were multiplied by the corresponding emission factors in Table 3.4 to quantify the emissions associated with the production of 10 Ps.

Table 3.4 Emissions from inputs.

Inputs	kg of CO_2	kg of CH_4	kg of C_2H_2
X	200	40	
Y		10	0.2
Z	5,000		

4. On the basis of Table 3.5, the emissions in the above table caused following impacts.
   a) Global warming impact (GWI)
   b) Photo-chemical smog
   Therefore, these emissions are classified as two separate environmental impacts.

Table 3.5 Emissions characterisation.

Inputs	Global warming impact kg CO_2 e-	Photo-chemical smog kg C_2H_2 e-
X	$200+40\times 28=1320$	$40\times 0.006=0.24$
Y	$10\times 28=280$	$10\times 0.006+0.2=0.26$
Z	5,000	0
Total	6,600 kg CO_2 e-	0.5 kg C_2H_2 e-

5. The global warming impact gases (i.e. CO_2, CH_4) will be converted to equivalent amount of CO_2 and photochemical smog causing gases (i.e. CH_4, C_2H_2) will be converted to the equivalent amount of C_2H_2.
6. Normalised value of GWI = 6,600*100/10,000 = 66%
   Normalised value of Smog = 0.5*100/2 = 25%
7. Eco-point value of GWI = 66% × 1 = 66 Eco-points
   Eco-point value of GWI = 25% × 5 = 125 Eco-points
   Total Eco-point is 191 Eco-points
8. The dominant impact is therefore photochemical smog
   X = (1,300 × 100 × 1)/10,000 + (0.24 × 100 × 5)/2 = 73.2 Eco-points; Y = (280 × 100 × 1)/10,000 + (0.26 × 100 × 5)/2 = 67.8 Eco-points; Z = (5,000 × 100 × 1)/10,000 = 50 Eco-points.
   The hotspot or the input causing the most impact is X. An engineering decision would be either to make X greener with less emissions or to look for alternative materials with reduced level of environmental impacts.

3.3.2.4 Step 4: Interpretation

Life cycle interpretation is the final step in the LCA study. It is a systematic approach where the outcomes of the LCI and the LCIA are classified, quantified, checked, and evaluated. LCIA enables us to identify the stage or process responsible for causing the most significant portion of impacts known as 'hotspots'. LCI is then used to find the input(s) or process causing this impact. Possible solutions can then be proposed. Interpretation of the results will result and appropriate conclusions and recommendations for engineering decision-making are then made.

3.4 Allocation Method

The allocation rule is used when valuable co-products are produced along with the main product. It means that the environmental burdens need to be shared by the co-products produced in a system. For example, fly ash is produced when coal is combusted to produce electricity. This fly ash, due to its cementitious property is sold as a continuous by-product to partially replace cement in concrete. Since this by-product has a market value (is considered a valuable item), the total emissions from the electricity generator need to be allocated to both electricity and fly ash production. There are broadly two major classes of allocation methods: the partitioning method and system expansion. The partitioning ratio in this case should reflect a physical or economic cause-and-effect relationship or causality between the inputs and outputs and the multiple products of the process in question.

In the system expansion method, co-products are usually substituted for equivalent products produced elsewhere. For example, a waste incineration service also produces energy. To isolate the waste treatment service, one must subtract the inputs and outputs required to produce an equivalent amount of energy elsewhere, which is avoided by the energy produced from the incineration process (Suh et al. 2010). System expansion considers the market mechanism under which a substitution between products takes place in reality.

In the case of partitioning, the inputs and outputs of the life cycle stages should be partitioned between their different co-products or co-services in a way that reflects the underlying physical relationships between them so that the values of inputs and outputs will change with the quantitative changes in the co-products or co-services delivered by the system. This is known as physical allocation. The economic partitioning factors are calculated as the share of the proceeds or sale of one co-product or co-service in the total outcome of the proceeds or sale of all products, which is expressed as follows:

$$A_i = \frac{n_i \times p_i}{\sum_{i=1}^{P} n_i \times p_i} \tag{3.2}$$

where A_i is the allocation factor of the ith co-product/co-service, n_i is the quantity of the ith co-product/co-service, and p_i is the price of the ith co-product/co-service. P is the total number of co-products or co-services. However, the applicability of the allocation methods is different for different circumstances. Economic allocation can be considered in most cases except for situations where physical casualty exists. In this case, the allocation by financial value can provide misleading results. According to the ISO 14041 (ISO 2007), allocation should reflect the physical relationship between the environmental burden and the functions; i.e. how the burdens are changed by quantitative changes in the functions delivered by the product system. Therefore, allocation can be based on the physical properties of the products, such as mass, volume, and energy, because data on the properties are generally available and easily interpreted. For example, Boustead et al. (1999) compared the physical allocation with the economic allocation for a mining operation with multiple outputs of lead, sulfur, zinc, cadmium, and copper and concluded that mass partitioning is more reliable, given that the only thing that has changed in all the cases is the economic value of the products as the physical process remains unchanged.

Physical allocation can be conducted based on the physical flow of mass or energy. For example, Hill et al. (2006) used mass and energy allocation methods in the study of corn. In the mass allocation method, the co-product credit was equal to the energy input of all the production steps leading to the creation of the co-product multiplied by the relative weight of the co-product. In the energy allocation method, the co-product credit was the amount of inherent energy (lower heating value) within each product assuming complete combustion at 90% boiler efficiency. The physical allocation method works well when there is a close correlation between the chosen physical property and the value of the co-products. The physical limitation of mass (weight) allocation is that it cannot be applied for energy services, for example, the output from a co-generation plant.

There are some cases where the physical relationship cannot be used to describe the effects of changing different functional units. Where the ratio of outputs is fixed, then the 'allocation' based on physical causality is impossible. Examples of this arise in the chemical industry, where the ratio of sodium hydroxide (NaOH) and chlorine (Cl_2) produced by electrolysing brine is fixed by stoichiometry, and in agricultural production, where ratios are defined by the physical and chemical structure of a plant crop (e.g. rapeseed oil and residue) or an animal (e.g. beef and leather). In these cases, economic allocation can be more suitable as the economic relationships reflect the socio-economic demands which cause the multiple-function systems to exist (Azapagic and Clift 1999). This is because the increase in demand of one of the co-products is actually increasing the pressure on the total product supply chain. Whilst the limitation of economic allocation can arise from the variability of prices and the low correlation between prices and physical flow, sometimes a price summarises complex attributes of product or service quality that cannot be easily measured by physical criteria. For example, bitumen is produced in petroleum refineries which usually produce other co-products such as gasoline, kerosene, fuel oil, and gas oil. If the refinery process yields 5% by mass of bitumen and 95% by mass of the other co-products (gasoline, kerosene, gas oil and fuel oil), it is necessary to investigate as to whether the variation in the ratio between the different co-products changes mass and energy data in the LCI. In this refinery example, the ratio between the mass of bitumen and the mass of the other co-products can only be varied in a small range which involves a significant change of the process parameters including energy consumption. Because of this lack of uniformity, economic allocation is more appropriate than other allocation methods.

When system expansion or allocation by physical quantities is not possible or not justifiable, the usual way is to allocate according to the economic value of the products (prices) (Niederl-Schmidinger and Narodoslawsky 2008). It is quite obvious that the market prices could fluctuate over time, which could also affect the validity and credibility of the LCA results (Marvuglia et al. 2010), but the price ratios of different co-products – which are used for revenue-based co-product allocation – are much more stable, particularly in the longer term, than the individual prices (Hischier et al. 2005; Guinée et al. 2004). Since economic allocation uses the price ratio rather than the absolute value, the price fluctuation effects can be overcome.

Like price fluctuations, the physical quantities of functional flows of a particular multi-functional process also fluctuate (for instance, the amounts of milk and wool produced in sheep breeding will fluctuate per year, as will the prices of milk and wool) and so the economic allocation method is the best generally applicable and consistent approach, although it is recommended to perform a sensitivity analysis with possible alternatives. Problems may arise in some exceptional situations, when markets (and therefore prices) are missing or when prices are distorted, as with imperfectly functioning markets (e.g. monopolies), or when there are government interventions (e.g. subsidies or compulsory requirements).

The economic allocation method can be considered straightforward when one or more co-products have markedly higher prices compared to others (Malmodin et al. 2001). For example, the manufacturing processes for electronic components consume much more energy per kilogram of product than other manufacturing processes, meaning that the energy consumption is more closely related to production cost than to weight. For electronics, cost may therefore be a better screening and allocation parameter than weight. Another advantage of economic allocation is that the results may be more rational in systems where large quantities of by-products with low economic value are produced. Most applications of economic allocation concern production of a 'main' product, which has a much higher price compared to other co-products, however, co-products may represent the major outputs in terms of mass or other physical quantities. In such situations, economic allocation can be applied, appealing to the 'economic causality' that is the basis of the process. Firms and manufacturers aim to produce the main product; co-products are only a side effect of production that they try to utilise as best as they can.

Both economic and physical partitioning/allocation can be expanded by expanding the product system to include the additional functions related to the co-products. This is known as system expansion. For example, iron ore is processed into steel and ground granular blast furnace slag (GGBFS) during the steelmaking process. Marginal slag (around 20%), which has cementitious properties displaces Ordinary Portland Cement (OPC). Hence, the system expansion of steel production involves the inclusion of the system of processes, including the production of its by-product (i.e. GGBFS) which is involved in OPC production. This enables the subtraction of the emissions associated with saved OPC production from the total emissions from steel and GGBFS production to find out the emissions of steel production.

3.5 Type of LCA

There are two main types of LCAs, attributional LCA (ALCA) and consequential LCA (CLCA) (Ekvall and Andrae 2006). The former one (state oriented) aims to illustrate the environmental impacts of a process, and the latter one (change oriented) deals with the impacts, resulting from the changes to a process.

Attributional LCA: ALCA provides an estimation of how much of the environmental impacts are contributed by a product throughout its life cycle. The examples so far provided in this chapter are attributional LCAs.

Consequential LCA: This LCA approach describes how environmental impacts will change in response to possible decisions or technological changes in the product supply change.

For example, district heating in a combined heat-and-power (CHP) plant in Sweden, generates emissions from the CHP plant but it reduces emissions in another plant in the electricity network by reducing its energy consumption. In this example, a CLCA includes both the emissions from the CHP plant and the reduction in emissions from another plant due to power generation from the CHP plant. In general, when a production process delivers more than one type of product, the CLCA should take into account how the process is affected by a change in the product investigated. If it affects the production of other products from the process, the system should be expanded to include the effect of that change. A more advanced CLCA can also include other types of consequences. An increased use of a material in the studied system can, for example, lead to less material being used in other systems.

ALCA is the most commonly used LCA method, and is suitable for presenting results for making policy decisions. For example, ALCAs of biofuels are routinely used to suggest that the use of biofuel as a replacement for conventional fuels will cause a certain percentage change in GHG emissions. However, because of several simplifications, ALCA cannot predict real-world environmental impacts. CLCA could conceptually be a superior approach, as it avoids many of the limitations of ALCA, by capturing consequences associated with the production of a product or delivering of a service. These limitations mean that even the best practical CLCAs cannot produce definitive quantitative estimates of actual environmental outcomes. Nevertheless, none of these approaches can provide absolute values and so a truncation error will exist in both approaches. Since these two approaches can only provide relative values, LCA is best used for comparative purposes. ALCA because of its simplistic approach is widely used by practitioners, industries, and researchers.

ALCA represents the environmental impacts resulting from a static system irrespective of economic or policy context. Although ALCA models impact without considering the function of changes in production, the result of an ALCA is commonly presented as an estimate of the effect of increasing or decreasing system outputs.

While ALCA is a statistical average, and context independent, CLCA ideally is dynamic, marginal, and context specific. ALCA is useful for comparing whether the effects or consequences of substitution need to be taken into account.

3.6 Uncertainty Analysis in LCA

There are uncertainties associated with the quality of data (i.e. inputs and output, and the emission factors) that are used to determine environmental impacts. Uncertainty is a measure of the knowledge of the magnitude of a parameter which is often quantified as a statistical distribution (Webster and Mackay 2003). The uncertainties result due to issues listed below Björklund (2002):

Data Inaccuracy: Due to variations in the values of the measurements used to derive the numerical values.

Data Gaps: Missing values in the model. When doing the LCA of biofuel, soil emissions associated with feedstock production cannot be ignored.

Unrepresentative Data: Data from non-representative sources. For example the emission factor for the electricity mix in the USA cannot be used in Denmark.

Model Uncertainty: Due to simplification of aspects that cannot be properly modelled in LCA, e.g. temporal spatial characteristics. For example, building service life or the use stage is usually considered 50 years regardless of the type of materials used in building construction. However it underestimates or overestimates the environmental performance of buildings with service lives of more or less than 50 years.

Uncertainty Due to Choices: e.g. allocation rules, system boundaries, weighting methods, etc.

Spatial Variability: e.g. real-world geographical differences. For example, soil emissions vary across regions due to variation in soil and climate.

Temporal Variability: Variations over time e.g. dispersion of emissions.

Epistemological Uncertainty: Due to the lack of knowledge on system behaviour.

Mistakes: Easy to make, difficult to find. Sometimes LCA practitioners forget to convert the units of an input and this error then propagates throughout the large spreadsheet providing the results.

There are typically four approaches to treat the aforementioned uncertainties, including a 'scientific' approach (more research, better data from the field or through experiments), a 'social' approach where stakeholders come to an agreement on the quality of data, a 'legal' approach, which requires the involvement of authoritative bodies to authenticate data and a 'statistical' approach (Monte Carlo, confidence intervals). The reduction in the inaccuracy of LCA data requires careful data collection, which is often costly and time-consuming and in some cases not even practically possible (e.g. the number of people who will be physically disabled due to production of a notebook computer). The first three approaches actually reduce the uncertainty by increasing the reliability and quality of data. The fourth approach actually determines the level of uncertainty depending on which of the first three approaches has been followed to improve the quality of data to attain an acceptable level of uncertainty. Monte Carlo simulation (MCS) is installed in most LCA software for conducting the statistical analysis. MCS is usually carried out for each data point used in the LCA analysis (Clavreul et al. 2012). The simulation is an iterative process utilising an input value from a probability function to produce a distribution of all possible output values for a large number of iterations at a required confidence level, which is usually 95% (Goedkoop et al. 2013). Once the inputs and outputs of the LCI are linked to relevant emission factor databases in the software, MCS can be conducted for a particular functional unit. For example, once the inventory data of 1 m^2 of building area is entered into the software and then linked to the corresponding emission databases, MCS of 1 m^2 of building can be conducted for a certain number of iterations and a specific confidence level. When running an MCS, two types of information will be asked. One is the level of confidence (e.g. 95%) and the other on is the number of iterations (e.g. 1000). The number of iterations can be more than 10,000. The higher the number of iterations, the more time it will take to generate uncertainty results. However, these iterations definitely produce more precise results. Figure 3.6 shows the MCS for a 100 m road using (i) traditional and (ii) recycling approaches for a confidence interval of 95% and 1000 iterations.

The results of an uncertainty analysis also help determine the probability of a predicted impact reduction potential. For example, only about 1.5% of the total GHG emissions from the pre-construction stage of a pavement can be mitigated due to the replacement of crushed rock base with recycled concrete aggregates. The level of environmental improvement is small and so is the uncertainty of the mean values of the carbon footprint that were determined in order to decide whether this 1.5% difference could be ignored. The MCS was used to analyse the uncertainty in the LCA at the pre-construction stage, using both traditional and recycling approaches (Figure 3.6). The standard deviations were 3.3% and 3.2% of the mean values of the carbon footprints in the pre-construction stage for the

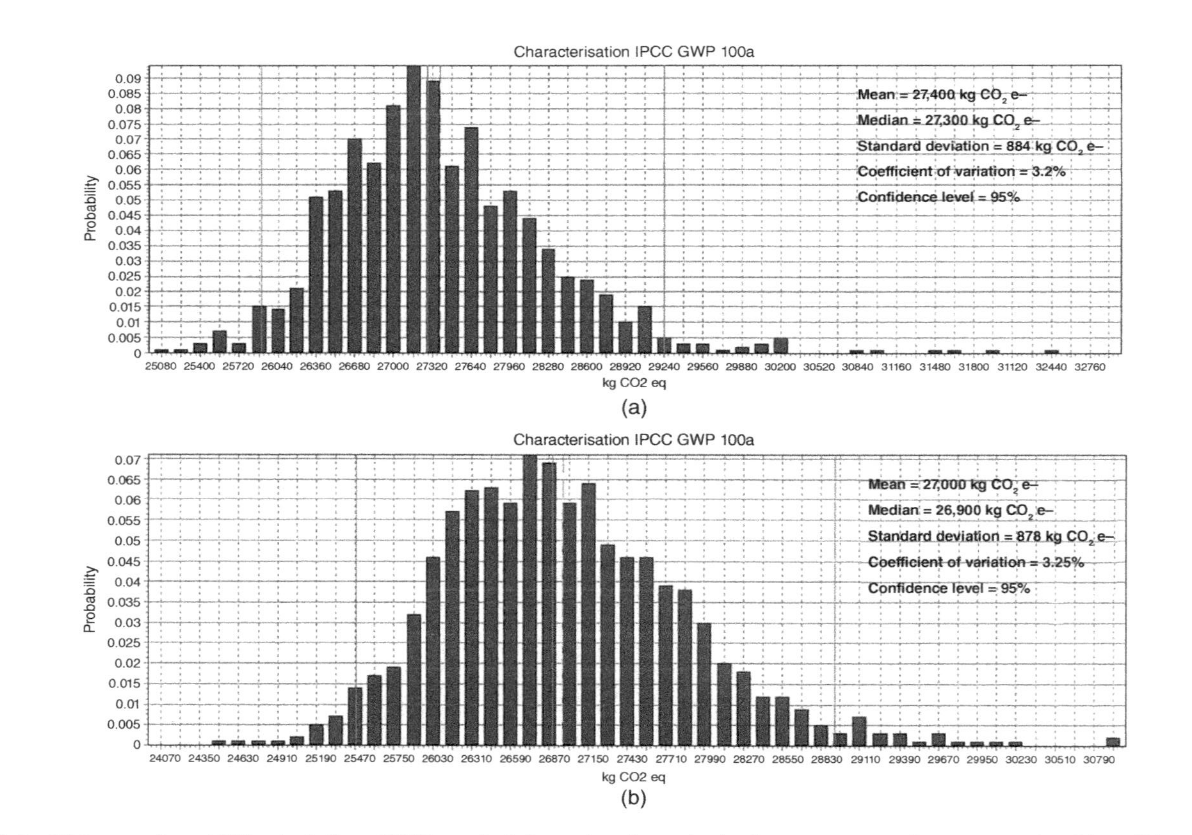

Figure 3.6 Histogram from MCS calculations (1000 runs) of the carbon footprint in the pre-construction stages using (a) traditional and (b) recycling approaches. Source: Based on Biswas (2014).

traditional and recycling approaches. These percentages were higher than the percentage differences (1.5%) between carbon footprints when using the traditional and recycling approaches. Therefore, the contribution of recycled crushed rubbles (RCR) to GHG mitigation was ignored.

The uncertainties sometimes depend on spatial and temporal variations and data/measuring uncertainty. The uncertainty of environmental impacts is caused by the parameter resolution in time and space. For example, there is considerable variability on a temporal and spatial level during normal crop production. The incorporation of local soil N_2O emission data in an LCA of the production and combustion of biodiesel in the semiarid climate of Western Australia found that the GHG emissions reduced from 63 kg of CO_2-e/GJ to 37 kg of CO_2-e/GJ, which was 41% less than the value estimated using IPCC default values. This demonstrates the significance of utilising regionally specific data when assessing GHG from the production and consumption of biodiesel.

In addition to data quality issues, the quality of emission databases of these inputs could cause the uncertainty of some impact results. For example, Table 3.6 shows the Monte Carlo simulation of 1 MWh of electricity generation performed for 1000 iterations using the SimaPro 8.3 software for a confidence interval of 95% (Arceo et al. 2019). The differences between the actual impact values and mean values calculated range from 0.02% to 83%. The calculated coefficient of variance (CV) of the global warming potential at 6% infers that the degree of uncertainty in the calculated impact is relatively small, which may be due to the fact that this category is commonly used and so there are more established emission databases for assessing this impact. Also this environmental impact has the lowest CV due to the fact that this is not directly affected by land use changes, variation in topography, water bodies, and vegetation coverage. However, discrepancies were found in more regional and local impact categories (i.e. acidification, eutrophication, human toxicity and eco-toxicities) due to the aforementioned reasons (Yoshida et al. 2013). Hung and Ma (2009) also noted that geographical location is important, which may be due to the fact that there are factors other than emissions, such as population, geography, locations (e.g. being next to the sea), which influence this impact and the LCA software cannot always handle this complexity.

3.7 Environmental Product Declaration

EPDs disclose the environmental performance of products based on an ELCA, which determines the environmental impacts for a product with pre-set categories of parameters based on the ISO 14040-44 series of standards. Having an EPD for

Table 3.6 Uncertainty analysis results using Monte Carlo simulation (FU = 1 MWh electricity supplied) (Arceo et al. 2019).

		Monte Carlo simulation results		
Environmental impacts	Unit	**Calculated value**	**Mean value**	**Coefficient of variance (CV)**
Global warming potential	kg CO_2 e-	295	284.3	5.6
Abiotic resource depletion (minerals)	kg Sb e-	5E−4	5.04E−4	13.0
Abiotic resource depletion (fossil fuels)	kg Sb e-	4.1	4.0	12.2
Land use and ecological diversity	ha	3.49E−4	3.36E−0	26.7
Water depletion	m^3 H_2O	1.5	1.5	12.9
Eutrophication	kg PO_4^{3-} e-	0.5	0.5	74.7
Acidification	kg SO_2 e-	2.7	2.6	59.0
Freshwater ecotoxicity	kg 1,4-DB e-	1.0	0.9	52.2
Marine ecotoxicity	kg 1,4-DB e-	1.1	0.8	55.3
Terrestrial ecotoxicity	kg 1,4-DB e-	1.37E−2	1.32E−2	32.6
Photochemical smog	kg NMVOC	4.2	4.1	77.5
Ozone depletion	kg CFC-11 e-	7.30E−5	7.30E−5	37.3
Ionising radiation	kBq U235 e-	2.1	1.4	118.5
Human toxicity	kg 1,4-DB e-	22.7	3.8	31.4
Respiratory inorganics	kg $PM_{2.5}$ e-	0.51	0.50	29.8

a product does not imply that the declared product is environmentally superior to alternatives – it is simply a transparent declaration of the life cycle environmental impacts of the product.

LCA is used to prepare EPDs which must meet the specific methodological requirements that are defined in product category rules (PCRs). The PCR varies from product to product due to the variation in the process, geographic areas, and environmental impacts associated their production. A PCR defines the criteria for a specific category of product and sets out the requirements that must be met when doing an LCA for developing an EPD for products under this category

Figure 3.7 Development and deployment of an Environmental Product Declaration.

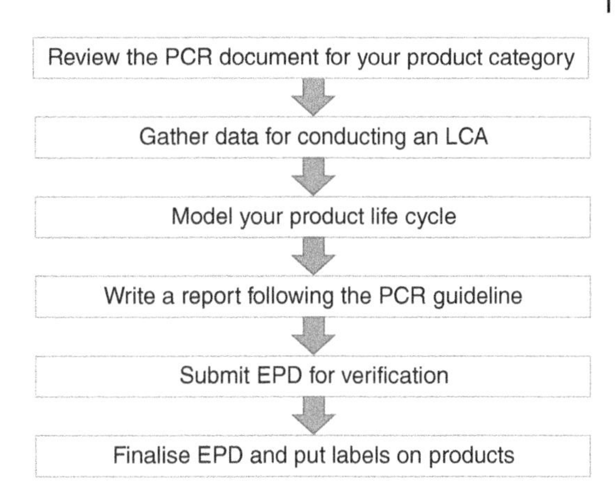

(Figure 3.7). PCRs guide the LCA practitioner on how to perform an evaluation of the environmental performance of a specific product. For example, rules for assessing the environmental performance of the manufacture of a chair will differ from those for assessing a desk. The key differences or similarities between the PCRs included differences in the purpose, standards, functional unit, scope and boundary, calculation rules/modelling, data quality rules, cut off rules, allocation, primary data requirements, secondary data sources, and impact categories. ISO 14025 (ISO 2006a) sets out a procedure for developing PCRs and the required content of a PCR, as well as the requirements for comparability.

PCRs offer some distinct benefits: greater consistency and comparability of assessments based on the same rules, modularity of assessment scope, transparency of requirements and in the development process, guidance, and clarity to users undertaking assessments within product sectors and the flexibility of use by any entity using ISO 14025. PCRs therefore provide the specific requirements for an EPD within a particular group and are critical to the transparency and comparability of EPDs. In addition to complying with PCRs, EPDs are also verified by third-party expert panels to ensure their accuracy prior to publication. The PCR defines the information that should be included in a life cycle assessment, helps minimise the practitioner's choice of assumptions and datasets, and defines the way the final data will be presented to the public. This protocol creates clear boundaries, much in the same way that a deed defines a parcel of land. Once the EPD is done, the environmental labels are put on products.

You need to have a program operator who is the individual or organisation that develops the PCRs or procedures that govern development of an EPD for a particular product or service in a region. If no PCR exists, LCA professionals need to be contacted to act as the program operator to develop the PCR which will need then to be submitted for verification. Typically, the PCRs are produced and owned by a third party, called the program operator.

In order to make the LCA assessment valuable, we need to assume that the LCAs are done in the same way so that an analysis of the same product will yield the same results, no matter who does the analysis. Environmental Product Declarations are owned by the companies that are selling the product that is being assessed by the EPD. The program operator is responsible for making sure that the rules are developed in a transparent and equitable fashion according to international standards, and that the validation of EPDs using the PCR is done in such a way that data and analysis quality is assured (Figure 3.7).

Case Study: The Gulf Organization for Research and Development (GORD) was the operator and owner of a PCR to administer the development of Gulf States concrete producers' EPDs to LCA practitioners and third party verifiers. In this case, the LCA practitioner was the Sustainable Engineering Group at Curtin University and VITO in Belgium was the verifier (Biswas et al. 2017). Following EN 15804:2012+A1:2013, the declared functional units are given below:

Declared Units	1 m^3 of ready mix concrete manufactured by SMEET, Qatar

The background LCA that had to be carried out to develop the EPD of ready mix and precast concrete followed Gulf Green Mark – EPD – Product Category Role (GGM-EPD PCR) (GORD 2015) using 100% local data from concrete manufacturing in Qatar.

System Boundary: The system boundaries of this LCA cover the modules A1 to A3 (Figure 3.8), and the LCA takes a '*cradle to gate*' system boundary:

- A1: Raw materials pre-processing and acquisition phase;
- A2: Transportation of all raw materials to SMEET;
- A3: Manufacturing of the different concrete products at SMEET factory

It must be noted that raw materials (A1), transport (A2), and manufacturing processes (A3) are included in the LCA calculation according to PCR rule EN 15804:2012+A1:2013, clause 6.3.4. Similar to '*cradle to gate*' approach for concrete has been followed in other recent LCAs of concrete (Biswas et al. 2017).

The following life cycle stages have been excluded due to the inability to predict how and where the materials are used during 'use' and at the 'end of life' (EoL) stages following manufacturing (i.e. A3):

- A4: Transportation of ready mix or precast concrete to the construction site;
- A5: Construction activities;

	Production stage			Construction process stage		Use stage							End of Life (EoL) stage				Resource recovery stage
	Raw materials	Transport	Manufacturing	Transport to construction site	Construction/installation	Use	Maintenance	Repair	Replacement	Refurbishment	Operational energy use	Operational water use	De-construction / demolition	Transport to waste processing	Waste processing	Final disposal	Reuse Recovery Recycling potential
Modules	A1	A2	A3	A4	A5	B1	B2	B3	B4	B5	B6	B7	C1	C2	C3	B4	D
Ready mix concrete	x	x	x	ND	ND	ND	ND	ND	ND	ND	ND	ND	ND	ND	ND	ND	ND
Precast concrete	x	x	x	ND	ND	ND	ND	ND	ND	ND	ND	ND	ND	ND	ND	ND	ND

X – Module declared; ND – Module not declared

Figure 3.8 Scope of declaration in EPDs: cradle to gate A1, A2, and A3.

- B1–B7: Building use and maintenance;
- C1–C4: End of life (i.e. dismantling and disposing of construction and demolition waste);
- D: Benefits and loads beyond the system boundaries, e.g. recovery of concrete materials.

A description of life cycle stages that were included in this LCA are as follows:

- *Raw material supply (A1)*: Extraction, handling, transportation, and processing of the raw materials including cementitious materials, aggregates, water, admixtures, and sand for use in the production of concrete (i.e. A3). The packaging of raw materials supplied to the batching plant was excluded as they are supplied in bulk instead of by industrial package.
- *Transportation of the raw materials from suppliers to batching and precast plants at SMEET (A2)*: This distance was multiplied by the weight of the materials to produce a declared unit of ton-kilometres (tkm). This is because the emission factors associated with transportation units are presented in LCA software as tkm values.
- *Manufacturing (A3)*: The amount of energy consumed to store, batch, mix, and distribute the concrete and operate the facility (ready mix concrete plant) and the amount of energy used in curing in the pre-cast plant. This stage also considers water use in mixing and distributing concrete. Maintenance of equipment is not included in the LCA, except for frequently consumed items. The packaging of manufactured products, including ready mix and precast concrete is not included in A3.

Since this LCA does not consider the impacts associated with the disposal of any degradable wastes/leachate into landfill, long-term emissions were not considered (Doka and Hischier 2005). Instead this LCA deals with short-term emissions including air emissions from combustion from the production and transportation of construction materials.

The LCI data for the processes of A1, A2, and A3 were collected from SMEET to develop the LCI overview for the different concrete classes studied.

The LCI which forms the basis of this LCA relates to the year 2016.

Once the scope has been defined, the LCA of concrete mixes was conducted following the procedure discussed in the previous section to calculate the environmental impacts in terms of eco-points. However, as the normalisation factor and weights vary from region to region, these values were provided in the PCR by the operator GORD. This PCR requires the reporting of a summary metric called 'GSAS eco-points' (Table 3.7). The metric is derived by normalising 11 impacts for a subset of the parameters from EN 15804:2012+A1:2013, then weighting them

Table 3.7 Parameters, normalisation factors and weightings for calculation of GSAS eco-points (GORD 2015).

Environmental impacts (EI)	Units	Normalisation factor (NF)	Weighting factor (WF)
Global warming potential (climate change) (GWP)	kg CO_2 e- per person per year	12,730	15.6
Ozone depletion potential (ODP)	kg CFC-11 e- per person per year	0.227	13.14
Acidification potential for soil and water (AP)	kg SO_2 e- per person per year	71.3	8.67
Eutrophication potential (EP)	kg PO_4 e- per person per year	33.4	5.04
Photochemical ozone creation potential (POCP)	kg C_2H_4 e- per person per year	21.5	7.71
Abiotic depletion potential – elements (ABPE)	kg Sb e- per person per year	39.1	6.31
Abiotic depletion potential – fossil fuels (ADPFF)	MJ per person per year	79,823	5.3
Net use of fresh water (FW)	m^3 per person per year	534	17.1
Hazardous waste disposed (HWD)	kg per person per year	183	8.4
Non-hazardous waste disposed (NWD)	kg per person per year	460	3.34
Radioactive waste disposed (high-level nuclear waste)(RWDHL)	kg per person per year	0.0110	9.4

according to factors that reflect their relative importance, then combining the normalised and weighted results (GORD 2015). Figure 3.9 shows the LCA output of one of the concrete mixes produced by the manufacturer.

Once the background report was developed by the Sustainable Engineering Group (SEG) at Curtin University, the critical review process of the LCA studies was commissioned by GORD. This was done between October and December 2016. After the receipt of the draft LCA background report from SEG (October 2016), VITO prepared a detailed list with review comments on methodological

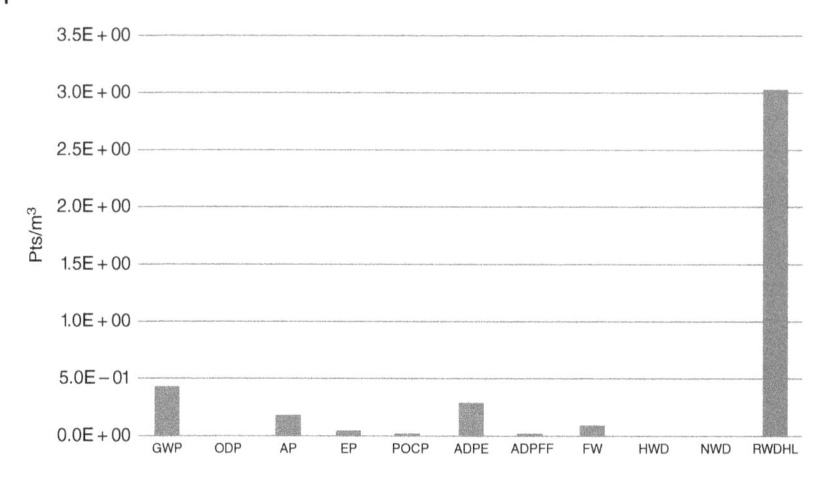

Figure 3.9 Eco-points of concrete ready mix C40. Source: Biswas et al. (2017) / with permission of Elsevier.

issues, assumptions made and data used. Furthermore, general questions to support the comprehensibility of the report and specific recommendations for improvements of the LCA studies were included. SEG and VITO then discussed the issues raised in the comments as well as suggestions made by VITO in several rounds. Based on the results of these discussions, SEG compiled a draft final background report.

VITO then confirmed in their critical review statement that the study 'Life Cycle Assessment of ready mix and precast concrete for environmental production declarations in Qatar based on Gulf EPD PCR' is based on the main guiding principles defined in the international standard series [ISO 14040, 2006] and [ISO 14044, 2006], the European norm EN 15804:2013 A1 and the GORD PCR for Gulf Green Mark EPDs. Further confirmations included:

- Methods used are scientifically and technically valid for the given goal and scope of the study;
- Data used are appropriate, sufficient, and reasonable in respect to the goal and scope of the study;
- Conclusions drawn reflect the goal and scope of the study and the limitations identified;
- Reports are transparent and consistent.

Once the background report had been finally approved, the EPD of the concrete mixes were produced, including results with a small description on the background and methodology for the client or concrete manufacturers to use for commercial purposes.

References

ABC (2017). Australian canola for European biodiesel emits half the greenhouse gas of fossil fuels. 18 Dec. www.abc.net.au/news/rural/2017-12-18/australian-canola-approved-as-low-emission-fuel-for-europe/9269232#:~:text=Europe%20had%20tightened%20its%20Renewable,set%20at%2035%20per%20cent (accessed 20 February 2021).

Arceo, A., Biswas, W., and John, M. (2019). Eco-efficiency improvement of Western Australian remote area power supply. *Journal of Cleaner Production* 230: 820–834.

Arceo, A., Rosano, M., and Biswas, W.K. (2018). Eco-efficiency analysis for remote area power supply selection in Western Australia. *Clean Technologies and Environmental Policy* 20: 463–475.

Aye, L., Ngo, T., Crawford, R.H. et al. (2012). Life cycle greenhouse gas emissions and energy analysis of prefabricated reusable building modules. *Energy and Buildings* 47 (April): 159–168.

Azapagic, A. and Clift, R. (1999). Allocation of environmental burdens in co-product product-related burdens (part 1). *International Journal of Life Cycle Assessment* 4 (6): 357–369.

Balkau, F., Gemechu, E.D., and Sonnemann, G. (2015). Life Cycle Management Responsibilities and Procedures in the Value Chain. In: Sonnemann, G., Margni, M. (eds) Life Cycle Management. *LCA Compendium – The Complete World of Life Cycle Assessment*, 10–12 November 1999, 315–318. Springer, Dordrecht. https://doi.org/10.1007/978-94-017-7221-1_14.

Biswas, W.K. (2009). Life cycle assessment of seawater desalinization in Western Australia. *World Academy of Science, Engineering and Technology* 56: 369–375.

Biswas, W.K. (2014). Carbon footprint and embodied energy consumption assessment of building construction works in Western Australia. *International Journal of Sustainable Built Environment* 3: 179–186.

Biswas, W.K. et al. (2017). Life cycle assessment for environmental product declaration of concrete in the Gulf States. *Sustainable Cities and Society* 35: 36–46.

Björklund, A.E. (2002). Survey of approaches to improve reliability in LCA. *International Journal of Life Cycle Assessment* 7: 64. https://doi.org/10.1007/BF02978849.

Boustead, I., Dove, W.T., Halada, K., and Matsuno, Y. (1999). Primary metal industry eco-profile calculations: a discussion of allocation methods. In: *Proceedings of the Fourth International Conference on Ecomaterials*, 10–12 November 1999, 315–318. Gifu, Japan: http://www.boustead-consulting.co.uk/download/gifu99.pdf (25 March 2021).

Clavreul, J., Guyonnet, D., and Christensen, T.H. (2012). Quantifying uncertainty in LCA-modelling of waste management systems. *Waste Management* 32: 2482–2495. https://doi.org/10.1016/J.WASMAN.2012.07.008.

Doka, G. and Hischier, R. (2005). Waste treatment and assessment of long-term emissions. *International Journal of Life Cycle Assessment* 10: 77–84. https://doi.org/10.1065/lca2004.12.181.9.

Ekvall, T. and Andrae, A. (2006). Attributional and consequential environmental assessment of the shift to lead-free solders. *International Journal of Life Cycle Assessment* 11: 344–353. https://doi.org/10.1065/lca2005.05.208.

European Environmental Agency (2019). EMEP/EEA air pollutant emission inventory guidebook 2019. https://www.eea.europa.eu/publications/emep-eea-guidebook-2019 (accessed 20 January 2020).

European Union (2010). *A Guide for Business and Policy Makers to Life Cycle Thinking and Assessment*. Luxembourg: Publications Office of the European Union.

Frischknecht, R., Althaus, H.J., Bauer, C. et al. (2007). The environmental relevance of capital goods in life cycle assessments of products and services. *International Journal of Life Cycle Assessment* 12: 7–17.

Goedkoop, M., Oele, M., Leijting, J. et al. (2013). *Introduction to LCAwith SimaPro*. Netherlands: PRé.

Guinée, J.B., Heijungs, R., and Huppes, G. (2004). Economic allocation: examples and derived decision tree. *International Journal of Life Cycle Assessment* 9 (1): 23–33.

Gulf Organization for Research and Development (2015). Gulf Green Mark Environmental Product Declaration 2015, Product Category Rules to EN 15804:2012+A1:2013. GORD, Doha.

Helling, R. (2015). Driving innovation through life-cycle thinking. *Clean Technologies and Environmental Policy* 17: 1769–1779. https://doi.org/10.1007/s10098-015-0928-7.

Hill, J., Nelson, E., Tilman, D., Polasky, S., Tiffany, D. (2006) Environmental, economic, and energetic costs and benefits of biodiesel and ethanol biofuels. *Proceedings of the National Academy of Sciences of the United States of America*, 103 (30), 11206–11210; DOI: https://doi.org/10.1073/pnas.0604600103.

Hischier, R., Althaus, H.J., and Werner, F. (2005). Developments in wood and packaging materials life cycle inventories in ecoinvent. *International Journal of Life Cycle Assessment* 10 (1): 50–58.

Hung, M. and Ma, H. (2009). Quantifying system uncertainty of life cycle assessment based on Monte Carlo simulation. *International Journal of Life Cycle Assessment*, 14, 19–27. DOI https://doi.org/10.1007/s11367-008-0034-8.

ISO (International Organization for Standardization) (2006). ISO 14040:2006. http://www.iso.org/iso/iso_catalogue/catalogue_tc/catalogue_detail.htm?csnumber=37456 (accessed 5 January 2021).

ISO (2006a). Environmental labels and declarations—type IIIenvironmental declarations—principles and procedures. European Standard EN ISO 14025. International Organization for Standardization.

ISO (2007). Environmental management—Life cycle assessment—Principles and framework. ISO 14040, International Organization for Standardization (ISO), Geneva.

Klöpffer, W. (2003). Life-cycle based methods for sustainable product development. *The International Journal of Life Cycle Assessment* 8 (3): 157–159.

Malmodin, J., Oliv, L., and Bergmark, P. (2001). Life cycle assessment of third generation (3G) wireless telecommunication systems at Ericsson. *Proceedings of Ecodesign 01: Second International Symposium on Environmentally Conscious Design and Inverse Manufacturing*, 11–15 December 2001, Tokyo, Japan. http://ieeexplore.ieee.org/stamp/stamp.jsp?tp=&arnumber=992305 (31 March 2021).

Marvuglia, A., Cellura, M., and Heijungs, R. (2010). Toward a solution of allocation in life cycle inventories: the use of least-squares techniques. *Int J Life Cycle Assess* 15: 1020–1040. https://doi.org/10.1007/s11367-010-0214-1.

McConville, J.R. (2006). *Applying Life Cycle Thinking to International Water and Sanitation Development Projects: An assessment tool for project managers in sustainable development work*. Michigan Technological University.

McDougall, F.R., White, P.R., Franke, M., and Hindle, P. (2001). *Integrated Solid Waste Management: A Life Cycle Inventory*, 2e, 1. USA: Blackwell Science Inc.

Niederl-Schmidinger, A. and Narodoslawsky, M. (2008). Life cycle assessment as an engineer's tool? *Journal of Cleaner Production* 16 (2): 245–252.

Suh, S., Weidema, B., Hoejrup, J. and Heijungs, S.R. (2010). Generalized Make and Use Framework for Allocation in Life Cycle Assessment. *Journal of Industrial Ecology* 14 (12): 335–353.

UNEP (2012). Towards a Life Cycle Sustainability Assessment: Making Informed Choices on Products. https://www.lifecycleinitiative.org/wp-content/uploads/2012/12/2011%20-%20Towards%20LCSA.pdf.

Van der Harst, E. and Potting, J. (2013). A critical comparison of ten disposable cup LCAs. *Environmental Impact Assessment Review* 43: 86–96.

Viere, T., Amor, B., Berger, N. et al. (2020). Teaching life cycle assessment in higher education. *International Journal of Life Cycle Assessment* 26: 511–527. https://doi.org/10.1007/s11367-020-01844-3.

Weber, C.L. and Matthews, H.S. (2008). Food-miles and the relative climate impacts of food choices in the United States. *Environmental Science & Technology* 42 (10): 3508–3513.

Webster, E. and Mackay, D. (2003). Defining Uncertainty and Variability in Environmental Fate Models. Canadian Environmental Modelling Centre Trent University Peterborough, Ontario, Canada.

Yoshida, H., Christensen, T.H., and Scheutz, C. (2013). Life cycle assessment of sewage sludge management: a review. *Waste Management and Research* 13: 1083–1101.

4

Economic and Social Life Cycle Assessment

4.1 Economic and Social Life Cycle Assessment

A product can be environmentally friendly but it may not be economically or socially viable. Therefore, using the same system boundary, goal, and functional unit as environmental life cycle assessment (ELCA) as discussed in Chapter 3, both life cycle costing (LCC) and social life cycle assessment (SLCA) can be additionally conducted to determine the economic and social performance of the same product to ensure that it fulfils the triple bottom line objectives of sustainability. All inputs that have been determined or quantified to develop the life cycle inventory for the ELCA are used in LCC of the same product. The values of inputs in terms of kg, l, kWh, and MJ are converted to a monitory value (e.g. Dollar) using unit prices in conducting a LCC analysis. Some output values such as wastes and GHG emissions that were determined during the life cycle inventory development process can also be considered in the LCC analysis for determining environmental costs or saving. Some input values which are not determined in the life cycle inventory (LCI) for ELCA, but are important to consider need to be included in the LCC analysis, such as labour cost and wages.

In the case of SLCA, the stages of ELCA are replaced with stakeholders. For example, mining, processing, manufacturing, use and end of life stages are replaced with miners, processors, manufacturers, users and recyclers, respectively, and may also include surrounding communities. Inputs like chemicals and energy are replaced with the perceptions/opinions/feedback from the stakeholders in order to determine the exposure to chemicals, wastes, and emissions generated in the product supply chain but also including working conditions, workforce management, and land use changes which may affect their well social being directly or indirectly.

Engineering for Sustainable Development: Theory and Practice, First Edition.
Wahidul K. Biswas and Michele John.

4.2 Life Cycle Costing

Economic factors are important in any engineering decision-making. The process of identifying and documenting all the costs spent over the life of a product or service is known as life-cycle costing (LCC). This is typically a cash flow oriented cost accounting system without having cause and effect relationships meaning that it cannot capture the changes in benefits associated with changes in the cost components. LCC is widely used in making investment decisions in engineering projects. It is an economic cousin of ELCA and if combined with ELCA, it can help in achieving the optimal and most cost-effective environmental solutions. For example, there are a number of ways, including renewable energy, recycling, energy efficiency measures, to reduce GHG emissions by 20% for electronics industry by 2030. LCC will determine the least best cost option which delivers the required level of environmental outcome. This assists decision makers and investors in achieving an environmental target in a cost-competitive manner.

LCC is defined as a technique which enables comparative cost assessments to be made over a specified period of time, taking into account all relevant economic factors, both in terms of initial costs and future operational costs (Heralova 2017). This is aligned with ELCA and is the sum of acquisition (capital) cost and ownership (operational) cost of a product or service over its life cycle from mining to material production, manufacturing, distribution, usage, maintenance, and the end of life disposal stages (Australian Standard AS/NZS 4536:1999).

Because of its ability to estimate or anticipate the future costs for cash flow analysis, LCC tools are also considered an effective technique for forecasting and evaluating the cost performance of a product or service. This requires analysis to assume the future costs on the basis of current and past trends. This creates a limitation in this methodology as it is based on the estimation and valuation of uncertain future events and outcomes, and therefore, subjective factors or value judgements are involved in the process that may affect the results (Ristimäki et al. 2013). For example, annual energy indexation is an uncertain factor due to fluctuations in energy prices, which could potentially impact on a life cycle cost outcome. Future cost figures may change due to changes in technology, market, life style, and political discussions.

Even though LCC appears not to be completely theoretically accurate, it has many benefits. For example, the analysis provides an indication of what strategic options and aspects to consider more seriously. Secondly, the results of the LCC analysis are presented with a common unit (currency) allowing the comparison between competing engineering solution options for an investment decision. Thirdly, an LCC analysis processes and simplifies a significant amount of information and provides a valuable life cycle perspective on the different alternative options considered.

The key incentive for applying an LCC analysis is to increase the possibility of a cost reduction during the operational phase, even if an additional increase in the initial investment is necessary. The LCC approaches help to compare capital cost with operating cost for competing design options and to find the optimised better option that offers long-term benefits. For example, the additional cost associated with the replacement of a compact fluorescent lamp (CFL) which has a life time of 4 years with a LED lamp with a 8-year life time would be $10, which is far less than the annual operational saving of $30 (or $240 over the entire life) meaning that the latter can offer long-term economic benefits. In addition, LCC allows better resource management due to longer-term cost visibility, the identification of high-cost functional stages and provides cost saving opportunities by helping us determine least cost options.

By applying an LCC perspective in the early design phase, engineers are able to obtain a deeper understanding of total costs during the life cycle to assist engineering design strategies. Buildings are a long-term investment as they are used for at least 50 years or more. Both ELCA and LCC need to be conducted for different building specifications (e.g. brick wall building, concrete wall building, terracotta building) in order to select the least cost specification with reduced levels of environmental impact. LCC could confirm that a little more expenditure during the pre-construction and construction stages could offer significant financial savings throughout the life of the building.

While performing LCC for a engineering project, functional and technical requirements of alternative engineering strategies need to specified for comparative assessment. This is because capital costs and operation costs vary with technological options. For example, the capital and operational costs of solar photovoltaic panels are different to diesel generators for generating electricity given the variation in technological characteristics and fuels used. Secondly, the relevant economic criteria discount rate (DR), analysis period, escalation rates, component replacement frequency, and maintenance frequency of alternative strategies need to be identified as they can also vary with location and product. Thirdly, significant costs need to be categorised or grouped to facilitate the differentiation process and to highlight the items increasing the overall LCC. For example, labour and fuel costs are operational costs but they need to be separated, otherwise they could underestimate or overestimate the cost of labour due to the use of labour intensive or automated processes. Fourthly, a systematic sensitivity approach needs to be carried out to reduce the overall uncertainty, so that it is clear which cost reduction policy instrument (e.g. tax, subsidy) or cost component (e.g. maintenance) is the influencing factor. Any small change in the variation of one item could result in significant variation in the LCC results known as the sensitive item in the LCC analysis and this allows investors or project managers to make the right strategies to reduce the cost of the sensitive item in order to reduce the overall LCC or to enable precautionary actions to avoid 'sensitive item' impact.

4.2.1 Discounted Cash Flow Analysis

The time value of money in LCC analysis is expressed as a discount rate which depends upon capital cost, inflation rate (IR), and social behaviour. The item which costs $1 now will cost more than $1 in the next year and the value will gradually increase in the subsequent years. This discount factor is used to convert all these gradually increasing future costs to a present value (PV) for conducting LCC. After considering the inflation rate and discount factor, the present value of any future cost of a product or service can then be calculated using the following equation:

$$\mathrm{PV} = \sum_{i=0}^{i=n} \frac{C \times (1 + \mathrm{IR})^i}{(1 + \mathrm{DR})^i} \tag{4.1}$$

PV = Present value

i = 1, 2, …, n; year value till end of life of the product or service

C = Present cost ($)

IR = Inflation rate (%)

DR = Discount rate (%)

The 'present cost' consists of capital costs, operational and maintenance costs, end of life disposal costs, and residual costs. The capital costs are usually the costs of fixed assets such as machinery and infrastructure required to convert inputs into useful products or services. The life of any project depends on the life of these assets. In the case of LCC analysis, the investment in these capital items is made in year 0. Therefore, there is no future cost of these fixed cost items. If $i = 0$ in Eq. (4.1) for capital cost, the value of C will remain unchanged. Once we have started to use these assets for producing products, operational costs will begin. These operational and maintenance costs begin from year 1. Operational cost items are required to produce products directly using the fixed assets. For example, we use materials/feedstock, energy and chemicals in machines in a factory to build or produce a product. On the other hand, maintenance costs are required for servicing machinery and infrastructure to produce products at the required level or standard without causing a loss in productivity. The replacement costs take place during the use or operational stage of a product life cycle and does not take place every year. For example, in solar electricity project of 20 years, the batteries need to be replaced every 4 or 5 years. Similarly, the membranes of a water desalination plant have to be changed every 4 or 5 years depending on the quality and demand of water. The costs associated with decommissioning, demolition of old infrastructure, and disposal or recycling occur at the end of life of the project. In some cases, the end of life (EoL) products still have salvage value, which is the amount that an asset is estimated to be worth at the end of its useful life. For example, building construction and demolition wastes can be used as recycled aggregates in concrete while complying with structural performance guidelines. This is known as residual or salvage value. A compressor after its end of life can also be remanufactured and so capital equipment can also have a salvage value.

The present values or PVs of costs calculated using Eq. (4.1) are the basis for quantifying the following economic indicators in the economic sustainability assessment of products or services.

$$\begin{aligned}\text{Life cycle cost} &= \text{capital cost} + \text{PV of replacement cost}\\ &\quad + \text{PV of operational cost} + \text{PV of maintenance cost}\\ &\quad + \text{PV of end of life disposal cost} - \text{PV of residual cost}\end{aligned}$$

Net present value (NPV): It is the difference between the present value of cash inflows into the industry (e.g. income, revenue) and the present value of cash outflows (e.g. material, energy, transport costs, administrative costs). The option having the highest NPV is considered the best option from an economic perspective.

Payback period (PP): It is the number of years it takes to recover the investment or incremental cost. The option having the shortest payback period is considered as attractive or acceptable. When the summation of PVs of net benefit associated with the replacement of an efficient product is the same as the incremental cost, the payback period is then over.

Benefit cost ratio (BCR): It is the ratio of the present value of net benefits to the present value of investment or incremental cost. If the BCR of any option is greater than 1, this indicates that the present value of benefits outweighs the present value of the investment or incremental costs.

Internal rate of return (IRR): It is the 'rate of return' that makes the net present value of all cash flows (both positive and negative) from a particular investment and project life equal to zero. The higher the IRR, the better the performance of the investment. This means that the profitability of an engineering project is possible and then to achieve a profit within a short period of time to quickly recover the incremental costs.

Annualised Life Cycle Cost (ALCC): ALCC is the LCC expressed in terms of a constant cost per year. It is the annual expenditure required to pay for the system over its life time and includes the costs of repayments on borrowed capital. ALCC is determined by multiplying the LCC with the capital recovery factor (CRF)

$$\text{ALCC} = \text{LCC} \times \text{CRF} \tag{4.2}$$

where CRF is the factor used to calculate the amount of regular payments needed to recover a present value at a given interest rate over a specified time period.

$$\text{CRF} = \frac{(1+d)^N d}{(1+d)^N - 1} \tag{4.3}$$

For a project life of 20 years and discount factor of 10%, the CRF is 0.12.

Levelised Cost or Unit Cost: The levelised cost is the amount equally spread throughout the project or product life time and is probably the most useful figure in comparing technologies. It expresses the average cost of generating each useful unit of energy during the life time of the system. This also known as unit cost and is expressed as,

$$\text{Unit cost} = \frac{\text{ALCC}}{\text{Annual outputs}} \tag{4.4}$$

For example, the total energy produced is 50,000 kWh and the ALCC is $100,000, the unit energy cost is,

$$\text{Unit cost} = \$100{,}000/50{,}000 = \$2 \text{ per kWh}$$

Practice Example 1: LCC, Payback period (PP) and benefit cost ratio (BCR) for alternative lamp options

Compact fluorescent lamps (CFLs) and light emitting diode lamps (LED) have been identified as alternatives to conventional incandescent lamps (CILs).

Compare the life cycle global warming potential (GWP) of a CIL, CFL, and LED lamp using following data.

Determine which is the best economic option. Assume that the whole quantity of lamps to meet the functional requirement has to be purchased at the beginning. The data for this exercise has been provided in Box 4.1.

Box 4.1 Data for practice example 1

	CIL	CFL	LED
Rating (W)	60	15	10
Life time (h)	2,000	8,000	16,000
Purchase price ($)	2	10	30
Electricity price ($)	$0.4/kWh		
Interest rate (%)	3		
Discount rate (%)	5		
Annual usages of lamps	2000		

Solution

Functional unit = 16,000 h

Life time of study = 16,000 h/2000 h = 8 yr

Number of lamps required to meet this requirement

CIL = 16,000 h/2000 h = 8

CFL = 16,000 h/8000 h = 2

LED = 16,000 h/16,000 h = 1

Capital costs (lamps to be purchased at beginning)

CIL = 8 × \$2 = \$16

CFL = 2 × \$10 = \$20, Additional investment cost = \$20 – \$16 = \$4

LED = 1 × \$30 = \$30, Additional investment cost = \$30 – \$16 = \$14

Annual cost of electricity consumption

CIL = 60 W × 2000 h = 120,000 Wh/1000 = 120 kWh × \$0.4/kWh = \$48

CFL = 15 W × 2000 h = 30,000 Wh/1000 = 30 kWh × \$0.4/kWh = \$12

LED = 10 W × 2000 h = 20,000 Wh/1000 = 20 kWh × \$0.4/kWh = \$8

Present values of operational costs (i.e. electricity), using Eq. 4.1

CIL (annual cost as per current price \$48)

$$= \frac{\$48x(1+3\%)^1}{(1+5\%)^1} + \frac{\$48x(1+3\%)^2}{(1+5\%)^2} + \frac{\$48x(1+3\%)^3}{(1+5\%)^3} + \frac{\$48x(1+3\%)^4}{(1+5\%)^4} + \frac{\$48x(1+3\%)^5}{(1+5\%)^5} + \frac{\$48x(1+3\%)^6}{(1+5\%)^6} + \frac{\$48x(1+3\%)^7}{(1+5\%)^7} + \frac{\$48x(1+3\%)^8}{(1+5\%)^8}$$

$$= \$47.09 + \$46.19 + \$45.31 + \$44.45 + \$43.6 + \$42.77 + \$41.95 + \$41.16$$

$$= \$352.51$$

CFL (annual cost as per current price \$12)

$$= \frac{\$12x(1+3\%)^1}{(1+5\%)^1} + \frac{\$12x(1+3\%)^2}{(1+5\%)^2} + \frac{\$12x(1+3\%)^3}{(1+5\%)^3} + \frac{\$12x(1+3\%)^4}{(1+5\%)^4} + \frac{\$12x(1+3\%)^5}{(1+5\%)^5} + \frac{\$12x(1+3\%)^6}{(1+5\%)^6} + \frac{\$12x(1+3\%)^7}{(1+5\%)^7} + \frac{\$12x(1+3\%)^8}{(1+5\%)^8}$$

$$= \$11.77 + \$11.55 + \$11.33 + \$11.11 + \$10.9 + \$10.69 + \$10.49 + \$10.29$$

$$= \$88.13$$

LED (annual cost as per current price \$8)

$$= \frac{\$8x(1+3\%)^1}{(1+5\%)^1} + \frac{\$8x(1+3\%)^2}{(1+5\%)^2} + \frac{\$8x(1+3\%)^3}{(1+5\%)^3} + \frac{\$8x(1+3\%)^4}{(1+5\%)^4}$$

$$+ \frac{\$8x(1+3\%)^5}{(1+5\%)^5} + \frac{\$8x(1+3\%)^6}{(1+5\%)^6} + \frac{\$8x(1+3\%)^7}{(1+5\%)^7} + \frac{\$8x(1+3\%)^8}{(1+5\%)^8}$$

$$= \$7.85 + \$7.7 + \$7.55 + \$7.4 + \$7.27 + \$7.13 + \$6.99 + \$6.86 = \$58.75$$

LCC = Capital cost + present value of operational cost

CIL = \$16 + \$352.51 = \$368.51

CFL = \$20 + \$88.13 = \$108.13

LED = \$30 + \$58.75 = \$88.75

Based on LCC, LED has the lowest LCC
Payback period (PP) of three lamps have been calculated as follows:

CFL = additional cost of investment is \$4 and life cycle operational cost saving is \$264.38
Additional cost can be recovered within first year's saving of \$47.09 – \$11.77 = \$35.31
PP = 4/35.31 = 0.11 years
LED = additional cost of investment is \$14 and life cycle operational cost saving is \$293.76
Additional cost can be recovered within first year's saving of \$47.09 – \$7.85 = \$39.24
PP = 14/39.24 = 0.36 years

Benefit cost ratio (BCR)

CFL = additional cost of investment is \$4 and life cycle operational cost saving is \$264.38 (i.e., 352.51 – 88.13)
BCR = \$264.38/\$4 = 66.095
LED = additional cost of investment is \$14 and life cycle operational cost saving is \$293.76
BCR = \$293.76/14 = 20.98

Based on PP and BCR, CFL lamp has the lowest PP and highest BCR.

Practice Example 2: Top loader vs front loader

Compare the life cycle costs of a top loading and front loading washing machine using the following data. The capacity of both loaders is the same (i.e. 10 kg) and they have the same life cycle of 5 years. The rate of inflation is 3%, while the discount rate is 10%. Data for LCC analysis has been provided in Box 4.2.

Box 4.2 Data for practice example 2

	Top loader	Front loader
Purchase price	\$700	\$1200
Wash cycles per year	250 wash cycles/year	
Water consumption	80 l/wash cycle	60 litre/wash cycle
Electricity consumption	600 kWh/year	320 kWh/year
Detergent consumption	80 g/wash cycle	40 g/wash cycle
Current price of detergent	\$ 8/kg	\$10/kg
Current price of electricity	\$0.4/kWh	
Current price of potable water	\$2.5/kilolitre	
End of life disposal cost	\$50	\$80

Solution

Cost of electricity consumption per year

$$\text{Top loader} = 600\ \text{kWh} \times \$0.4/\text{kWh} = \$240$$

$$\text{Front loader} = 320\ \text{kWh} \times \$0.4/\text{kWh} = \$128$$

Cost of water consumption per year

$$\text{Top loader} = 80\ \text{l/wash cycle} \times 250\ \text{wash cycles/year} \times \$2.5/1000\ \text{l} = \$50$$

$$\text{Front loader} = 60\ \text{l/wash cycle} \times 250\ \text{wash cycles/year} \times \$2.5/1000\ \text{l} = \$37.5$$

Cost of detergent consumption per year

$$\text{Top loader} = 80\ \text{g/wash cycle} \times 250\ \text{wash cycles/year} \times \$8/1000\ \text{g} = \$160$$

$$\text{Front loader} = 40\text{g/wash cycle} \times 250\ \text{wash cycles/year} \times \$10/1000\ \text{g} = \$100$$

Operational costs per year (based on current price)

$$\text{Top loader} = \$240 + \$50 + \$160 = \$450$$

$$\text{Front loader} = \$128 + \$37.5 + \$100 = \$265.5$$

End of life disposal costs (based on current price)

Top loader = \$50

Front loader = \$80

Present values of operational costs during life cycle of 5 year

$$\text{Present value} = \sum_{i=1}^{i=n} \frac{\text{PC}x(1+\text{IR})^i}{(1+\text{DR})^i} \quad \text{where } i = 1 \text{ to } 5, \text{IR} = 3\%, \text{DR} = 10\%$$

PV of life cycle operational cost of a top loader

$$= \frac{\$450x(1+3\%)^1}{(1+10\%)^1} + \frac{\$450x(1+3\%)^2}{(1+10\%)^2} + \frac{\$450x(1+3\%)^3}{(1+10\%)^3} + \frac{\$450x(1+3\%)^4}{(1+10\%)^4} + \frac{\$450x(1+3\%)^5}{(1+10\%)^5}$$

$$= \$421.36 + \$394.55 + \$369.44 + \$345.93 + \$323.92 = \$1855.21$$

PV of life cycle operational cost of a front loader

$$= \frac{\$265.5x(1+3\%)^1}{(1+10\%)^1} + \frac{\$265.5x(1+3\%)^2}{(1+10\%)^2} + \frac{\$265.5x(1+3\%)^3}{(1+10\%)^3} + \frac{\$265.5x(1+3\%)^4}{(1+10\%)^4} + \frac{\$265.5x(1+3\%)^5}{(1+10\%)^5}$$

$$= \$248.60 + \$232.78 + \$217.97 + \$204.10 + \$191.11 = \$1094.57$$

Present values of end of life costs

PV of end of life disposal cost of a top loader

$$= \frac{\$50x(1+3\%)^5}{(1+10\%)^5} = \$36$$

PV of end of life disposal cost of a front loader

$$= \frac{\$80x(1+3\%)^5}{(1+10\%)^5} = \$57.58$$

	Top Loader	**Front loader**
Capital cost (same as purchase price)	\$700	\$1200
Life cycle operational cost (present value)	\$1855.21	\$1094.57
End of life disposal cost (present value)	\$36	\$57.58
Life cycle cost	\$2591.21	\$2352.15

Payback period (PP)

Additional capital cost for front loading washing machine

= \$1200 – \$700 = \$500

Recovery of additional cost from first year operational cost saving

$$= \$421.36 - \$248.60 = \$172.76$$

$$\text{Remaining capital cost} = \$500 - \$172.76 = \$327.24$$

Recovery of additional cost from second year operational cost saving

$$= \$394.55 - \$232.78 = \$161.77$$

$$\text{Remaining capital cost} = \$327.24 - \$161.77 = \$165.47$$

Recovery of additional cost from third year operational cost saving

$$= \$369.44 - \$217.97 = \$151.47$$

$$\text{Remaining capital cost} = \$165.47 - \$151.47 = \$14$$

Recovery of additional cost from fourth year operational cost saving

$$= \$345.93 - \$204.10 = \$141.83$$

$$\text{Remaining capital cost} = \text{nil}$$

The additional capital cost will be recovered within $3 + (14/141.83) = 3.1$ years.

Analysis: Here the capital cost is increased due to the use of more efficient technology to mitigate energy and water consumption which means that the external costs or environmental costs have been internalised or avoided by using a top loader washing machine and LED lamps. If the government imposes a carbon tax or environmental taxes (to mitigate environmental impacts), this means that there could be financial savings in reducing both environmental impacts and the operational costs with the application of resource efficient/conservation measures. In the case of the LED lamps, it is better than the CFL lamps in terms of LCC but not in terms of BCR and payback period. However, if the company or a user is going ahead with the LED lamps, this decision could result in the conservation of huge amounts of resources for future generations than what would have been possible by using CFL lamps. Secondly, the use of LED is not making any loss. Whilst the BCR of the LED lamp is lower than CFL, the benefits with the use of the former are still higher than the cost. The company or a user will be regarded as a good corporate citizen for deciding to use of more environmentally friendly LED.

4.2.2 Internalisation of External Costs

External cost or externalities happen when the social or economic activities of producers or manufacturers or industrialists have an impact on the surrounding environment and community, and when the first group fails to fully account for these impacts' (e.g. pollutions) and do not apply any mitigation measures (European Commission 1994). This external cost can be internalised by considering pollution mitigation costs part of the operational costs (producers costs), which can then be included in the life cycle costs of the product to make the environment as important as the other cost items used in the LCC analysis (e.g. capital, labour, resources, technology and other factors of production).

Externalities can be classified according to their benefits or costs in two main categories:

Environmental Externalities: These are classified as local, regional, or global environmental impacts. For example, climate change caused by emission of CO_2 or the destruction of the ozone layer by emissions of CFCs are global environmental impacts. More specifically, sulfur dioxide (SO_2) is produced by many companies across the world. SO_2 emissions have given rise to local and international conflicts over 'acid rain', well-known as 'transboundary pollution'. For example, the death of fish in the lakes in Scandinavia was the result of acid rain caused by the emissions from industrial processes in the UK and Germany (Schreiber and Newman 1988). Additionally, trans-boundary pollution was also caused between southern Brazil and Uruguay because of a coal power station at Candiota in Brazil (Martinez-Alier 2001). In a large country like, the United States, acid rain is created by (i.e. New England) pollution in the western states (Martinez-Alier 2001). However, SO_2 can considerably be reduced by installing scrubbers, or by changing the fuel used in power stations. The money spent on making this change by the polluting industries is known as internalising the external cost of pollution control.

Human Health Externalities: Valuation procedures are needed for putting a value on a person becoming ill due to pollution, or future climate change damage caused by a tonne of CO_2. Such evaluation of externalities has uncertainties due to the various assumptions, methods, risks, and moral dilemmas. It is sometimes a very challenging task to fully implement the internalisation of externalities through policy measures and instruments (for example emission standards, tradable permits, subsidies, taxes, liability rules and voluntary schemes). Nevertheless, they provide an opportunity for politicians to improve the internalisation of the external costs from energy markets.

The question then arises whether the internalisation of externalities in the pricing mechanism for competing technologies is necessary. For example, a substantial difference in the external costs of two competing electricity generating technologies may result in a situation where the least-cost technology (where only internal costs are considered) may turn out to be the highest-cost solution to society if all costs (internal and external) are taken into account. Therefore, the inclusion of regulatory measures and market mechanisms will enable the environmental technology to become more cost-effective. Box 4.3 provides a detailed mechanism for internalising the external costs for an engineering project.

Once the industry has internalised this external cost, the LCC equation can be updated as follows:

$$\text{Life cycle cost} = \text{Capital cost} + \text{PV of replacement costs}$$
$$+ \text{PV of operational cost} + \text{PV of maintenance cost}$$
$$+ \text{PV of end of life disposal cost} - \text{PV of residual cost}$$
$$+ \text{PV of abatement cost}$$

Box 4.3 Mechanism for internalising the external costs

The below Figure explains the mechanism of internalising the external cost. A is a point of equilibrium, at the intersection between the marginal production cost (MPC) incurred by the manufacturer and marginal social benefit (MSB) in terms of meeting society's demand for goods or services. MPC is the cost associated with the production of an additional unit to supply or provide services to society. MSB represents utilities provided to meet societal demand and hence forms a demand curve. At point A, the price is optimum (i.e. P_3) as both producers and consumers benefit due to the production of Q number of products. The consumer's and producer's surplus are represented by the triangular area P_5AP_3 and P_1AP_3, respectively. However, the equilibrium point at A does not take into account the amount of environmental damage and resource scarcity caused by the production of Q number of products to meet the demand. The environmental damage and resource scarcity associated with the production of Q number of products incurs external costs include health, and the environmental management costs for society and the surrounding communities where these manufacturing activities are taking place. This cost is known as an external cost. When this marginal external cost (MEC) associated with the manufacturing of an additional product is added to the MPC, it is known as marginal social cost (MSC) and the point of equilibrium will move to B with a new optimum price at P_4. With this increased cost, the producer has to reduce the production from Q to Q1 as the MSB cannot be less than the MSC. If MSC > MSB, or if the position of MSC is at point C, it means that the additional quantity (Q−Q1) that will be produced by the manufacturer will remain unsold or be wasted resulting in a socially inefficient production. However, in a free market economy with no external costs, the producer continues to produce Q number of products without internalising the external costs. Failing to comply with this moral obligation results in market failure as presented by a triangular area ABC in the diagram, which is also known as a welfare loss of economy. Instead of society bearing this external cost, the producer should either spend money on pollution control or abatement technologies to prevent the externality or pay environmental taxes which ultimately generates revenue for the government to take environmental protection measures.

(Continued)

Box 4.3 (Continued)

In below Figure, if the manufacturer is paying taxes, then the consumer's and producer's surplus will be P_5BP_4 and P_2DP_1, respectively, and the government revenue that will be generated will be equal to a rectangular area P_4BDP_2. Instead of paying taxes, when the industry invest in abatement technology to internalise the external cost, the producer surplus will change to P_4BP_2. The abatement option appears to be more effective given the fact that industry's direct action will reduce the impact on the surrounding community and the environment. Cleaner production strategies which are discussed in the next chapter are important to consider as they help to abate pollution or internalise the external cost in an economically feasible way.

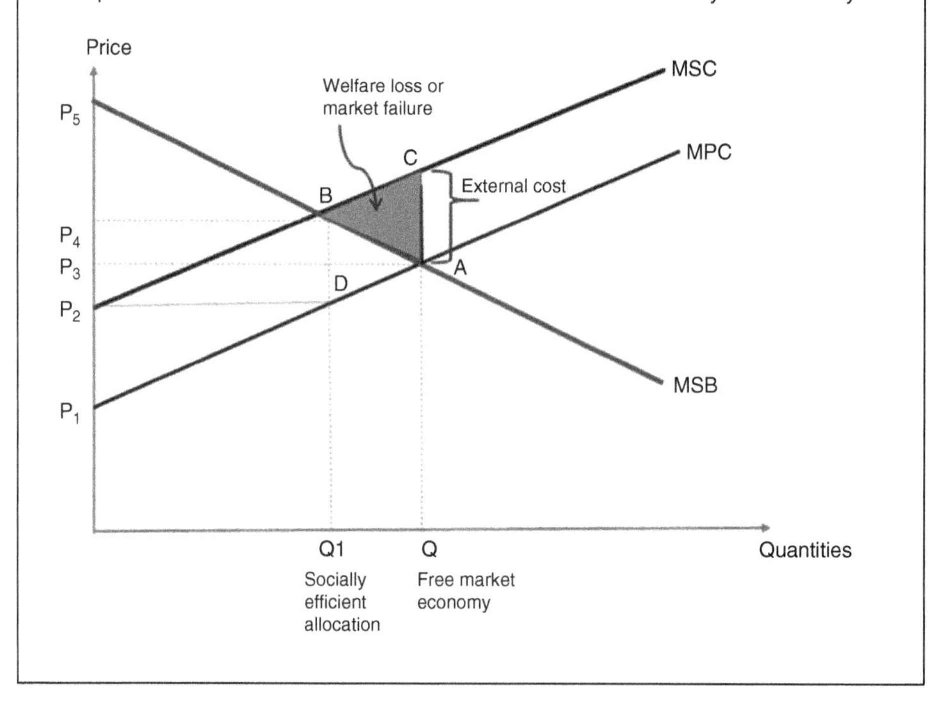

4.3 Social Life Cycle Assessment

As per United Nations Environment Programme (UNEP), SLCA is a technique that aims to assess the social and sociological aspects of products or services and their potential positive or negative impacts along their life cycle (UNEP 2009). The SCLA concept emerged long after ELCA and LCC. There is an increasing awareness of SLCA and new frameworks/methods have recently been developed at an increasing rate. SLCA contributes to the improvement of social performance assessment at different stages of a life cycle by identifying the 'social hotspots'. Measuring the social dimension of sustainability is a very complex matter as it may involve mixed methods for calculating indicators and there is no common unit for assessment like ELCA and LCC. In the case of ELCA, all environmental indicators are calculated on the basis of a functional unit, while all indicators are presented in terms of $ in the case of LCC analysis. In addition, there is often a lack of data and the data collection stage is quite expensive and time-consuming. Like ELCA, as discussed in Chapter 3, SLCA follows the same ISO guideline. The four steps of SLCA as per ISO14040-44:2006 (ISO 2006) are as follows:

4.3.1 Step 1: Goal and Scope Definition

The first step is to define the functional unit and product or service utility. For example, like ELCA, the functional unit could be one GJ equivalent amount of biodiesel production. The SLCA will work out the amount of social impacts (SIs) created during the life cycle of the production of a product. Once the functional unit has been determined, some types of social impacts based on the quantitative data can be determined (e.g. number of jobs created, accessibility, affordability). For some social impacts, the functional unit is not necessary as they are mainly dependent on people's personal judgement or the level of satisfaction. For example, working conditions and human rights, cultural heritage cannot be quantified directly or based on a functional unit, as they are entirely based on stakeholders' value judgements. Their value judgement or perception can be converted to a quantitative value using a Likert scale which was discussed in Chapter 2 as well as using an example below. The scope involves the determination of appropriate social indicators of a product or service by involving experts which has also been discussed in Chapter 2.

These social indicators will vary from place to place and product to product due to variation in socio-economic condition, geographical locations, and stakeholders. For example, stakeholders in the fertiliser supply chain (i.e. input suppliers, fertiliser producer, fertiliser distributers and farmers) are different to stakeholders in the iron supply chain (i.e. miners, processors, manufacturers). Furthermore, these industries are location specific, deploy different types of technologies and create different levels of income under different working conditions. Also the indicators for the same product will vary with location and geography as the socio-economic situation, resource utilisation pattern, energy mix, availability of resources (i.e. transportation) all vary. Indicators need to be selected in a way that these are relevant to the product or service. It is always wise to choose a key and limited number of indicators quite explicitly representing the social characteristics of the product or service.

A participatory approach involves a consensus survey that can be considered either face to face, online or via a postal survey to engage with stakeholders and area experts in the selection of indicators for improving the creditability of the sustainability assessment. Prior to engaging stakeholders in the consultation process, a thorough literature review of international journals, government documents, and reports published by recognized organisations (e.g. World Bank, Dell Computers) and a large number of industry sustainability reports need to be carried out as a preliminary selection of the social, economic, and environmental indicators.

A questionnaire for the survey should be designed based on the preliminarily selected indicators to collect feedback and opinion from the stakeholders and area experts on the relevance and importance of these indicators.

Following Eqs. (2.1)–(2.3) and an example in Box 2.2 in Chapter 2, the weight of the social indicators will be determined. Also, the example below provides a set of equations to be used for calculating the weights of social indicators. Once the weights have been calculated, stakeholders who are directly and indirectly affected by the production of a product or delivery of a service need to be interviewed to gauge their social perception for each criteria. It should be notable that the experts and stakeholders are two different groups where the former assist in the selection and weighting of indicators and the latter consists of people who are directly and indirectly affected by the production of a product during different stages of the product supply chain.

Stakeholder involvement needs to be identified prior to lunching a stakeholder survey. Examples of some stakeholder categories and subcategories/indicators are listed below.

Workers: Freedom of association, discrimination, child labour, fair salary, working hours, forced labour, health and safety, and social benefits.

Local community: Respect of indigenous rights, net migration rate, safe, secure, and healthy living conditions, local employment, cultural heritage, community engagement, and access to material and immaterial resources.

Society: Contribution to economic development, corruption, technology development, prevention of armed conflicts, public commitments to sustainability issues.

Consumers: Health, safety and transparency, consumer privacy, end of life responsibility.

Value chain actors: Fair competition, promoting corporate social responsibility, supplier relationship, respect to intellectual property rights.

As mentioned earlier, some of these subcategories can be calculated based on the functional unit and some of them have to be based on the qualitative judgement of the stakeholders in the productsupply chain.

4.3.2 Step 2: Life Cycle Inventory

Data collection steps and methods for inventory development vary and data may be both quantitative and qualitative. The information on sub-categories like child labour, fair salary, working hours, and access to material and immaterial resources can be obtained in numerical form for a particular functional unit. For example, the number of workers as child labour, which is usually a widely used indicator for human rights, used in different stages of building construction can be determined in terms of person-hours.

In the case of sub-categories like freedom of association, the subjective judgement of the relevant stakeholder in the supply chain has to be collected. Once the stakeholders have been determined, relevant questionnaires for different stakeholders are developed to collect their feedback or level of satisfaction. The qualitative feedback that is received or collected from the stakeholders forms the social life cycle inventory. It should be noted that different social indicators are different for different stakeholders. For example, the question related to human rights is more relevant to workers, while the questions in relation to cultural heritage are more relevant to the local community where the production/activities are going on. Social perceptions could be gauged using a five or multiple-point Likert scale. The highest point on the scale (e.g. it will be '7' on a seven-point Likert scale) is regarded as the highest level of satisfaction that a stakeholders can offer. Using the scale from 1 to 7, where 1 means totally disagree and 7 means totally agree or expected score.

4.3.3 Step 3: Life Cycle Social Impact

This is a positive or negative pressure on the well-being of stakeholders. ISO 14044 (2006) describes this step as follows: 'The calculation of indicator results involves the conversion of LCI results to common units and the aggregation of the converted results within the same impact category. The outcome of the calculation is a numerical indicator result' (ISO 1999). This is actually a basic aggregation step, bringing text or qualitative inventory information together into a single summary, or summing up quantitative social and socio-economic inventory data within a category. Final stakeholders' perspectives are appraised by determining the gaps between expected and perceived quality of each criterion, which is the difference between social perception and expectation (i.e. 7). If the gap equals zero, the actual state of social impact exactly matches the stakeholders' expectation. If the gap is negative, the actual state of social aspect is not meeting stakeholders' expectations.

The examples of some social impacts (SI) are human rights, working conditions, health and safety, cultural heritage, governance, socio-economic repercussions. These impacts are scored in terms of gaps between their actual values and the maximum value to achieve sustainability performance. The gaps calculated from the responses received from all respondents for a certain criteria are averaged. In most of the cases, a social impact may consist of a number of criterion (C). In this case, the formula for calculating the social impact is as follows:

$$\mathrm{SI} = \sum_{C=1}^{C} \mathrm{Gap}_c \times W_c \tag{4.5}$$

The overall score of the assessed system, which is the sum of the product of the gap and weight of each criterion. Using the same boundary and indicators, SLCA of competing products will be conducted and the product with the reduced level of gap is considered to offer better social performance. Once these values are determined, they are compared with the threshold or optimum values on a Likert scale to find the sustainability gap. The examples with calculations are shown in Chapter 2.

Each of these SI consists of a number of criteria (C). For example, human right SI for palm oil based biodiesel production could be based on these criteria, including free from the child labour employment, free from the employment of forced labour, equal opportunities, and free from discrimination.

4.3.4 Step 4: Interpretation

Interpretation assists in the identification of significant issues so that appropriate recommendations can be provided to overcome negative social consequences. The social impact with the largest gap is known as the social hotspot. Accordingly, all criteria under this impact category or SI need to be investigated. Each of these criteria depends on people's or stakeholders' feedback or the qualitative data collected from the field. The affected stakeholders need to be consulted further to obtain their suggestions to satisfy the criteria. Once these criteria are satisfied by incorporating social improvement measures, the social sustainability performance of the given product can be improved.

Practice Example 2: Social life cycle assessment

Box 4.4 Example ofa SLCA

Your company is planning to import fuel from country Y in order to run a transportation business. The fuel company in country Y uses modern equipment to increase its production and offers good working conditions for its workers. However, the fuel refinery was built on an archaeological site which has in fact reduced the number of visitors for educational purposes. They recruit workers from the surrounding areas. Whilst they are contributing to local employment, they are reluctant to take care of flora and fauna in the surrounding area as their activities sometimes cause deforestation and water pollution.

According to government policy, your transportation company is required by law to provide a social life cycle assessment of the fuel imported. Because of your background in sustainable engineering, your company has assigned you to travel to country Y to conduct this assessment.

The social impact categories that are found relevant to this fuel production are human rights, working conditions, cultural heritage, socio-economic repercussion, and governance.

The first thing that you need to do in order to carry out this SLCA is to give weight to each social impact category. Therefore, you need to find experts whom you will ask to rank five impact categories from 1 to 5, in which the most important impact is marked as 5 and the least important as 1. These experts are neither directly nor indirectly associated with the fuel company. After conducting this survey, you have obtained data that is presented in below Table.

Social Impact Categories	Expert1	Expert2	Expert3	Expert4	Expert5
Human rights	5	4	3	4	4
Working conditions	3	5	4	3	5
Cultural heritage	1	3	1	5	1
Socio-economic repercussion	2	1	2	2	3
Governance	4	2	5	1	2

a. *Calculate the weight of each social impact category*

Once you have ascertained the weights of these five impact categories, you now conduct a survey of the people who are involved in the fuel production as well as affected by fuel production, including management, workers, local community, government officials, and researchers. Using the scale from 1 to 7, where 1 means totally disagree and 7 means totally agree, you have asked respondents how they would rate these impact categories. The responses that you had received from 5 respondents are presented in below Table.

Social Impact Categories	R1	R2	R3	R4	R5
Human rights	3	2	4	3	5
Working conditions	6	5	5	7	5
Cultural heritage	3	2	4	5	1
Socio-economic repercussion	6	4	5	4	5
Governance	3	2	4	3	5

(Continued)

Box 4.4 (Continued)

b. *Calculate the sustainability gap for each impact category.*
c. *Draw a radar chart to determine the social hotspot.*

Solution

Table 4.1 shows the formulae for calculating weights of social indicators (Manik et al. 2013).

Table 4.1 Weighting process (Manik et al. 2013).

Expert → Impact category or criterion ↓	E_1	E_2	…	E_n	Weight (W)
C_1	R_{11}	R_{12}	…	R_{1n}	$\left(\sum_{j=1}^{n} R_{1j}\right) / \left(\sum_{i=1}^{n}\sum_{j=1}^{m} R_{ij}\right)$
C_2	R_{21}	R_{22}	…	R_{2n}	$\left(\sum_{j=1}^{n} R_{2j}\right) / \left(\sum_{i=1}^{n}\sum_{j=1}^{m} R_{ij}\right)$
…	…	…	⋱	⋮	…
C_m	R_{m1}	R_{m2}	…	R_{mn}	$\left(\sum_{j=1}^{n} R_{mj}\right) / \left(\sum_{i=1}^{n}\sum_{j=1}^{m} R_{ij}\right)$

Note – i = Criterion, 1, 2, 3, …….. m; j = experts, 1, 2, 3, ……….. n

The steps used for calculating the weights of social impacts using above equations are given below:

The scores provided by the experts were incorporated into the above equations to calculate weights of social impacts (Table 4.2).

The second step is to sum up the ranks (R) provided by 'n' number of experts to calculate the total score is as follows:

$$\left(\sum_{j=1}^{n} R_{1j}\right) = 5 + 4 + 3 + 4 + 5 = 20 \tag{4.6}$$

The summation of the total scores of m criterion is as follows:

$$\left(\sum_{i=1}^{n}\sum_{j=1}^{m} R_{ij}\right) = 20 + 20 + 11 + 10 + 14 = 75 \tag{4.7}$$

Table 4.2 Calculation of weights of social impact categories.

Expert → Impact category or criterion ↓	E1	E2	E3	E4	E5	Total score	Weight (W)
Working condition (C2)	5	4	3	4	4	20	0.27
Cultural heritage (C3)	3	5	4	3	5	20	0.27
Socio-economic repercussion (C4)	1	3	1	5	1	11	0.15
Governance (C5)	2	1	2	2	3	10	0.13
	4	2	5	1	2	14	0.19
						75	1.00

Weight of criterion 1 (C_1), which is human right, is,

$$\frac{\left(\sum_{j=1}^{n} R_{1j}\right)}{\left(\sum_{i=1}^{n}\sum_{j=1}^{m} R_{ij}\right)} = \frac{20}{75} = 0.27 \tag{4.8}$$

The third step is to determine the gap. Seven is the highest point on a Likert scale. The scores given by the respondents (R) on a Likert scale are deducted from 7 to determine the Gap (G). The gaps for all responses for a social impact are averaged to determine G′ (Table 4.3).

The fourth step is to use an excel spreadsheet to draw a radar chart (Figure 4.1). It shows that the human rights with the highest gap is the dominant social hotspot. Alternatively, the students can draw the same graph on a piece of graph paper.

This is a simplified example for SLCA. In an actual case, each of these SIs consists of a set of criterion. In this case, Eq. (4.5) needs to be calculated the same for each SI.

Table 4.3 Determination of gaps.

Social impact categories	R1	R2	R3	R4	R5	G'	PG = $G' \times W$	Score, (7 − PG)
Human rights	−4	−5	−3	−4	−2	−3.6	−1.0	6.0
Working conditions	−1	−2	−2	0	−2	−1.4	−0.4	6.6
Cultural heritage	−4	−5	−3	−2	−6	−4	−0.6	6.4
Socio-economic repercussion	−1	−3	−2	−3	−2	−2.2	−0.3	6.7
Governance	−4	−5	−3	−4	−2	−3.6	−0.7	6.3

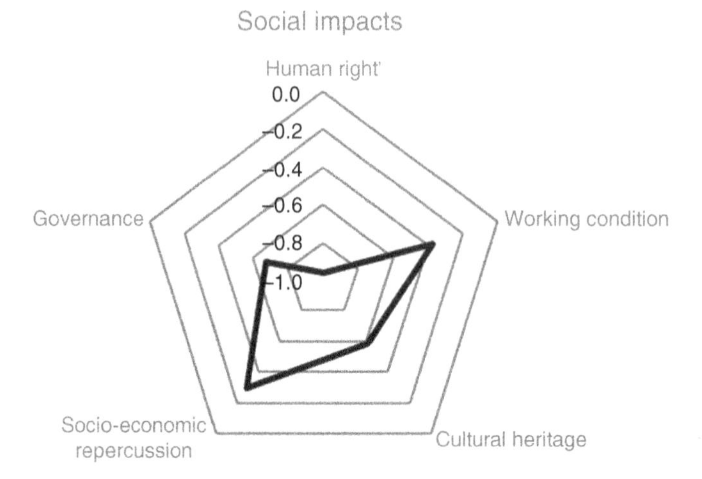

Figure 4.1 Rader chart of social impacts.

4.4 Life Cycle Sustainability Assessment

Life cycle sustainability assessment (LCSA) involves the use of ELCA, LCC, and SLCA tools to calculate the environmental, economic, and social indicators of any product or service. This allowed the integration of all indicators into a single score and also identified which of the three objectives of sustainability require more attention. Secondly, it helps compare the overall sustainability performance of different versions of the same product or different approaches to achieve a sustainable solution.

LCSA will provide a measure of current sustainability performance, a clear statement of what might be achieved in terms of future performance targets and a yardstick for measurement of progress along the way. The selection of the right performance indicators, covering validity, relevance, sensitivity and measurability is important for any sustainability assessment of a project. Selecting key indicators for TBL objectives is often a brainstorming task, but it makes the assessment process more objective, cheaper, and more time efficient, while a larger set of indicators increases complexity and makes the assessment a time-consuming, expensive, and data-intensive process. Therefore, an optimum number of key indicators that are aligned with sustainability objectives and standards, should be determined. LCSA can be based on the analytical hierarchy process (AHP) of multi-criteria decision analysis to assess sustainability using indicators. System boundaries in LCSA studies vary with the scope of the study. However, the three LCSA objectives, i.e. the environmental, social, and economic factors, should be studied using the same system boundaries. An LCSA framework should be robust enough to analyse the TBL objectives simultaneously and explain the interdependencies among the environmental, social, and economic aspects being impacted.

SLCA and LCC are added to life cycle assessment (LCA) to make an equation for LCSA. Kloepffer (2008) put the LCSA framework into the conceptual formula in 2007 [13]

$$\text{LCSA} = \text{LCA} + \text{LCC} + \text{SLCA} \tag{4.9}$$

LCSA = Life cycle sustainability assessment

LCA = Environmental life cycle assessment

LCC = LCA-type life cycle costing

SLCA = Social life cycle assessment

Guinée et al. (2016) called it a framework rather than a method in itself. Firstly, it helps to broaden the level of analysis from product to sector- and then to economy-wide questions and analyses. Secondly, it deepens the analysis by including physical, economic, and behavioural relations other than just technological considerations.

Since SLCA and LCC follow the LCA approach, the terminologies of Eq. 9 has been slightly modified as follows:

$$\text{LCSA} = \text{ELCA} + \text{LCC} + \text{SLCA} \tag{4.10}$$

Using the procedure in Chapter 3, ELCA will be used to calculate environmental indicators. LCC and SCLA procedures discussed in this chapter will be used to calculate economic and social indicators, respectively. All these indicators were integrated into one single sustainability score by using the procedure explained in Section 2.8 in Chapter 2. Once environmental, economic, and social indicators have been calculated, they were compared with the corresponding threshold values to calculate the gaps. These gaps were multiplied by the corresponding weights of the indicator. Then Eqs. (2.4)–(2.7) can be used to calculate the overall sustainability score. The equations also help show the breakdown of the overall score in terms of TBL objectives (i.e. Eq. (2.6)) and the breakdown of objectives in terms key performance indicators, which are the lowest units of the hierarchy of indicators (i.e. Eqs. (2.4)–(2.5)). The further breakdown is important to find the TBL hotspot(s) specifically or predominantly responsible for the overall sustainability score.

TBL indicator selection should be based on the factors in the region studied, and the LCSA framework should be flexible enough to handle the variation in region-specific impact indicators.

References

Guinée, J. (2016). Life cycle sustainability assessment: what is it and what are its challenges? In: *Taking Stock of Industrial Ecology* (ed. R. Clift and A. Druckman). Springer, Cham: https://doi.org/10.1007/978-3-319-20571-7_3.

ISO (International Standard Organization) (1999). Environmental management – Life cycle assessment – Life cycle impact assessment. ISOIFDIS 14042: 1999(E). ISO copyright office, Case postale 56 0 CH-1211 Geneva 20.

Klöpffer, W. (2008). Life cycle sustainability assessment of products. *International Journal of Life Cycle Assessment* 13 (2): 89–95.

Manik, Y., Leahy, J., and Halog, A. (2013). Social life cycle assessment of palm oil biodiesel: a case study in Jambi Province of Indonesia. *International Journal of Life Cycle Assessment* 18: 1386–1392. https://doi.org/10.1007/s11367-013-0581-5.

Martinez-Alier, J. (2001). Mining conflicts, environmental justice, and valuation. *Journal of Hazardous Materials* 86: 153–170.

Ristimäki, M., Säynäjoki, A., Heinonen, J., and Junnila, S. (2013). Combining life cycle costing and life cycle assessment for an analysis of a new residential district energy system design. *Energy* 63: 168–179.

Heralova, R.S. (2017). Life cycle costing as an important contribution to feasibility study in construction projects. *Creative Construction Conference* 2017, CCC 2017, 19-22 June 2017, Primosten, Croatia.

Schreiber, R.K. and Newman, J.R. (1988). Acid precipitation effects on forest habitats: implications for wildlife. *Conservation Biology* 2 (3): 249–259.

United Nation Environmental Programme (UNEP) (2009). Guidelines for social life cycle assessment of products. UNEP/SETAC Life Cycle Initiative at UNEP, CIRAIG, FAQDD and the Belgium Federal Public Planning Service Sustainable Development.

Part III

Sustainable Engineering Solutions

5

Sustainable Engineering Strategies

5.1 Engineering Strategies for Sustainable Development

As discussed in Chapters 3 and 4, environmental life cycle assessment (LCA), social life cycle assessment, and life cycle costing can help identify environmental, social, and economic hotspots to enable engineers to make sustainable decisions. This chapter discusses cleaner production (CP) strategies that can potentially be incorporated into the product life cycle to treat environmental and economic hotspots. There can be both direct and indirect social implications of the application of CP strategies as environmental and economic improvement can also enhance or reduce social well-being.

Sustainable development which is built on three pillars such as economic growth, ecological balance, and social equity, can be achieved through CP strategies. Economic growth, without taking into account the environmental concerns will not be sustainable in the long term. Similarly, environmental protection which comes at the expense and sacrifice of basic human needs may also be considered unacceptable. An ambitious social program which prevents a responsible wealth-creating mechanism could ultimately be self-defeating.

The social objective of sustainable development is not explicitly addressed in CP. A CP strategy could lead to intergenerational equity in terms of resources conservation, clean air, and affordability. The objective of CP is to achieve eco-efficiency at micro or farm level by doing more with less or producing more products with less cost and less environmental impacts (EI).

These strategies are integrated into environmental management systems (EMSs), an ISO14001-4 guideline, proposed by International Standard Organization to achieve the environmental targets of industries (ISO 2004). Applying an EMS in an organisation and constantly using cleaner production strategy (CPS), leads to better risk management and a reduced negative impact on the environment, ultimately resulting in the reduction of energy and material costs

Engineering for Sustainable Development: Theory and Practice, First Edition.
Wahidul K. Biswas and Michele John.

as well as waste management and pollution control costs which assists industries in developing competitive advantage.

5.2 Cleaner Production Strategies

The true challenge of sustainable development lies in applying sustainability theory in practice. CP offers a practical approach in attaining sustainable development. CPS allows the producers of goods as well as the providers of services to produce more with less environmental impacts and costs – less raw material, less energy, less waste generation, whilst reducing environmental impacts and enhancing conservation of resources for future generations. CP investigates the root causes of the impact, rather than just the symptoms or impacts. This is the highest tier of environmental management. First tier is passive, which ignores pollution, then comes reactive (i.e. dilution and dispersion), then comes constructive which is typically an end of pipe pollution control and then finally proactive or CP strategy.

Cleaner production is considered a continuous application of an integrated, preventive environmental strategy towards processes, products, and services in order to increase overall efficiency and reduce damage and risks for humans and the environment. In summary, cleaner production also applies to processes in conserving raw materials and energy, eliminating toxic materials, and reducing the quantity and toxicity of associated emissions and wastes before they leave the process. CP is also known as pollution prevention or waste minimisation.

CP is a logical extension of our conventional environmental management in reducing material consumption and waste generation. This is a preventative approach rather than a pollution control approach. Prevention is better than cure in a way that it is applied during the processing and manufacturing activities of the production of goods and services, which not only avoids the cost of pollution control but also reduces operational costs by reducing chemical and energy consumption (Figure 5.1). CPS requires a person to examine what they are performing in an industrial operation and to look for better, more efficient ways or strategies to do it – ways that result in higher productivity, reduced resource inputs, reduced waste and most importantly reduced risk of environmental impacts (e.g. by reducing harmful emissions, e.g. SO_x from refinery, ppm from the fertiliser industry).

Cleaner production is not just limited to manufacturing processes, it can be applied across the product life cycle. Like eco-efficiency, CPS can improve the environmental performance of processes, products, and services. CP assessment allows companies to identify business opportunities and efficiency gains along the entire supply chain. Since cleaner production can be applied to LCA to restructure the

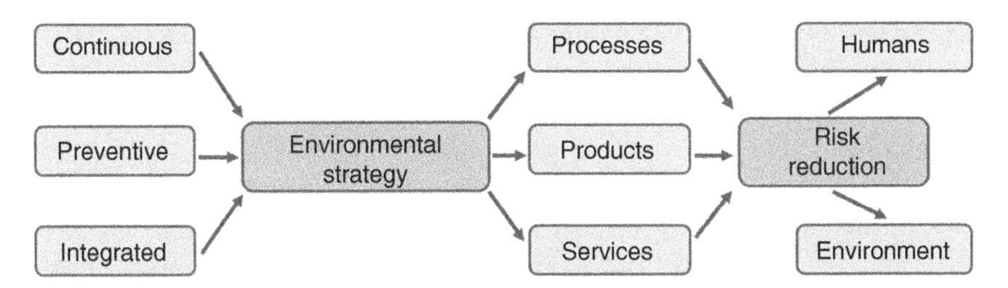

Figure 5.1 Basic principles of cleaner production strategies.

environmental supply chain of products or service, it also applies to products for reducing the environmental impact during the product life cycle, from raw materials extraction to product disposal. This is also known as design for the environment or life cycle design.

There are five types of CP strategies, classified into two tiers. The first tier includes good housekeeping, input substitution, production modification, and technology modification and the second tier is focused on-site recycling (IVAM 2008).

Cleaner production also refers to managerial and operational actions as it is not only to prevent leaks and spills but also re-enforce existing operational instructions. This is an important category because of its strong effects usually without any need for major investment. Human behaviour is often the main reason for inefficient production, and internal policy instruments are often needed to make this behavioural change.

The five CP strategies are now described:

5.2.1 Good Housekeeping

The good housekeeping strategy is considered as a low hanging fruit as they do not require innovation, large investment, or long-term research and development to improve efficiency and processes. However, the need for behavioural change is important to implement this strategy.

Good operational practices in an industry begin with efficient production planning. For example, in a dairy plant with multi-functional production and bottling equipment, producing regular milk and then the milk based products (custards, yoghurt and chocolate or fruit desserts) should reduce the maintenance costs and wastewater generation by reducing the equipment cleaning needs in order to prevent cross-contamination and avoids separate production lines for incompatible products.

Energy management is another important part of a good housekeeping strategy. Costs can be reduced by avoiding high-peak energy consumption and also avoiding the need to install high load equipment. An example is seen in a cooling system where a glycol solution is pumped from chillers that surround hundreds of balls that are frozen during the night when electricity rates are lower. (ARN 2010). During the day, the system pumps the glycol solution surrounding the ice balls through a heat exchanger, which provides cool air to the data center. This reduces the need to run the chillers during the day, when rates are higher. EnergyAustralia estimates standby electricity wastage can contribute up to 10% of an electricity bill – money which customers rather see in their pockets than their retailer's bank account (Canstar Blue 2016).

Staff have to bring into practice the procedures and instructions to the work floor on a day-to-day basis. Proper maintenance and cleaning of equipment should be a day-to-day routine, resulting in a clean and tidy workplace and avoiding health risks and hazards.

The application of good housekeeping strategies usually help to reduce the following wasteful practices in a production system:

- Spill and leak prevention
- Better inventory management
- Improved production planning
- Enhanced operating and maintenance practices

Box 5.1 shows how a Western Australian processed chicken company was able to reduce waste management costs and energy costs by fixing leaks, recycling cardboards by implementing CPS measures.

Box 5.1 Good House Keeping

Canon Food in WA is a manufacturer and supplier of cooked chicken products to remote mining areas in Western Australia. This food factory educates employees in CPS to reduce waste to floor drains, and leaks, and to increase cardboard recycling and energy efficiency. All these tasks are considered low hanging fruits as they do not require research and development and significant capital investment. As a result, there are environmental benefits associated with the reduction of wastes and bio-chemical oxygen demand (BOD) to the sewer. The economic benefits resulting from this CPS strategy include no need for capital investment. There is a net saving of $500 a month in sewer discharge costs, the grease waste disposal costs reduced to $800, and the productivity of the factory increased using aforementioned good housekeeping strategies.

5.2.2 Input Substitution

This CP strategy is aimed at reducing the environmental impact of a product or process or service through the use of alternative materials (e.g. bio-cement instead of normal ordinary portland cement [ODC]), process (e.g. use of compressed air as a replacement for lubricating oil for cooling purposes), auxiliaries, product raw materials (e.g. use of recycled steel instead of virgin steel), and consumables (e.g. renewable or bio-degradable based detergent instead of chemical detergent), replacement of toxic or harmful materials with less toxic materials, use of renewable materials (e.g. use of biodiesel produced from waste cooking oil), and the use of materials with increased durability, recyclability, recoverability, and are biodegradable. It may require new technology or process design to be able to use these alternative materials and an example is provided in Box 5.2.

Box 5.2 Input substitution

Ford Australia introduced a cleaning process involving a high pressure water jet for removing paints from skids and then separating paint from waste water, which in fact eliminated the need for a traditional process requiring the use of caustic soda, and also conserved energy. The total cost to implement this was $120,000 with a net annual savings of $300,000.

Another approach by Ford Australia is the use of plastic media as a new paint stripping process, avoiding the use of caustic soda and organic strippers. Once the paint has been removed from the car body, the plastic media was recovered and reused. This process was quicker than the traditional one. The investment was $140,000 but the annual savings were more than $100,000, resulting in a payback period of 2–3 years.

5.2.3 Technology Modification

This strategy mainly concerns efficiency improvement and can include lower and higher cost options. The lower-cost efficiency options include better process instrumentation, control provisions (e.g. building energy management system), equipment insulation (e.g. sandwich walls avoid more heat loss than a double brick house), and the layout of equipment (e.g. properly aligned machine will reduce vibration and increase the machine life). In the case of higher cost options, the use of more efficient process equipment, including integrated recovery processes (e.g. use of heat exchangers to recover waste heat for use in another process), cleaner process technology (e.g. modifying engine and power train to convert conventional cars to a hybrid car) and chemistry (e.g. the reaction efficiency can be increased by using enzymes as they can replace a good number of chemical reactions), and associated energy and chemical consumption. Box 5.3 shows that the use of Clipsal C-Bus lighting control systems has optimised energy consumption and reduced associated GHG emissions.

Box 5.3 Technology modification

In Coca Cola Amatil in Kewdale, Western Australia, lighting accounts for a reasonable portion (i.e. 15–20%) of the total energy consumption. The installation of Installed Clipsal C-Bus lighting control systems involved three lighting controls providing the right amount of illumination in the right place at the right time, including

- high daylight – only lights on in critical areas,
- low daylight – proportion of lights on in areas being used, and
- night – all lights on in areas.

Using this technology modification resulted in lighting energy savings of 30–40%, GHG emission reduction ~400 ton CO_2 e-, and cost savings of $45,000 with a payback period of only 2 years.

5.2.4 Product Modification

Sustainable products need to be redesigned to reduce net environmental impact over the product life cycle. They need to be designed in a way that it will help achieve environmental efficiency, conserve scarce resources, use renewable or sufficiently available materials, increase product durability, and designed for reuse, recycling, and disassembling, minimise harmful substances and environmental impacts during production and use, and involve environmentally friendly packaging and logistics (e.g. H_2 fuel). Box 5.4 shows an example from the experience of a company Fulleon Ltd., which shows that a fire alarm can be designed in a way where the material consumption and associated GHG emissions can be reduced.

Box 5.4 Product modification

Fulleon Ltd. manufactures and sales parts for fire alarm systems. They have modified the product design in a way that reduced number of parts by 35%, plastic consumption by 27%, assembling time by 35%, and packaging costs by 24%. The product functionality also improved at the same time. As a result, some environmental benefits resulted, including the reduction of raw material consumption, energy use (e.g. machines run for less time), use of cardboard packaging, and the transport costs for materials and products. Less waste was generated at the end of the product's life. The associated economic benefit was the saving of manufacturing costs of £100 K/year.

5.2.5 On Site Recovery/Recycling

This CP strategy reduces environmental impact through on site recovery, including reuse and recycling of heat recovery (e.g. heat exchangers, super heaters), water recycling (e.g. use of reverse osmosis to produce process and cooling water), and materials recovery (e.g. on site application and off site application). This strategy will enhance waste management through knowing the amount and the type of materials wasted, waste management costs, and what could be some relevant strategies to reduce this waste. Box 5.5 shows that the use of a remanufactured compressor offers both environmental and economic benefits while maintaining the same level of functionality and durability as a new compressor.

Box 5.5 On-site recycling

The Sustainable Engineering Group at Curtin University conducted an LCA research for Recom Engineering (Western Australia) to compare the global warming performance of a remanufacturing compressor with an Original Equipment (OEM) compressor. This LCA showed that there are environmental benefits with remanufacturing a compressor which reduced CO_2 emissions by almost 1.5 ton/unit, which is equivalent to taking a small car off the road. Additional economic benefits could be the reduction of Carbon Tax penalties. For example, if the carbon price was set at $50 per ton of CO_2 a new (OEM) compressor would face a cost of $79.50 and a remanufactured compressor a cost of $5.85. Also the price of this remanufactured compressor was one third of the price of a new compressor while offering the same service life and durability warranty.

5.3 Fuji Xerox Case Study – Integration of Five CPS

The integration of all five CP strategies by Fuji Xerox helped resource conservation through dematerialisation. They produce a multi-functional printer, fax machine, scanner, and photocopier as one machine. The energy consumption due to the integration of these four devices into one device reduced energy consumption from 1400 to 700 kWh during use (Fuji Xerox 2008). In 2001–2002, the plant saved $23.5 million due to this product modification strategy. Most importantly, there was 99% resource recovery from the end of life products which reduced the requirements for raw materials by 17,400 tons, between 1995 and 2007 (Fuji Xerox 2008).

By embracing extended producer responsibility, Fuji Xerox were able to take back EoL equipment from the customers. Fuji Xerox do not finish their relationship with the customers after selling their products. Instead, they lease their equipment and then at the end of life, they take it back for disassembling and decommissioning. This is how this company achieved economic and environmental efficiency, and its leadership in corporate sustainability provides a benchmark for other organisations, particularly those in the electronic industry.

5.4 Business Case Benefits of Cleaner Production

There are financial and other business benefits associated with the application of CP strategies. CP strategies save money by avoiding both direct and indirect costs during the production process. The direct costs that are avoided include external costs such as environmental taxes, waste disposal costs, energy, and material costs. Indirect costs that are avoided included internal costs such as the avoidance of lost value/depreciation costs, waste handling costs, income loss, unit cost due to improved productivity, and increased sales due to improved product quality. This also helps in risk management by reducing product liability, complying with regulations, enhancing the supplier of choice and obtaining license from the regulatory authority to operate. Other avoided costs are associated with the environmental, health, and safety benefits of CP strategies. CP typically saves costs on raw material, energy, and water use which makes companies more profitable and competitive with reduced operational costs.

CP also improves the company image and worker's health and safety conditions, and reduces waste treatment and disposal costs. These strategies can be integrated within the business EMSs. The purpose of an EMS, which will be discussed later in this chapter, is to help industries to achieve environmental targets by integrating CP to reduce environmental impacts in a cost-competitive manner and also to establish their identity as a good corporate citizen.

5.5 Cleaner Production Assessment

When CP is applied at micro level, it focuses on the production processes of a particular industry or a service delivered by an industry. A systematic approach for applying CP strategies, is known as CP assessment (as presented in Figure 5.2).

Four steps of the CP assessment framework have been discussed as follows:

5.5.1 Planning and Organisation

In order to conduct any CP assessment, it is important to obtain management commitment as it involves investment in assessment, measurements, and monitoring as well as the changes in processes, technologies, and human resources required to reduce energy, material consumption, emissions and wastes generation. Planning is strongly grounded on company's organisation structure and information systems, current status of plant equipment and technology. There could be an initial challenge in the planning process, as the organisation has to perform additional responsibilities like production data reviews, emissions, and wastes measurements and making staff aware of these CP practices.

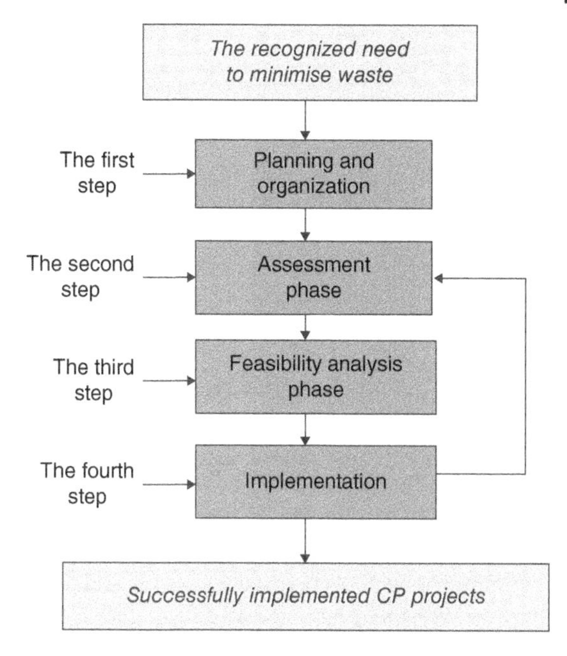

Figure 5.2 Steps for conducting cleaner production assessment (UNEP 2015).

Secondly, the prevailing concepts and attitudes can potentially delay the assessment process which can be overcome by conducting training and awareness programs. The planning stage also requires the participation of external stakeholders like regulatory bodies, suppliers of inputs and consumers to determine the type of CPS needed to be considered to avoid waste and emissions in a particular industry. The use of chemicals, processes, and product design varies with industry and so the type of emissions and wastes produced will vary. Stakeholders in both upstream and downstream supply chains can potentially collaborate with other industries in implementing CPS. Once industry specific emissions and wastes are identified, a plant-wide goal for environmental management can determine the internal efficiency standards, environmental policies and agreements. The company needs to form a balanced project team, including a management representative, a production manager, an environmental officer, a motivated supervisor, a motivated technician or operator, an external advisor capable of the assessment and measurement of the pollution and wastes created, and the implementation of CP strategies to achieve the targets. Finally, the team needs to be empowered with all the information necessary and be able to access data from the organisation.

5.5.2 Assessment

Once industry specific environmental issues are identified and an assessment team has been developed, a material and energy balance needs to be conducted to estimate the inputs in terms of chemicals and energy used in the different processes as well as to estimate the outputs in terms of the wastes and emissions produced in production. This is similar to developing a life cycle inventory for an industry using a mass balance approach which was done for an LCA discussed in Chapter 3. However, in this case, it does not consider all the stages of the product life cycle, but focuses more on micro-level production considering only where the products are produced or the service is delivered. The steps for developing an inventory for CP assessment are as follows:

Step 1: Specify all processes including production, materials handling and storage, utilities, etc.
Step 2: Identify inputs and outputs including at least raw materials, auxiliaries, energy, water, waste, waste water, and air emissions.
Step 3: Link unit operations by means of the material, energy, and water flows.

This inventory will help identify sources/processes producing more wastes and emissions so that a management approach can be taken. Like LCA, it is not required to convert these emissions and wastes directly into environmental impacts. Possible causes for waste generation will be identified to determine the right CPS for reducing the wastes. Figure 5.3 shows how the cause diagnosis can be done in order to find the problematic areas that will help recommend CP strategies. For example, in this paint spraying and baking unit of the car workshop, the problems are identified using the following standard questions:

How does technology choice and equipment design affect waste generation? – The overspray associated with the use of a spray gun generated waste. Also the efficiency of the baking unit is low meaning that it consumes significant electricity and produces emissions.
How do choice and quality of input materials affect waste generation? – The paint used is enriched with solvent and toxic dyestuffs. The disposal of the solvent and toxic dyestuffs to water bodies and soil can cause severe health effects and destroy the ecosystem.
How does equipment operation and maintenance practice affect waste generation? – As mentioned earlier, excess spraying during operation.
How does product specification affect waste generation? This question is not relevant here as it has nothing to do with car design.
As can be seen, all CP strategies may not necessarily be applied in every particular industry.

The third part of the assessment is to generate ideas or possible options to solve the problems or apply standard 'prevention practices' to all waste and emissions using the five CP strategies discussed in the previous section, including

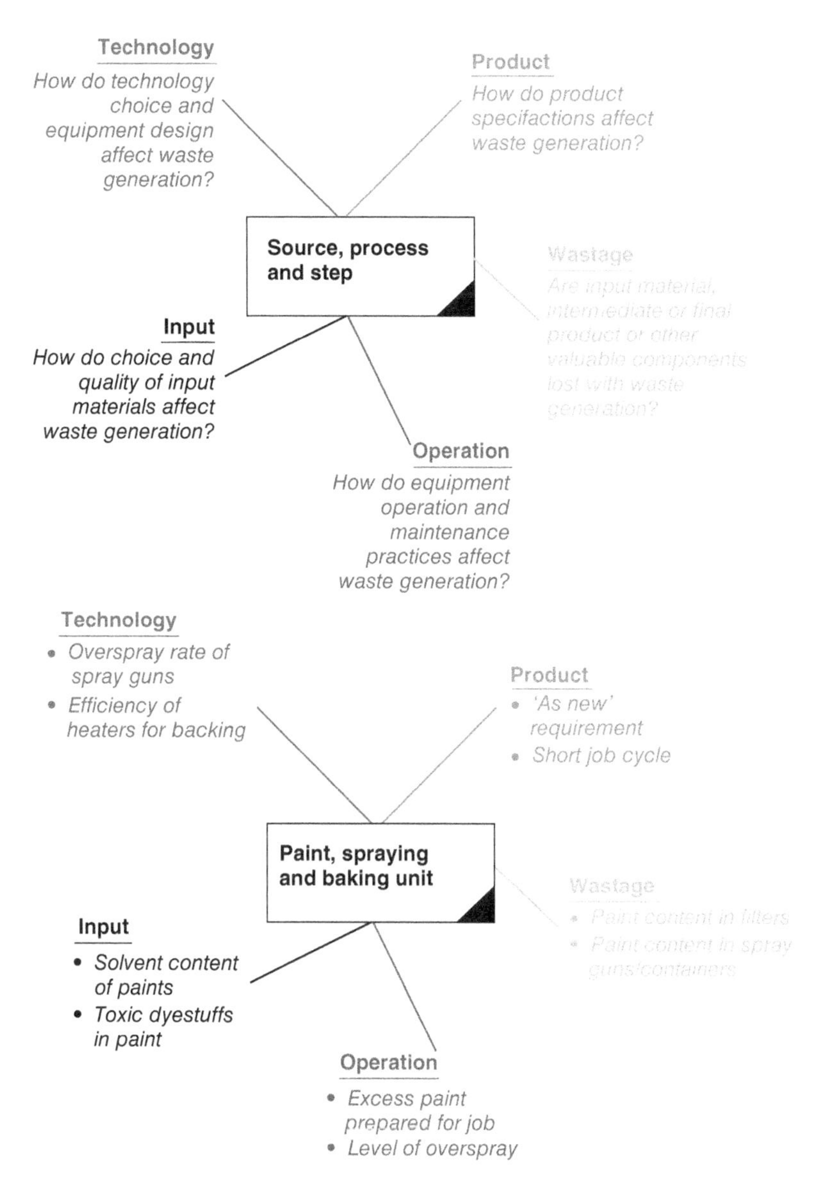

Figure 5.3 Cause diagnosis.

product modification, input substitution, technology modification, good housekeeping, and on-site recycling. Specified criteria need to be designed to select solutions/ideas. A standard cause categories/diagnosis approach can be used to assess the impact on process efficiency and waste generation. The design criteria includes product specifications, choice and quality of input materials, selection of technology, design and lay-out of equipment, operation and maintenance of equipment and waste parameters.

In the case of the car workshop example, Figure 5.4 shows how options can be generated to apply the five different CP strategies to reduce waste and emissions in a car workshop. The questions that need to be asked prior to the selection of the appropriate CP options are listed below.

Technological Modification: What changes in technology and equipment are needed to reduce waste generation? Improve heating efficiency and use low overspray gun.

Input Substitution: How do input material specifications have to change to reduce waste generation? Use high solid or water based paints.

Good Housekeeping: How can equipment operation and maintenance practices be improved? Proper job estimation practices, staff training on overspray reduction.

Recycling: How can valuable components be reused or recycled or recovered? Cleaning and reuse of filters, recovery of washed solvents.

Product Modification: What changes in product modification are conceivable to reduce waste generation? Not applicable to this particular case study.

Screening of ideas/options: Select all ideas/options that may be implemented immediately. The remaining options/ideas should then be divided into three categories:

- Interesting options but more analysis is needed
- Information source only
- Brainstorm further in project teams to overcome obstacles, and encourage innovative thinking

Solicit ideas outside the project team to encourage participation from all areas in the company. In order to do this, provide example options databases, manuals, conduct technology surveys and benchmarks, check applicability of all five prevention practices and CP assessment practices, followed by an internal brain storming workshop to select the best options.

5.5.3 Feasibility Studies

Once the CP options have been considered, preliminary, technical, economic, and environmental implications need to be conducted.

Preliminary evaluation: This evaluates the CP options as to whether the necessary technology can easily implement the option and whether it can implement the option within a reasonable timeframe without disrupting production.

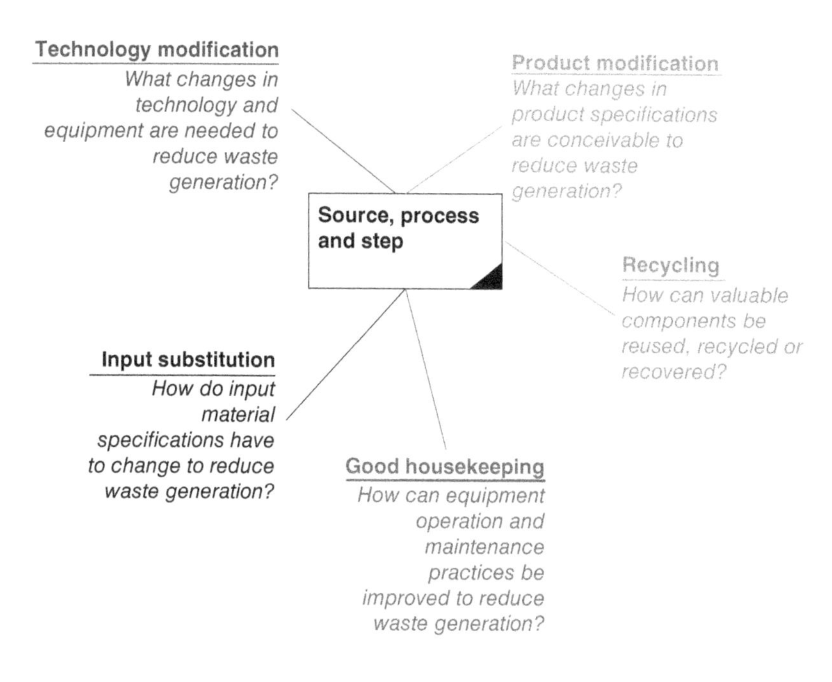

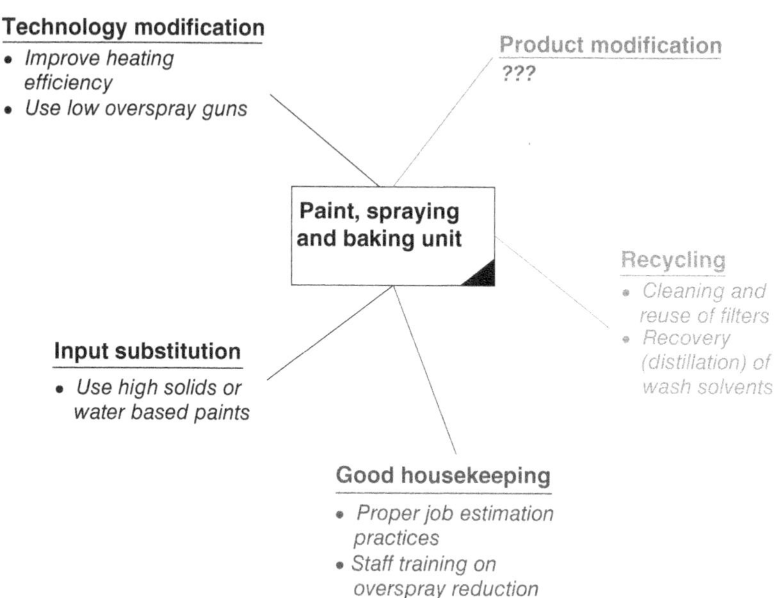

Figure 5.4 Option generation.

Technical evaluation should identify the main benefits to be gained by implementing this option. Firstly, it will identify and evaluate necessary changes due to the use of existing plant infrastructure (Will it work? What do we have to do to make it work?). This will then lead to an engineering design for the project, so that the investment can be costed. The assessment of the impact of the respective option on material and energy balance is then conducted (e.g. What will we gain?). This will enable a before and after comparison, providing the basis for the quantification of operational benefits (cost savings and revenue gains).

Economic evaluation should determine whether the option is cost-effective, which requires an in depth feasibility assessment.

Major steps in financial analysis on the basis of discounted cash flows for economic evaluation are as follows:

i) Establish the baseline
   Gather existing information
ii) Calculate the costs
   Determine the capital or investment cost of the project
   i) *Establish the lifetime of the equipment and compute annual depreciation*
iii) Calculate the benefits
   Determine revenue implications of the project
   Estimate any changes in operating costs
iv) Calculate incremental cash flows
v) Assess the project's financial viability using various decision rules

Calculation of the Costs

i) Firstly, the capital cost of new equipment or modification of the infrastructure for installation of CP options have to be determined by remembering that these expenditures took place before the commissioning of the project (or in Year 0)
ii) Secondly, the lifetime of the aforementioned capital assets and their depreciation have to be estimated. Annual depreciation is calculated as follows: [capital cost – salvage value]/lifetime of the equipment.
iii) Thirdly, the incremental benefits associated with reduced energy and water saving costs can also be calculated due to the use of CPS. The cost values of the net savings of inputs and outputs annually are incremental benefits from the use of CP options/strategies.
iv) Fourthly, operating costs including additional costs (e.g. operating and maintenance, and hidden costs (e.g. monitoring, reporting, etc.) also have to be calculated.

Table 5.1 Format for calculating incremental operational benefit.

	Unit cost ($) (P)	Rate before cleaner production (B)	Rate after cleaner production (A)	Difference D = B – A	Incremental benefit D*P
Process inputs					
Materials					
✓ A					
✓ B					
Energy					
✓ A					
✓ B					
Water					
Labour					
Product process outputs					
Product					
By-product					
Non product process outputs					
Waste					
Waste water					
Air emissions					
		Incremental operational benefits ($/year)			

Table 5.1 can be used to help determine the incremental benefits. Since processes, inputs and outputs (i.e. emissions and wastes) vary with products, strategies, it is not mandatory to obtain all of the information noted in this worksheet.

Following the work of Shrestha et al. (1998) on the energy efficient technology assessment, net annual benefit (NAB) has also been considered as an indicator for finding cost-effective options using Eq. (5.1).

$$IC + [AAC_i - AAC_0] + [AOMC_i - AOMC_0] \quad (5.1)$$

where

AAC_i and AAC_0 = annual annualised costs of the CPS and existing options, respectively

$AOMC_i$ and AOM_0 = annual operational costs of the CPS and inefficient existing options appliance, respectively

The procedure for calculating the annualised cost has been shown in Chapter 4. The following equation has been used to calculate NAB using Eq. (5.2).

$$\text{NAB} = \text{IOB} - \text{IC} \tag{5.2}$$

where

IOB = incremental operational benefit due to the use of the efficient appliances = annual savings × energy price

Environmental evaluation: Cost-effective or economically feasible CP options are to be assessed on the basis of their environmental performance. The evaluation of environmental improvements includes the reduction in pollutants generated, toxicity of emissions or waste generated, energy material and water consumption, and the total pollutant load of the product.

Selection of feasible options: Integrate the results of technical, economic, and environmental evaluation. Firstly, rule out options that do not meet the basic hurdle rate set or have doubtful technical and environmental impact. Secondly, select among feasible options with highest NAB.

Sort options to identify additional information requirements such as being people or equipment based, simple or complex, and cheap or expensive. Then use results to guide the feasibility studies. People based options most often do not require detailed technical evaluation. Cheap options typically require simplified economic evaluation.

5.5.4 Implementation and Continuation

Once the options have been selected, a plan needs to be developed to implement selected CP strategies. Table 5.2 shows an example of this plan where tasks are assigned to staff to meet the milestone to achieve the environmental target.

There is also a need for preparation of preventative maintenance schedules and to arrange supply of critical spare parts, draft work instructions, update quality manuals and plant project initiation documents (PIDs) and train operators, supervisors, and technicians. Once the CP options have been introduced to the industry, the CP progress needs to be monitored by recording the change in waste quantity, resource consumption, and profitability by taking into account the changes in total production output, shifts between product ranges, and the introduction of new products.

Practice Example

Students can follow the assessment procedure discussed above. Select a factory, refinery or a chemical processing plant producing a certain product. Once you have selected this case study, then undertake the following tasks:

Table 5.2 Implementation plan.

Option	Responsible person	Deadline for implementation	'Milestones'
Rinse solvent reuse	Supervisor	2 mo	Monitoring programme (6 mo duration)
Insulation (steam and condensate)	Chief technician	3 mo	
Reuse cooling water for process bath make up	Technical supervisor	2 mo	Modification of working instructions (4 mo)

Mass Balance

i) Draw a detailed process flow chart showing how the main feedstock is converted to a product through chemical processing. The flow chart should show a schematic diagram consisting of all components in symbolic forms (e.g. reactor, separator, generator, heat exchanger, distiller, precipitator, etc.) which are required for the production of the product.
ii) Estimate the amount of chemicals, energy, and water associated with the production of a certain unit of product. This unit of product is known as a functional unit, which is required for conducting the mass balance. For example, one million dollar equivalent amount of iron ore transported to China is a functional unit. Accordingly, you need to work out the amount of inputs in the form of chemicals, energy, and water associated with the production of a million dollar equivalent of iron ore transported China.
iii) In addition to inputs, estimate the amount of outputs in the form of waste and emissions generated in different processes due to the conversion of feedstock into a product.
iv) You will then draw a similar type of flow chart and provide units for all inputs utilised and outputs generated.

Identification of Environmental Improvement Opportunities

v) Once the flow diagram showing detailed inputs and outputs has been drawn, then determine the processes creating waste and emissions, known as the hotspot(s). Select top two hotspots from the flow chart.
vi) Generate CP options for reducing these waste and emissions for each of the two selected hotspots. The CP options are product modification, input substitution, technology modification, good housekeeping, and on-site recycling.

Cost-Effective Mitigation Strategies

vii) Once CP options have been generated, then estimate the amount of waste and emissions that can be mitigated due to the use of these options. You need to develop a table consisting of five columns with one column for processes creating the waste or emission, one column for writing down the name of the relevant CP option, one column for the technical characteristics/details of the option, one column for the justification for choosing these CP options and the final column for estimating the mitigation potential (e.g. 10 ton of residue avoided per ton of alumina production).

viii) Estimate the costs, including capital and operational, for mitigating wastes/emissions with CP options and also estimate the operational cost savings (e.g. energy, chemicals and water), if any, associated with the application of CP options. Consider $50/ton for any type of solid waste disposed to landfill and $25 per ton for any type of gaseous emission. Use a discounted cash flow analysis to carry out an economic analysis, where the capital cost of CP option, operational costs (maintenance), replacement costs (if any), and benefits (savings in operational cost and environmental cost) have been utilised.

ix) Conduct a benefit cost analysis to determine the economically feasible CP options. An example of a benefit cost analysis has been provided in Chapter 4.

5.6 Eco-efficiency

Eco-efficiency is a strategy that creates more value with less environmental impact and costs, focuses on dematerialisation and the decoupling of resource use and emissions from economic growth. This can be simply gained when firms reduce energy and resource consumption and/or reduce waste generation by changing their products, internal processes, and infrastructure. Eco-efficiency can be considered as other side of the coin to CP strategies as the latter is an operational strategy to achieve the former by reducing impacts, costs, risks, and liabilities. This is in fact a management strategy that creates more value with less impact, through delinking goods and services from the use of natural resources. According to the World Business Council for Sustainable Development (WBCSD 2000), eco-efficiency is the delivery of competitively priced goods and services that satisfy human needs and bring quality of life, while progressively reducing ecological impact and resource intensity throughout the life cycle, to a level at least in line with the Earth's estimated carrying capacity.

The challenges that can be addressed with eco-efficiency include the growth of world population combined with unsustainable production and consumption

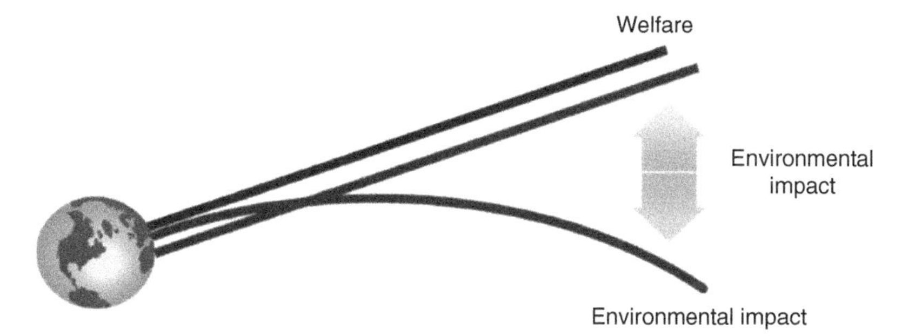

Figure 5.5 Delinking welfare from the use of nature.

patterns that place increasing stress on the population-supporting capacities of our planet. The three key objectives of eco-efficiency to address these challenges are:

- To provide more value with less environmental impacts, by reducing material and energy intensity, and toxic dispersion, while enhancing material recyclability, use of renewable resources, product durability, and service intensity.
- To de-link the growth of human welfare from the use of nature (Figure 5.5). For example, in the case of Sweden, GDP grew by 58%, while CO_2 emission reduced by 23% during the period 1990–2013 (GoS 2015). In Germany, GDP increased by 28% and CO_2 emissions decreased by 22% during 1991–2012 (Morris and Pehnt 2014).
- To improve both economic and ecological efficiency, environmental performance needs to be improved in a cost-effective or cost-competitive manner.

Eco-efficiency ratio: The eco-efficiency ratios at micro and macro levels using Eqs. (5.3) and (5.4) are expressed as follows:

- On the micro-level (company):

$$\text{Eco-efficiency} = \frac{\text{product/service value}}{\text{environmental influence}} \tag{5.3}$$

- On a macro-level (government):

$$\text{Resource produtivity} = \frac{\text{more welfare}}{\text{less resource use}} \tag{5.4}$$

Micro-level efficiency focuses on a particular technology or a process. For example, in the case of a refrigerator, the capacity increased from 150 to 300 l over 10 years ago, while the power consumption was decreased from 1000 to 200 kWh due to a massive technological improvement measures by using variable operating conditions, selection of the most efficient refrigeration cycle and components, use of design effective control systems and the selection of the refrigerant. The emission factor of the electricity is 900 kg of CO_2 e- per MWh assuming that the electricity mix remains the same over these years (Morris and Pehnt 2014).

Using above information, the eco-efficiency of the refrigerator 10 years ago would be 150 l/(1000 kWh × 0.9 kg CO_2 e-/kWh) or 0.17 l/kg of CO_2 e-. After 10 years, the eco-efficiency is 0.83 l/kg of CO_2 e-. This is an example for micro-level eco-efficiency as it is happening at the product level.

At a macro-level, eco-efficiency indicators are ratios of welfare indicators (mainly GDP or gross value added) and environmental indicators (use of nature). This 'ecological efficiency' ratio expresses how much benefit or welfare is achieved from one unit of 'nature'. On the macro-economic level, GDP is the most often used indicator of 'welfare'. However, some alternatives have been proposed including the United Nations Environment Program (UNEP's) human development index (HDI) or the index of sustainable economic welfare (ISEW). For the 'use of nature', we can choose from a wide range of environmental indicators. The denominator for nature is of two types, input based and output based (i.e. emissions and wastes resulted from any production process). The input based denominators are typically raw-material input, gross inland energy consumption, land-use, and water consumption. Output based denominators are greenhouse effects, acidification, ozone depletion, and hazardous waste chemicals.

5.6.1 Key Outcomes of Eco-efficiency

Key outcomes of eco-efficiency analysis are as follows:

Optimised Processes: Moving the industry from costly end of pipe solution to approaches that prevent pollution in the first place.

Eco-innovation: Manufacturing smarter by using new knowledge to make old products more resource efficient to produce and use.

Networks/Virtual Organisations: Shared resources increase the effective use of physical assets.

New Services: e.g. Leasing products not selling them, spurring an increased focus on product durability and recycling.

Waste Recycling: Using wastes or by-products of one industry as potential raw materials and resources for another industry – moving towards zero waste.

5.6.2 Eco-efficiency Portfolio Analysis in Choosing Eco-efficient Options

A number of options can do the same job or solve the same problem. However, it is important to identify the option that can deliver the best environmental performance. The cost can increase due to the additional investment in environmentally friendly technology for improving the environmental performance or vice versa. Kicherer et al. (2007) developed a procedure for estimating eco-efficiency portfolio positions for a number of possible options or solutions for a particular task or project as a basis for selecting the most eco-efficient option(s). The economic and environmental performance of a product or service are balanced in the eco-efficiency framework. An eco-efficient option needs to demonstrate acceptable environmental and economic performance. Environmental life cycle assessment (ELCA) explained in Chapter 3 is used to calculate the life cycle environmental impacts (LCEI) of a product and life cycle costing, which is discussed in Chapter 4 is used to calculate the life cycle cost of the same product using the same system boundary. Eco-efficiency portfolio analysis helps combine LCC and ELCA results to determine the eco-efficiency portfolio position of a product/option/solution.

The first step is to normalise the environmental impacts in the form of 'inhabitant equivalents' (Kicherer, et al. 2007). The normalised value of each environmental impact is calculated using Eq. (5.5).

$$NV_i = \frac{LCEI_i}{GDEI_i}\ [Inh] \tag{5.5}$$

NV_i = The normalised value of the environmental impact
$LCEI_i$ = The life cycle environmental impact over all life cycle stages
$GDEI_i$ = The gross domestic environmental impact
Inh = The net capita of inhabitants
i = Refers to the ith impact category. This impact categories vary with products across regions as different products have different production processes. The impact categories of the same product vary with countries due to geographical, climatic, and resource utilisation differences. This means that the impacts considered in a country or region will be different in another country for the same product.

Prior to aggregating the individual environmental impacts into a single impact, the impact categories are weighted to reflect the relative degree of importance of the system boundary conditions. The weighting factors of impacts vary across regions due to variations in climate and resource utilisation patterns. The total environmental impact (EI) was determined by weighting and aggregating using Eq. (5.6).

$$EI = \sum NV_i \times WF_i \tag{5.6}$$

EI = The total normalised environmental impact
WF_i = The weighting factor. The summation of each WF_i must add to 100%.

In order to combine life cycle costs with environmental impacts, an LCC needs to be normalised in a similar method by using the GDP of the same region to reflect the costs in the same units of inhabitants per year (Kicherer et al. 2007). The equation used to calculate the normalised cost is shown in Eq. (5.7).

$$\mathrm{NC} = \frac{\mathrm{LCC}}{\mathrm{GDP}_{\mathrm{cap}}}\ [\mathrm{Inh}] \tag{5.7}$$

NC = The normalised total cost
LCC = The life cycle costing over all life cycle stages
$\mathrm{GDP}_{\mathrm{cap}}$ = Gross domestic product per capita of the region

The next step is to calculate the eco-efficiency portfolio as presented in Figure 5.6 by plotting NC on the abscissa and EI on the ordinate (Kicherer et al. 2007). It should be noted that the options that are not eco-efficient will be located where the highest values of each axis tend toward the bottom left and the the most eco-efficient options will be located towards the top-right of the portfolio. A diagonal line which runs through the origin is used to visualise the eco-efficiency. Any options that are above this line are deemed eco-efficient. The most eco-efficient option is that which has the largest perpendicular distance above the diagonal line.

To determine the preliminary portfolio position of each option, Eqs. (5.8) and (5.9) were used.

$$\mathrm{PP}_{\mathrm{e},n} = \frac{\mathrm{EI}_n}{\left(\frac{\mathrm{EI}}{j}\right)} \tag{5.8}$$

$$\mathrm{PP}_{\mathrm{c},n} = \frac{\mathrm{NC}_n}{\left(\frac{\mathrm{NC}}{j}\right)} \tag{5.9}$$

$\mathrm{PP}_{\mathrm{e},n}$ = environmental impact preliminary portfolio position for option n
$\mathrm{PP}_{\mathrm{c},n}$ = cost preliminary portfolio position for option n
n = refers to the nth eco-efficiency option

These preliminary positions (PP) are then improved by the relevance factor ($R_{\mathrm{e,c}}$) factor, in order to get a new position (PP′) in which a balance between environmental impacts and costs exists. The relevance factor is determined by the ratio of the average environmental impact of all options to the average cost of all options as shown in Eq. (5.10).

$$R_{\mathrm{e,c}} = \frac{\left(\frac{\sum \mathrm{EI}}{j}\right)}{\left(\frac{\sum \mathrm{NC}}{j}\right)} \tag{5.10}$$

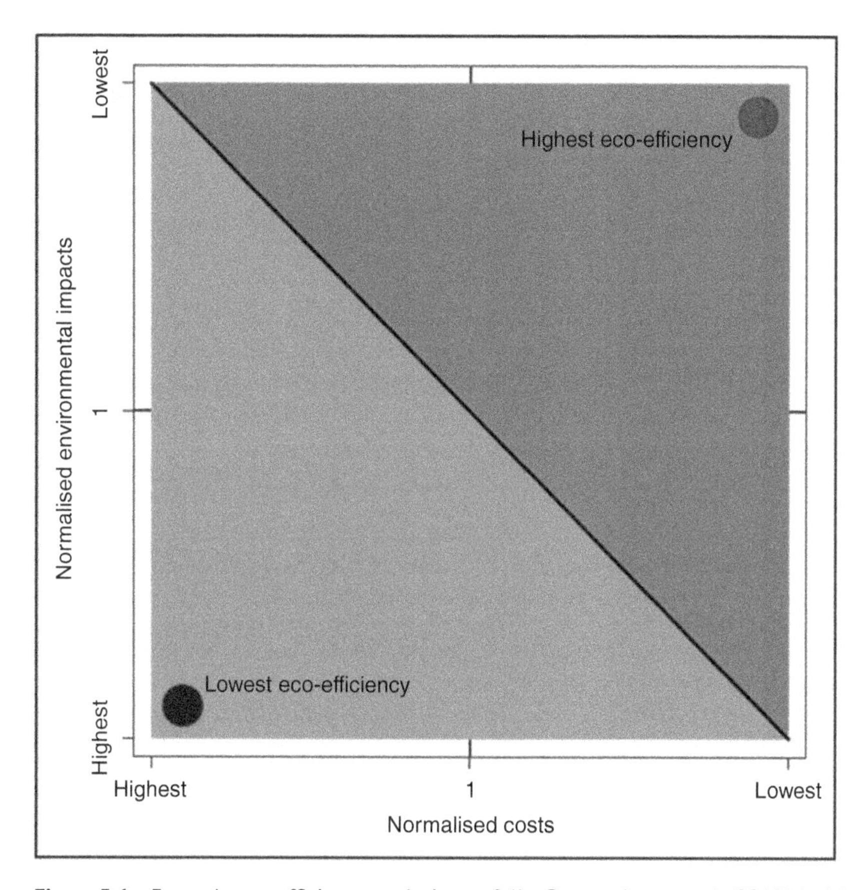

Figure 5.6 Example eco-efficiency analysis portfolio. Source: Arceo et al. (2019) / with permission of Elsevier.

$R_{e,c}$ = The relevance factor of environmental impact to the cost of all options

j = The number of eco-efficient options to be considered in this portfolio

To determine the final portfolio position of each option, Eqs. (5.11) and (5.12) are used.

$$PP'_{e,n} = \frac{\left[\frac{PP_{e,n}}{j} + \left[PP_{e,n} - \left(\frac{PP_{e,n}}{j}\right)\right] \times \sqrt{R_{e,c}}\right]}{\left[\frac{PP_{e,n}}{j}\right]} \tag{5.11}$$

$$\mathrm{PP}'_{\mathrm{c},n} = \frac{\left[\frac{\mathrm{PP}_{\mathrm{c},n}}{j} + \frac{\left[\mathrm{PP}_{\mathrm{c},n} - \left(\frac{\mathrm{PP}_{\mathrm{c},n}}{j}\right)\right]}{\left[\sqrt{R_{\mathrm{e,c}}}\right]}\right]}{\left[\frac{\mathrm{PP}_{\mathrm{c},n}}{j}\right]} \tag{5.12}$$

$\mathrm{PP}'_{\mathrm{e},n}$ = adjusted environmental portfolio position of eco-efficiency option n
$\mathrm{PP}'_{\mathrm{c},n}$ = adjusted cost portfolio position of eco-efficiency option n

The adjusted portfolio positions are influenced by the relevance factor resulting in a balanced position between the environmental and economic factors (Kicherer et al. 2007). The final positions are plotted graphically to visually assess the potential eco-efficient options. The resulting portfolio is ideal for comparison with multiple eco-efficient options, a major goal of this assessment.

Box 5.6 provides practice example for determining the eco-efficiency portfolio.

Box 5.6 Eco-efficiency performance of the use of C&D wastes and industrial by-products in high compressive strength concrete applications

The experimental tests conducted confirmed that the following concrete classes utilising recycled aggregates are structurally sound:

1) MIC = 100 NA + 100 OPC (control)
2) M2 = 100 RA + 100 OPC
3) M3 = 50 RA + 50 NA + 90 OPC + 10 SF
4) M4 = 50 RA + 50 NA + 60 OPC + 30 FA + 10 SF

(OPC = ordinary portland cement; NA = natural aggregate; RA = recycled aggregates; SF = silica fumes; FA = fly ash.)

The alternative concrete mixes could offer the same structural performance as conventional concrete. The task is to compare the eco-efficiency performance of concrete mixes using both natural and recycled aggregates, OPC, and by-product based cementitious materials with conventional concrete using natural aggregates and OPC. Table 5.3 shows the environmental impacts created by the production of 1 m^3 of both alternative and conventional concrete mixes (i.e. MIC, M2, M3, and M4) and these impacts are particularly considered for building materials in Australia. These impacts were normalised by dividing by the corresponding normalisation factors and then these normalised values are multiplied by the corresponding weights. Both NF and weights of these impacts were determined for building materials by the Green Building Council of Australia. In the case of cost, no operational cost is involved as the system followed a cradle to gate approach and also the system boundary was limited to the production of 1 m^3 of concrete mix. Therefore, a discounted cash flow analysis was not conducted. It included the total amount of materials, labour and energy involved in the production of these concrete mixes. The costs of four concrete mixes are normalised by dividing by the per capita GDP of Australia in year 2019. Tables 5.4 and 5.5 show how the initial portfolio (PP) and final portfolio (PP') positions were calculated to compare the eco-efficiency performance of the four concrete mixes.

Solution

Table 5.3 shows how Eqs. (5.5) and (5.6) are used to calculate impacts in terms of per inhabitant. Table 5.4 shows how Eq. (5.7) is used to calculate the normalised value of the cost. This table also shows how Eqs. (5.8)–(5.10) were used to calculate the relevance factor, and the preliminary portfolio positions for environmental and economic performance. Table 5.5 shows how Eqs. (5.11) and (5.12) was used to calculate the adjusted portfolio positions PP′e and PP′c. Finally, Figure 5.7 shows the portfolio positions of four different types of concrete mixes and concludes that M4 (50 RA + 50 NA + 60 OPC + 30 FA + 10 SF) is the only eco-efficient mix.

5.7 Environmental Management Systems

EMS follow ISO 14001-4 guidelines to assist industries to meet environmental targets using CP strategies and to comply with the environmental regulations.

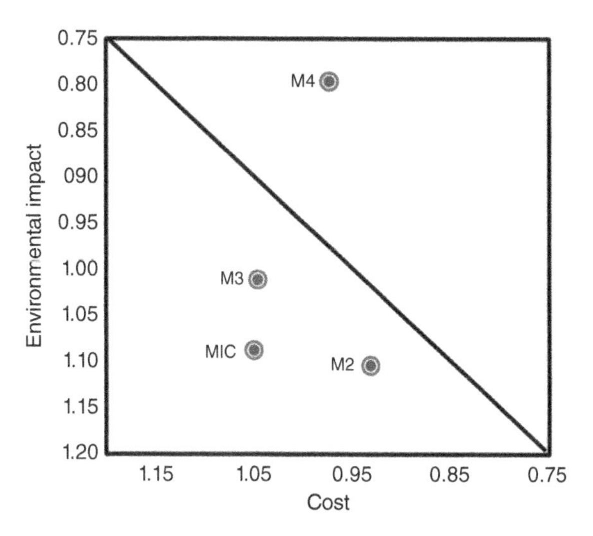

Figure 5.7 Eco-efficiency portfolio positions for 4 concrete mixes.

Table 5.3 Conversion of impacts to inhabitant equivalent [Inh].

		Characterised values						Inh (using Eqs. (5.5) and (5.6))			
Impacts	**Units**	**MIC**	**M2**	**M3**	**M4**	**$GDEI_i$ per Inh**	**WF**	**MIC**	**M2**	**M3**	**M4**
Global warming	kg CO_2	5.29E+02	5.29E+02	4.85E+02	3.72E+02	28,690	20%	2.97E−03	2.97E−03	2.73E−03	2.12E−03
Eutrophication	kg PO_4 e-	2.52E−01	2.53E−01	2.32E−01	1.79E−01	19	3%	3.83E−04	3.83E−04	3.52E−04	2.70E−04
Land use	Ha a	1.23E−03	9.61E−03	5.34E−03	5.11E−03	26	21%	7.40E−06	7.49E−05	4.05E−05	3.86E−05
Water use	M3 H_2O	2.02E+00	2.06E+00	1.99E+00	1.87E+00	930	6%	1.35E−04	1.37E−04	1.33E−04	1.25E−04
Human toxicity	DALY	1.12E−06	1.17E−06	1.05E−06	8.19E−07	3,216	3%	9.60E−12	9.64E−12	8.90E−12	6.94E−12
Fresh water ecotoxicity	DALY	9.65E−11	1.02E−10	9.11E−11	7.16E−11	172	7%	3.86E−14	4.00E−14	3.65E−14	2.83E−14
Marine ecotoxicity	DALY	4.86E−08	5.00E−08	4.62E−08	3.93E−08	12,117,106	8%	3.04E−16	3.02E−16	2.90E−16	2.46E−16
Terrestrial ecotoxicity	DALY	2.04E−11	2.07E−11	1.92E−11	1.57E−11	88	10%	2.37E−14	2.40E−14	2.24E−14	1.82E−14
Ozone depletion	kg CFC-11 e-	3.53E−06	3.98E−06	3.45E−06	2.94E−06	0.002	4%	6.92E−05	7.70E−05	6.79E−05	5.71E−05
Terrestrial acidification	kg SO_2 e-	1.96E+00	1.97E+00	1.81E+00	1.42E+00	123	3%	4.89E−04	4.91E−04	4.51E−04	3.54E−04
Photochemical smog	kg NMVOC	1.91E+00	1.90E+00	1.75E+00	1.36E+00	75	3%	7.09E−04	7.06E−04	6.49E−04	5.05E−04
Ionising radiation	kBq U235 e-	2.31E+00	2.18E+00	2.08E+00	1.63E+00	1,306	2%	3.33E−05	3.17E−05	3.00E−05	2.36E−05
Respiratory effects	kg PM2.5 e-	1.09E−01	1.14E−01	1.02E−01	7.96E−02	45	3%	7.38E−05	7.69E−05	6.91E−05	5.40E−05
Abiotic depletion	kg Sb e-	1.42E−04	1.55E−04	1.36E−04	1.13E−04	300	8%	3.96E−08	4.27E−08	3.76E−08	3.08E−08
						Total environmental impact (EI) in Inh		4.87E−03	4.95E−03	4.52E−03	3.55E−03

PO_4 = Phosphate; Ha = Hectare; DALY = disability-adjusted life year; CFC = chlorofluorocarbon; NMVOC = Non-methane volatile organic compounds; U235 = Uranium-235; PM2.5 = Particulate Matter less than 2.5 microns in diameter; Sb = Antimony.

Table 5.4 Calculation of initial portfolio (PP) positions of concrete mixes.

EI [Inh] (from Table (5.4))	Costs	Costs [Inh] (using Eq. (5.7))	Using Eq. (5.10)	PP_e (using Eq. (5.8))	PP_c (using Eq. (5.9))
EI_{MIC} = 4.87E−03	$Cost_{MIC}$ = 351	NC_{MIC} = 4.88E−03	REC = 9.63E−01	$PP_{e,MIC}$ = 1.09E+00	$PP_{c,MIC}$ = 1.05E+00
EI_2 = 4.95E−03	$Cost_2$ = 311	NC_{C2} = 4.32E−03		PP_{e2} = 1.11E+00	PP_{c2} = 9.31E-01
EI_3 = 4.52E−03	$Cost_3$ = 350	NC_{C3} = 4.86E−03		PP_{e3} = 1.01E+00	PP_{c3} = 1.05E+00
EI_4 = 3.55E−03	$Cost_4$ = 325	NC_{C4} = 4.52E−03		PP_{e4} = 7.93E−01	PP_{c4} = 9.73E−01

Table 5.5 Calculation of final portfolio positions (PP′).

PP'_e (using Eq. (5.11))	PP'_c (using Eq. (5.12))
$PP'_{e,MIC}$ = 1.09E+00	PP'_{c1} = 1.05E+00
PP'_{e2} = 1.10E+00	PP'_{c2} = 9.32E−01
PP'_{e3} = 1.01E+00	PP'_{c3} = 1.05E+00
PP'_{e4} = 7.97E−01	PP'_{c4} = 9.73E−01

Once industries have an EMS in place, they are certified by the ISO (International Standard Organization) meaning that the industries are able to export their products, meet World Trade Organisation requirements and improve their good corporate stewardship. According to ISO 14001, an EMS includes the organisational structure, planning activities, responsibilities, practices, procedures, processes, and resources for developing, implementing, achieving, reviewing, and achieving environmental policy requirements (ISO 2016). The organisational structure means the involvement of company people ranging from the Chief Executive Officer down to shop floor workers, and external agents who can be both directly and indirectly affected due to externalities caused by the industry. Each plays different roles in developing actions to implement CP strategies using available human and natural resources and technologies to reduce pollution and wastes, energy and material consumption to achieve environmental targets, and to comply with the environmental policy. EMS is a continual management improvement process as monitoring and measurement activities are involved to ensure that the CP strategies used are actually reducing emissions, wastes, energy, and material consumption to reduce impacts over a certain period of time, known as a target period.

5.7.1 Aims of an EMS

The key aims are as follows:

Identification and Control of Aspects, Impacts, and Risks: The aspect in this case is the pollution associated with human activities such as the combustion of fossil fuel for power generation or chemical processing, with different emissions creating different impacts (e.g. global warming impact, acidification) and then different impacts causing different effects (e.g. human health, damage of an ecosystem) or risks.

Establishing and Achieving Environmental Policy, Objectives, and Targets, Including Compliance with Legislation: Different industries create different types of emissions and wastes and therefore, their environmental aspects, impacts, mitigation strategies, and targets will be different. For example, petroleum refineries release gases like SO_x and NO_x into the atmosphere, which are aspects that need to be separated during the refining process, while mining and mineral processing industries have different aspects which include the release of dust during these operations. The emissions of SO_x and NO_x creates impacts such as acid rain and eutrophication, which are different to the impacts caused by dust emissions such as visibility loss and human eco-toxicity. Pollution prevention strategies or environmental policy and targets therefore vary with industry, and the targets for mitigating pollution vary with the policies imposed by different regions and countries.

Identifying Environmental Opportunities: This requires a detailed process with data enabling management or EMS team to identify the inputs in terms of energy, chemicals, processes and technologies that cause the most impact as determined by the regulatory authority. CP strategies can then help to mitigate these impacts to an acceptable level.

Monitoring and Continual Improvement of Environmental Performance: Once the mitigation strategies are applied, measurements of inputs and outputs are conducted as part of an EMS process to monitor either on a daily or monthly basis to observe the progress in terms of the inputs and outputs reduction achieving the environmental target. This will depend on the policies, availability of technology and the required skills to operate and maintain these mitigation measures. If the progress does not happen as anticipated, the EMS is flexible enough to incorporate changes in the mitigation measures that further help achieve the environmental target. Also this target may change with changes in policy, technology, and socio-economic considerations, and therefore, there may be a need to change these mitigation measures initially selected. This is known as a continual improvement process.

5.7.2 A Basic EMS Framework: Plan, Do Check, Review

Plan: This involves the issuing of a policy statement, signed by the facility manager that must commit to continual improvement, pollution prevention, and environmental compliance. Planning identifies aspects and impacts from facility activities, products, and services, and then reviews the legal requirements, for setting objectives, for achieving targets, and then establishing a formal EMS program. This is usually the most expensive component of an EMS as it involves the collection and measurement of chemical and energy use data, pollution and waste data, and the involvement of external agents to help facilitate the planning process.

Do: The first task is to allocate roles and responsibilities to people to implement an EMS, providing EMS training, establish internal and external communication mechanisms, documenting control systems, offering operational controls, and then their integration in operations emergency procedures to avoid risks and environmental hazards.

Check: This involves the periodic monitoring of environmental performance, identifying root causes of problems and conducting corrective and preventive actions, maintaining environmental records, and conducting periodic EMS audits.

Review: Reviewing, includes progress reviews and acting to make necessary changes/improvements.

5.7.3 Interested Parties

The development of an EMS involves all company staff and the CEO, the local community, Department of Environment, consultants, political leaders, environmentalists, and local businesses (Figure 5.8). Pressure groups and local community participation help prevent potentially negative impact by the industry and the surrounding ecosystem. The regulatory authority or the Department of Environment makes the CEO aware of impacts detrimental to the environment. The CEO works with engineers, procurement officers, finance officers, and the human resource officers to procure the right technology within company's budget and to operate technology with the required level of technical skill to mitigate emissions and wastes to achieve industry's environmental mitigation target. Local businesses should also participate with the local community in the discussion process. The participation of suppliers of inputs like chemicals and fuels is important as they can provide industries with inputs with less environmental impact. For example, a local utility company can supply electricity that is produced from solar photovoltaic panels instead of from a diesel fuel power plant. Finally, the involvement of

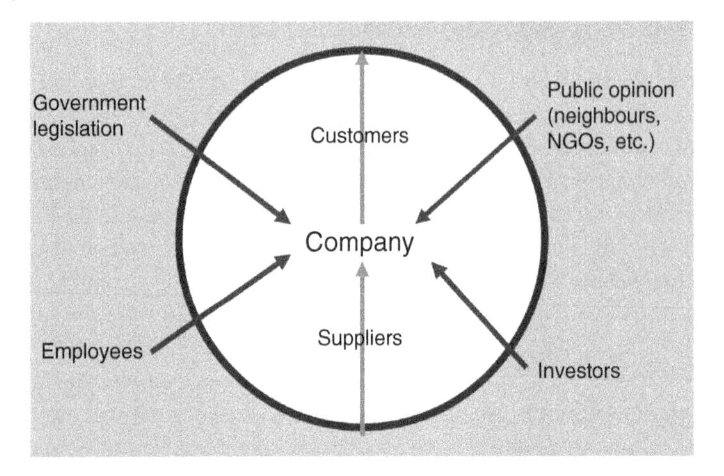

Figure 5.8 Direct and indirect stakeholders in the EMS development process.

customers is important as a change in product specification due to environmental improvements could increase the price or make the product unaffordable. Therefore, a number of consultations may be required as a part of an EMS process, which enables the industry to meet its environmental target.

5.7.4 Benefits of an EMS

Using an EMS, an industry can demonstrate that it is a good corporate citizen. Industries certified by ISO14001-4 can export their products as the World Trade Organization (WTO) is increasingly focussing on trade rules that relate to environmental protection policies (IISD 2020). Along with Canada, the United States, and Europe, the UNEP, the World Bank and World Conservation Union (IU CN) are becoming more active in the environment and trade debate (WWF 2021).

Secondly, an EMS prioritises, coordinates, and focuses activities on achieving its goals and helps anticipate problems, and take preventative rather than corrective action. The EMS allows the application of CP strategies to take competitive advantages of environmental management opportunities.

Thirdly, an EMS enables industry to demonstrate 'due diligence' by taking reasonable steps in advance or to eliminate or minimise risks to the organisation and to the environment.

Fourthly, an EMS requires industry to use efficient technologies and processes for production which reduce chemical and energy use and reduce pollution control costs. This ultimately reduces operational costs and helps industries to gain a competitive advantage.

Finally, an EMS helps industry to maintain good organisational morale by considering community and the surrounding environment in their production decision. It helps maintain an open interface between the industry and environment.

Box 5.7 shows how the use of EMS guidelines helped a cable manufacturer in Ghana to achieve their environmental target, gain market access, and improve corporate reputation. This was based on the work conducted by Botchway Nancy and Gbedemah Shine Francis in Ghana (Nancy and Shine Francis 2018).

Box 5.7 Drivers and implications of an EMS in Cable Manufacturing Company in Tema, Ghana

The core business of this company is the design and production of cabling systems for three major growth markets in Ghana; mining, oil and gas, and building sectors that require ISO 14001 EMS certification prior to any business. In order to exist in this growth market, this company has developed used ISO14001-4 to obtain a certification from the ISO. This certification process did not only help the manufacturer to sustain the market but they were able to increase their resource efficiency. The reduction in potential environmental liability risks with the reduction in energy consumption was found to be the main competitive advantage of ISO 14001 EMS certification. The firm also saved a large amount of energy, which not only reduced the operating costs but also increased market access.

The company has appointed also trained internal auditors from various departments to audit activities based on the certified standards (ISO 14001, OHSAS 18001 and ISO 9001) for the continual improvement of the production process. In order to implement the EMS, some staff appointed Health, Safety and Environment Representatives for certain processes in reviewing the environmental and risk register, which helped document environmental and safety performance.

The company set an annual budget for energy. The manufacturer desires to maintain its presence at all levels of oil and gas production, providing a wide range of energy and telecom cables for onshore and offshore exploration, production, and distribution, as well as petrochemical infrastructure. Market share (25%) is the most important competitive advantage the firm derived from ISO 14001 certification. A significant competitive advantage is the advantage that organisations gain over their competitors, which then provides great value to the customers and a great benefit to the company (Lynch 2000).

After energy, water pollution was the second most important consideration as the Centre for Public Interest Law and Anor, on behalf of the people of Tema Manhean, once took the Tema Oil Refinery to court in 2007 for polluting a local lagoon. Along

(Continued)

Box 5.7 (Continued)

with oil refining, there are many industries including a cable metal manufacturing company situated in the industrial enclave. The best insurance against any future environmental liability is to get a certified EMS. The main area of environmental liability prevention for the firm is in the hazardous fuel and oil used in maintenance and production processes. The company installed underground fuel and oil separators for this storage area.

FM Global engineering personnel regularly visit the firm to evaluate hazards and recommend improvements to their work practices to reduce physical and financial risks if a loss occurs. The QHSE Assistant Manager explained that 'prior to the discharge of effluent (wastewater from the recycling plant), physicochemical and bacteriological analysis are done to confirm the water quality of the effluent'.

Corporate Commitment: Senior management clearly signalled to staff that improving energy use by the firm was a corporate goal that both plant managers and all staff should care about.

The main conclusion of this study was that environmental performance of the company in energy consumption needs continually improving even after the certification of ISO 14001 EMS because the EMS is implemented alongside key performance indicators and the commitment of top management.

5.8 Conclusions

This chapter identifies the potential role of CP initiatives in increasing the efficiency of a production system and associated supply chain management, whilst reducing their associated environmental impacts in a cost-competitive manner.

The substantial contribution by both heavy and small-medium sized industries to environmental emissions warrants investigation of CP strategies to assist manufacturers and industry to both reduce their ecological footprint and improve their production efficiency.

In addition, the product supply chain needs to evolve, or in some instances radically change, its production practices with the use of CP strategies with a consequential improvement of environmental performance at 'hotspots'. In order to assess the environmental performance of supply chains, each stakeholder should develop a system-based approach, to help lead the implementation of CP initiatives. Government policies and supporting initiatives, international trade requirements, carbon trading schemes, and capacity building have all been and will potentially continue to be, the major drivers for implementing EMS in production supply chains.

References

Arceo, A., Biswas, W., and John, M. (2019). Eco-efficiency improvement of Western Australian remote area power supply. *Journal of Cleaner Production* 230: 820–834.

ARN (2010). Ice balls help data center go green. https://www.arnnet.com.au/article/364219/ice_balls_help_data_center_go_green/ (accessed 10 April 2021).

Canstar Blue (2016). The cost of leaving appliances on standby. https://www.canstarblue.com.au/electricity/cost-leaving-appliances-standby/ (accessed 14 May 2021).

Fuji Xerox (2008). Fuji Xerox and environmental sustainability. Fuji Xerox Australia Pty Ltd. 101 Waterloo Rd, Macquarie Park NSW 2113. http://www.ecobuy.org.au/uploads/documents/FXA%20sustainability%20profile%20A4LR%20no%20crops.pdf.

GOS (Government Offices of Sweden) (2015). Sweden: Decoupling GDP growth from CO2 emissions is possible. Minister of Foreign Affairs, Government Offices of Sweden.

IISD (2020). Trade and Environment: Revisiting Past Debates and Weighing New Options. https://sdg.iisd.org/commentary/policy-briefs/trade-and-environment-revisiting-past-debates-and-weighing-new-options/ (accessed 25 May 2021).

ISO 14004:2016. Environmental Management Systems – General Guidelines on Implementation.

ISO (International Organization for Standardization) (2004). ISO 14001:2004 Environmental management systems — Requirements with guidance for use. ISO Central Secretariat Chemin de Blandonnet 8 CP 401–1214 Vernier, Geneva, Switzerland.

IVAM (2008). Cleaner Production Manual. ASIE/2006/122–578 Improving the living and working conditions of people in and around industrial clusters and zones. file:///C:/Users/230077I/Downloads/08-cleaner-production-manual.pdf.

Kicherer, A., Schaltegger, S., Tschochohei, H., and Pozo, B.F. (2007). Eco-efficiency. *International Journal of Life Cycle Assessment* 12 (7): 537–543.

Lynch, R. (2000). *Corporate Strategy*, 4e. Harlow, UK: Pearson Education Limited.

Morris, C. and Pehnt, M. (2014). Energy Transition The German Energiewende. An initiative of the Heinrich Böll Foundation January 2014, www.energytransition.de https://pl.boell.org/sites/default/files/german-energy-transition.pdf (accessed 5 July 2021).

Nancy, B. and Shine Francis, G. (2018). Corporate Environmental Management Systems and Outcomes: A Case Study of ISO 14001 Implementation in a Cable Manufacturing Company in Tema. *Ghana European Scientific Journal* 14 (31): 320–336.

Shrestha, R.M., Biswas, W.K., and Shrestha, R. (1998). The implications of efficient electrical appliances for CO2 mitigation and power generation: the case of Nepal. *International Journal of Environment and Pollution* 9 (2–3): 237–252.

UNEP (2015). *Introduction to Cleaner Production (cp) Concepts and Practice*. Institute of Environmental Engineering (APINI) Kaunas University of Technology, Lithuania for UNEP, Division of Technology, Industry, and Economics https://fliphtml5.com/ovpk/bicc/basic.

World Business Council for Sustainable Development. (2000, August). Eco-efficiency: Creating more with less. Retrieved from "Archived copy" (PDF). Archived from the original (PDF) on 2016-05-15. Retrieved 2013-02-12.

WWF (World Worldlife Organisation) (2021). History. https://www.worldwildlife.org/about/history.

6

Industrial Ecology

6.1 What Is Industrial Ecology?

If we further breakdown the title, 'Industrial' focuses on product design and manufacturing processes to deliver goods and services to meet societal demand, and it has the means (technology) to address environmental and economic challenges. In the case of 'Ecology', this represents non-human 'natural' systems that act as models for industrial activity and where, technological activity is placed in context with larger ecosystems, which act as sources and sinks for this activity. The conversion of industrial wastes to resources for neighbouring industries in an industrial ecosystem not only reduced residue area (i.e. sink) but also avoids the use of land in upstream activities to produce virgin materials (i.e. source). For example, the conversion of cow dung to electricity using a biogas generator in a dairy farm not only meets the energy demand of any neighbouring industries but also avoids upstream processes such as mining or exploration, processing, and the combustion of gas or coal to produce electricity. Reducing 'Source and Sink' wastes through technological innovation can enhance carrying capacity while reducing ecological footprint. This keeps the total throughput of the human economy small enough to avoid exceeding two critical physical limits of the ecosystem: its capacity to regenerate itself, and its capacity to absorb our wastes. Activities that enable nature to regenerate itself are often excluded in a conventional industrial production systems. For example, printed circuit boards (PCBs) are designed in a way where it is a challenging task to extract rare earth elements (REEs) from waste PCBs. These REEs are scarce in nature and so they are obtained by disturbing a large areas of land, using toxic inputs and finally leaving a large amounts of hazardous waste and abandoned areas. In fact, in an industrial eco-system, there is an open interface between an industry and the surrounding environment to avoid externalities through pollution prevention measures.

Engineering for Sustainable Development: Theory and Practice, First Edition.
Wahidul K. Biswas and Michele John.

6.2 Application of Industrial Ecology

Industrial ecology considers the whole supply chain of a product or a service, including upstream and downstream activities of an industry so that the stakeholders can collaborate, and apply the necessary tools to make the production process resource efficient. IE is, therefore, a market place for concepts, tools, methodologies, and practices for moving beyond micro-level strategies (e.g. cleaner production and eco-efficiency), including life cycle assessment, material flows accounting, regional synergies, green chemistry, eco-industrial parks, industrial symbiosis (IS), industrial metabolism, regional synergies, and biomimicry (ISIE 2021). The broad category actors of IE are industry, community, and the government. The applications of industrial ecology can be applied at firm, between firm and at regional levels in achieving sustainability in terms of resource conservation and pollution prevention.

Industrial Metabolism: This studies the physical processes that convert resources into finished products and wastes, which happens usually at the firm level.

Material Flow Analysis: This happens at regional level as mass flows within defined geographic boundaries

Industrial Symbiosis: This is also known as 'regional synergies', where there is an exchange of by-products, water, and energy between neighbouring industries. This can happen either between firms or at regional level.

Biomimicry: Studies nature's best ideas and imitates these to solve industrial problems. Industrial ecosystems can follow a natural eco-system to enhance recycling, reuse, recovery, and remanufacturing activities. In any natural eco-system nutrient cycling and hydrological cycling help to achieve a self-sustaining system.

Green Chemistry: This happens at a firm level or an industry level to design chemical products or processes to reduce or eliminate the use and generation of hazardous and toxic substances. Avoiding the toxification of land associated with industrial production processes, conserves land and natural resources for future generations.

Green Engineering: This also happens at a firm level by improving industrial processes by optimising process design and operating conditions such as the use of compressed air instead of lubricant as a coolant in machining operation. Process integration is the application of green engineering as it integrates a large number of chemical processes into single processes to increase resource efficiency in the production process.

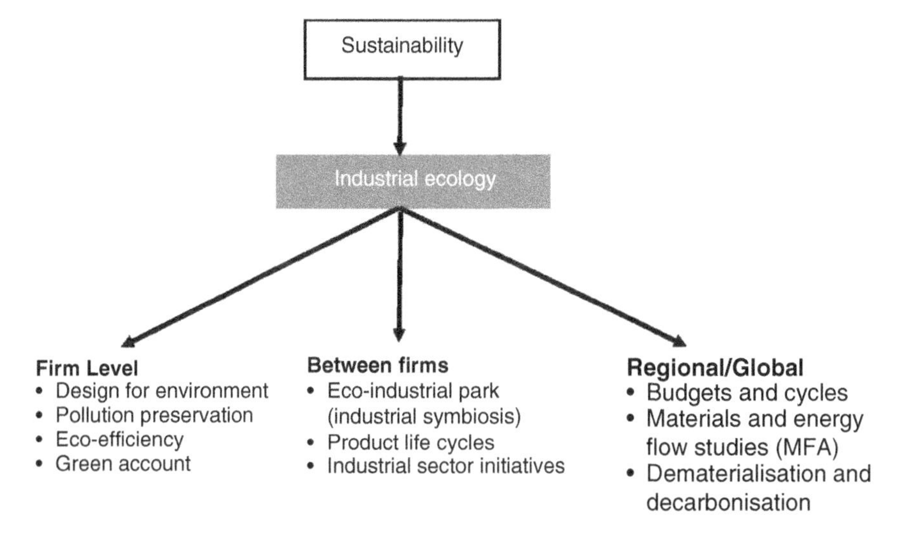

Figure 6.1 Spatial levels of industrial ecology. Source: Chertow (2000).

Figure 6.1 shows that the IE can happen at firm, between firms and regional levels. At a firm level, it mainly applies the principles of cleaner production strategies, designing products for disassembly and reuse, and taking preventative measures to enhance eco-efficiency performance. When two neighbouring firms are exchanging products, wastes and by-products with each other or as part of a supply chain (e.g. fly ash from a nearby coal power plant is used to make concrete in a neighbouring concrete plant) they form a mini industrial circular economy. Regional level interaction occurs in large industrial areas where there is at least more than 10 industries. The regional level usually consists of clusters of three to five different industries. The cluster is formed in a way that the industries within a cluster can exchange by-products in a technically and economically feasible manner. Industrial ecology therefore plays a pivotal role in conserving resources for future generations and is considered as a pathway to achieve sustainability.

6.3 Regional Synergies/Industrial Symbiosis

Industrial symbiosis is defined as the capture, recovery, and reuse of previously discarded or unrequired resources from one industrial operation by other, traditionally separate, industries operating in close proximity (Chertow 2000).

It involves traditionally separate industries in a collective approach to gain competitive advantage by conducting physical exchange of materials, energy, water, and/or by-products.

Industrial symbiosis uses metaphors drawn from natural ecosystems to suggest that industrial production can be reconfigured into an 'industrial ecosystem' where firms are interconnected through the exchange of wastes, by-products, wastewater, and energy between each other (Gibbs 2008). At the heart of the concept of industrial symbiosis is a deceptively simple argument that proposes a way to reduce or eliminate the negative impacts of economic development. As discussed in Chapter 5, saving the environment also means protecting your profit. By using an example of natural ecosystems, an industry can shift from the current wasteful linear model of production to a circular economy by mimicking nature, where natural resource inputs are reduced, wastes transformed into firm inputs and energy cascaded through the industrial ecosystem (i.e. energy loss from a process is recovered by heat exchangers for use in another process and it goes on like this) (Gibbs 2008).

Figure 6.2 shows how an industrial system can follow a natural ecosystem to become an eco-industrial ecosystem. In natural ecosystems, the energy and materials produced by one biomass (plants) is consumed by another biomass (microorganisms, animals) through nutrient cycles. Regeneration is known as the process of maintaining nutrients in soil for the growth of plants. By-products are either directly converted to organic nutrients (i.e. carbon compounds) or indirectly to inorganic nutrients (N, P, K). Microorganisms in soil convert organic materials, such as manure or plant waste, into inorganic forms of nitrogen that plants can absorb or assimilate. These processes helping sequester nutrients in soil, otherwise known as mobilisation. Enhanced sequestration of nutrients in soil sustains the growth of vegetation, thereby maintaining the supply of nutrients across the food chain (i.e. from herbivore to carnivore).

The aforementioned processes are not always found to exist in industrial systems, as surplus heat is commonly dissipated in the atmosphere, used water is introduced into nearby waterbodies and potentially recyclable by-products are often disposed of at landfill, which results in a linear flow system or linear economy. The concept of industrial symbiosis is able to demonstrate that processes and industries are interlinked, interdependent, and interacting with each other through waste recycling, remanufacturing, and by converting industrial wastes or by-products to reduce the dependence on virgin materials to help achieve a circular economy. The essence of industrial symbiosis is that there are interactions between, and interdependence of, industries in an industrial area, compared with the emphasis on independence and competitiveness as a more conventional view of industrial systems. This provides a basis for thinking about ways in which various waste producing industries can be connected into a

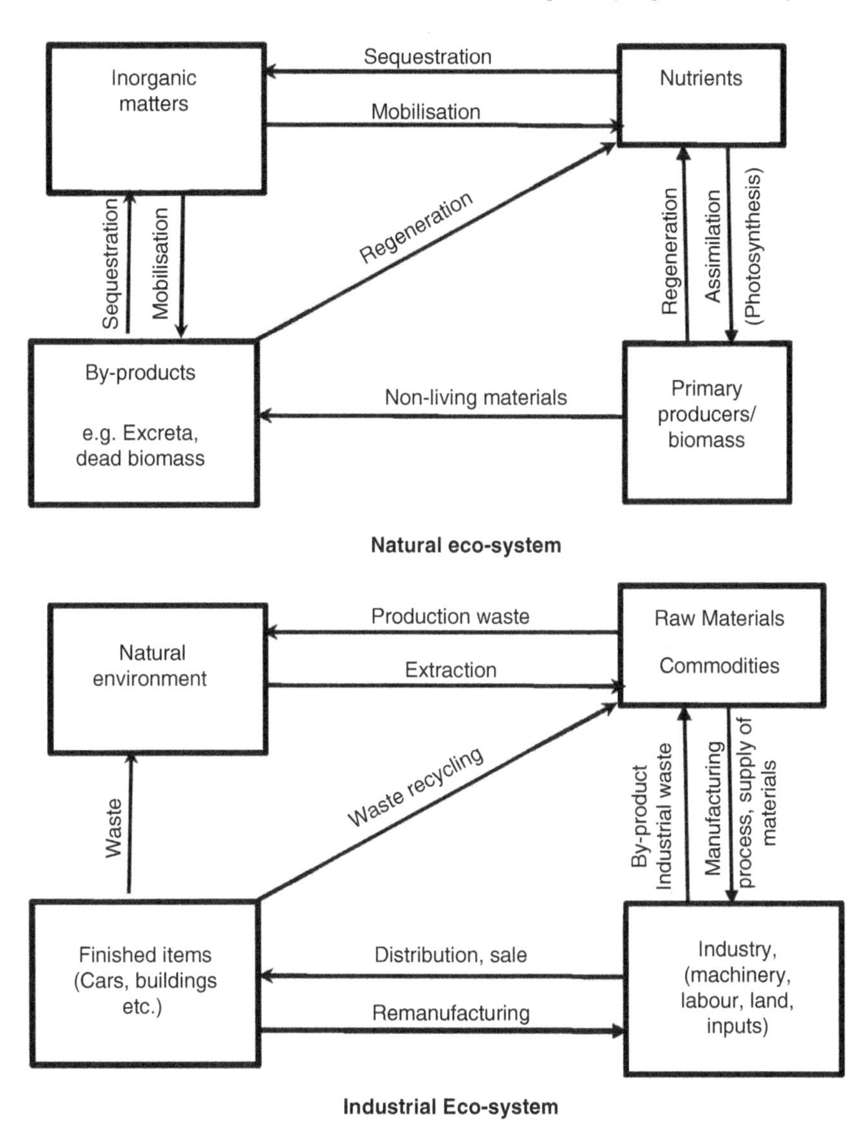

Figure 6.2 Mimicking a natural ecosystem to develop an industrial system.

network of exchange to minimise the amount of industrial wastes or by-products they produce and/or currently going to residue area or landfill (i.e. disposal sinks) or lost in intermediate processes. An industry can shift from the current wasteful linear model of production to a circular economy by mimicking food webs in natural systems, where natural resource inputs are reduced, wastes transformed into firm inputs and cascade waste heat recovery system (i.e. waste heat recovery and use continues until the point when no more recovery is possible). Similarly, wastewater and waste heat can be recovered from one industry for use by a neighbouring industry. In an industrial ecosystem, the firm or process impacts are addressed in the parallel concept of industrial metabolism, which is concerned with the efficiency of the metabolic processes within individual species (or in this case firms or processes). The higher the metabolic rate, the higher the conversion rate of feed stocks to products as well as the conversion of industrial wastes to resources. Therefore, unlike conventional approaches that emphasise waste minimisation from an individual industry or facility (i.e. pollution prevention, zero waste), industrial symbiosis focus is on waste minimisation of the larger system where the industries collaborate with each other in the waste exchange process. A better understanding of material flows is necessary to increase the resource efficiency of industrial metabolism by closing material flows into loops of recycling and reuse. This is known as the circular economy.

The main triple bottom line benefits of industrial symbiosis are resource conservation for future generations (i.e. inter-generational equity), employment creation by using locally available resources (i.e. intra-generational equity), avoiding deforestation by reducing residue area and reducing greenhouse-gas (GHG) emissions by avoiding upstream processes of virgin materials (i.e. environmental benefit), and gaining competitive advantage by reducing operational costs by avoiding the need for imported virgin materials (i.e. economic benefit).

Industrial symbiosis (IS) or synergy can also be termed as a 'round put' (opposite to throughput which is closing the loop) to distinguish it from linear forms of industrial production. IS typically is comprised of materials and water recycling, a focus on 6Rs and energy cascading to reduce the requirement and processing of virgin raw materials (Korhonen 2001; Lovins et al. 1999).

6.4 How Does It Happen?

The keys to industrial symbiosis are collaboration and the synergistic possibilities offered by geographic proximity (Chertow 2008). For example, the Kwinana Industrial Council (KIC) helps industries in the Kwinana Industrial Area (KIA) in Western Australia to exchange by-products, wastewater, and waste heat with each other. KIC in fact helped develop regional synergies or symbiotic relationships between industries through workshops and a number of independent consulting processes to help find technically and economically feasible symbiosis options. This process needs a facilitating structure to enable the collaboration. This is done by first identifying the key stakeholders, then by identifying the synergies through increased sharing of information on by-products between companies, and increasing their communication and collaboration. Next feasibility assessments and screening of options, and giving support in the development of symbiotic exchanges to the corresponding companies. This collaborative process has also been called the ABCDE model or A2E model consisting of Awareness (identification of synergies: information on others have and wants), Benefit recognition (development of knowledge of full benefits, sustainability evaluation tools, cost benefit analysis), Communication (communication systems where the facilitators are helping companies gain top management support) are processes within the company, Development (collaboration, negotiation, research and development on technology and processes) and Execution (societal license to operate resource synergies) (Harris 2007).

Once the symbiotic relationships have been commenced, the industries evaluate and communicate the success stories and build stronger relations with key neighbouring industries to increase further collaboration.

Last but not the least, there is often a need for operational and contractual arrangements between industries that enable commitment of the necessary resources for the implementation of industrial symbiosis. The key areas for discussion and negotiation are ownership of project assets, ownership and liabilities of traded resource, supply risks with availability of critical process input, demand risks with utilisation of the recovered by-product or utility generated, and pricing of the exchanged resource for each of the synergy partners.

6.5 Types of Industrial Symbiosis

There are a number of different synergy types:

- Inorganic by-product synergies
- Utility synergies
- Supply chain synergies
- Service synergies
- Gaseous synergies

  Inorganic by-products synergy: This is the exchange of solid inorganic by-products between industries. For example, the by-product gypsum produced by a fertiliser plant can be used for residue area amelioration at a neighbouring alumina refinery. This exchange could result in TBL benefits in terms of land and biodiversity conservation and the commercialisation of residue by-products.

A second example is a cement producer selling by-products of lime kiln dust to a nearby pig iron plant and pigment plant (van Berkel et al. 2006). This will assist the cement producers to reduce the amount of lime kiln dust (LKD) being landfilled (perhaps down to zero). It has been recently found that LKD can also be used as an asphalt filler. A hypothetical numerical calculation based problem for students' practice has been developed based on this case study (Box 6.1),

Box 6.1 LKD inorganic by-product synergy

This is a hypothetical example only and not based on real data. ABC Road Construction Co. uses hydrated lime (HL) as filler in asphalt mix, where HL accounts for 5% of asphalt mix by weight and the density of asphalt mix is 2.45 ton/m^3.

X Lime Co. produces hydrated lime for Alumina, Gold and Building Industries that has a market value of $150/ton. About 1.3 tons CO_2 equivalent are emitted due to production of 1 ton of HL and 1.27 ha of land is used for mining 1000 tons of limestone.

The research at Greenleaf University has found that lime kiln dust (LKD) can be a perfect substitute for HL. Jay Jay Cement produces LKD as a by-product which is currently being stored at a designated residue area. The requirement of additional land for storing LKD (i.e. 8.5 ha/ton of LKD) causes deforestation and land degradation.

Construction industries are now required to reduce GHG emissions to avoid carbon tax as the existing government of Country A has just introduced a carbon tax of $30/ton CO_2 e. This policy has created a market for LKD and Jay Jay Cement also have now begun to sell LKD at $25/ton.

ABC Road Construction Co has recently signed a contract to build a 10 km long and 7 m wide road which has a 50 mm thick asphalt wearing course. In the second week of your appointment to your new position at ABC Road Construction Co. you have been asked by your Engineering Manager to assess the economic and environmental implications of replacing HL with LKD in asphalt mix. Calculate the following parameters associated with the replacement of HL with LKD in asphalt for conducting this task.

(1) Economic benefits for ABC Road Construction Co.
(2) Carbon Tax reduction for X Lime Co.
(3) Land conservation for future generation (Ha)
  - Area of Land Conserved by Jay Jay Cement
  - Area of Land Conserved by X Lime Co.

Solution

Volume of asphalt $= 10\,\text{km} \times 1000\,\text{m/km} \times 7\,\text{m} \times 50\,\text{mm} \times \text{m}/100\,\text{mm} = 3500\,\text{m}^3$
Weight of asphalt $= 3500\text{m}^3 \times 2.45\,\text{ton/m}^3 = 8575\,\text{tons}$
Weight of HL $= 8575\,\text{tons} \times 0.05 = 429\,\text{tons}$

Economic benefits for ABC road cons co. = 429 × ($150 − $25) = $53,594
CO_2 mitigation = 1.3 tons of CO_2/ton of HL × 429 tons of HL = 557 tons of CO_2
C-tax avoided = $30/ton of CO_2 × 557 tons of CO_2 = $16,721
429 tons of HL × 8.5 Ha/ton of HL mined = 3644.4 Ha conserved.
429 tons of HL × 1.27 Ha/1000 tons of limestone = 0.544 Ha conserved.

Table 6.1 shows some examples of industrial by-products, which can be converted to resources for neighbouring industries.

Utility Synergy: This synergy is mainly of two types, including a wastewater synergy and utility synergy.

Water Utility Synergy: Wastewater from one industry can be treated to make process water for the neighbouring industry. For example, the Kwinana Water Reclamation Plant (KWRP) in Western Australia is a joint initiative of the Water Corporation and Kwinana industries to achieve the double benefit of greater overall water efficiency and reduced process water discharges into Cockburn Sound (Department of Water 2016). A micro filtration/reverse osmosis unit was built at a cost of Aus$ 25 million to take secondary treated effluent from the nearby Woodman Point wastewater treatment facility to produce processed water with a low total dissolved solids (Figure 6.3). This water is used by CSBP, Tiwest, Kwinana Cogeneration Plant and BP to replace potable water (6 GL/year, about 2–3% of the total potable water use in the drought-affected Perth metropolitan area). The low TDS will enable the process plants to cut

Table 6.1 Inorganic by-products reuse in Australia.

Inorganic by-product	Reuse
Iron and steelmaking slags	Accepted and used as a cement replacement
Fly and bottom ash	Cement blending, concrete addition, CSIRO Australia advanced construction material technology (with Blast Furnace slag and silica fume)
Lime kiln dust	Asphalt filler in pavement
Chemical gypsum (Flue Gas Desulfurisation and Phosphogypsum)	Pasminco Hobart Smelter Australia
Construction and demolition waste and recycled concrete aggregate	Some in New South Wales and Victoria Limited in Western Australia
Boiler slag	Cement

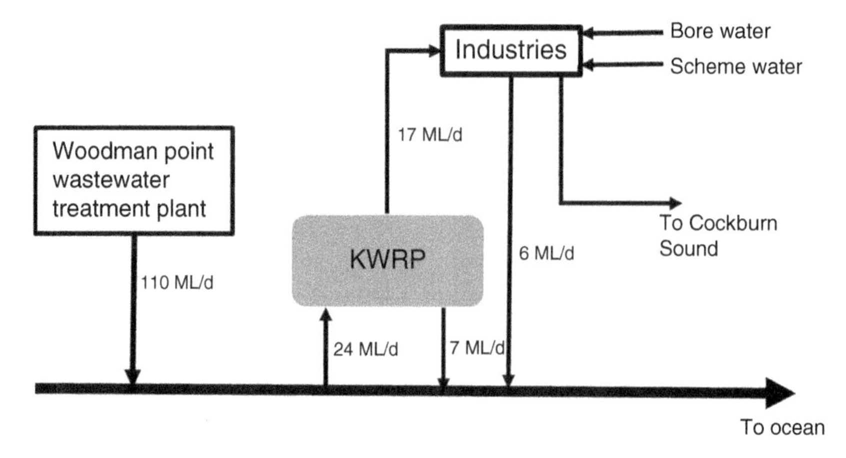

Figure 6.3 Iconic synergy project – KWRP.

their chemical use in cooling towers and other process applications, thereby reducing metal loads in their effluents. In exchange for taking water from the KWRP, the industries will be able to discharge their treated effluents into a deep ocean outfall through a Water Corporation wastewater pipeline, thereby eliminating the current discharges of treated process water into the environmentally sensitive Cockburn Sound (Oughton et al. 2021).

There is another Western Australian KIA example where an artificial wetland treatment was built at CSBP Chemical Plant (CSBP 2010). This is an innovative nutrient stripping wetland to further reduce the nitrogen discharges from other industries in the industrial are to the adjacent ocean inlet. The nutrient rich wetland is planted with sedges to sequester CO_2 emissions, thereby further providing carbon saving benefits to CSBP.

Energy Utility Synergy: In the case of utility synergy, energy is often recovered from flue or exhaust gases and process applications through heat exchangers and depending on the temperature of the recovered heat, neighbouring industries can use it for different applications. For example, if it is a high temperature recovered heat (i.e. >300 °C), it can be potentially considered for converting water to steam for electricity generation. Low temperature recovered heat can be used to preheat air or water in process applications. For example, hydrogen produced by oil refinery as a by-product from its refinery is piped to an industrial gas producer and supplier facility where it is purified and pressurised (van Beers and Biswas, 2008). The compressed hydrogen is then trucked to a bus depot and off-loaded to a refuelling facility, from which the hydrogen fuel cell buses are refuelled.

A second example is the Kwinana Cogeneration Plant (116 MW capacity), which is located on land at the BP Kwinana oil refinery, and produces all process steam for the refinery, and generates electricity for BP as well as the grid (van Beers and Biswas, 2008). The cogeneration plant is fired with natural gas supplemented with excess refinery gas. The total benefit has been estimated as a reduction of about 170,000 tons of carbon dioxide emissions per annum. Box 6.2 shows a hypothetical example for conducting numerical calculations.

Box 6.2 Hydrogen by-product synergy

Hydrogen is a by-product from an oil refinery and is piped to an industrial gas producer and supplier (BOL Gases) facility site next door. The BOL Gases separates, cleans, and pressurises hydrogen by-product for the hydrogen buses in a 'Green City'. The price of pure hydrogen gas is $2/m^3$. BOL use this price to sell hydrogen gas to Green City buses. The additional capital cost for BOL Gases for purifying the hydrogen is $10,000 per annum and operating cost is $5000 per annum. BOL receives about 150×10^3 m^3 of crude hydrogen annually, 80% of which is converted to purified hydrogen fuel for Green City buses. The Green City buses receive 70% of their hydrogen supply from BOL Gases and reduces 50 kg of CO_2 emissions/year with the use of 1 m^3 of hydrogen.

Determine the economic benefit for BOL from this industrial symbiosis process. Calculate the environmental benefit of one Green City bus.

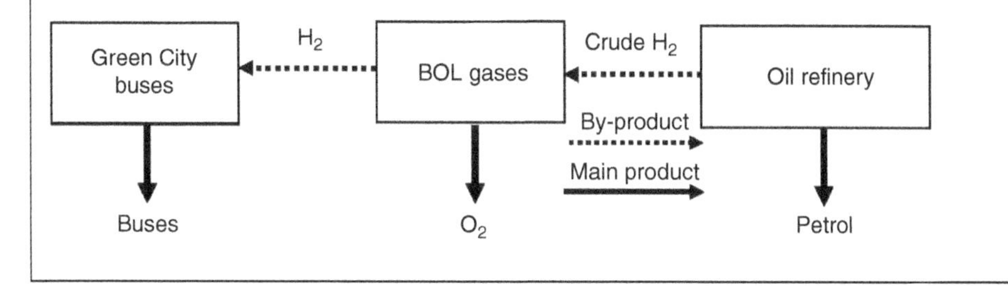

Solution

The gas is sold at $2/m^3$.

The annual conversion/processing cost $= \frac{\$10{,}000+\$5{,}000}{150 \times 10^3\ \text{m}^3} = \$0.1/\text{m}^3$.

Net profit $= \$2 - \$0.1 = \$1.9/\text{m}^3$.

80% of the crude H_2 is converted to pure H_2.

Therefore, the amount of H_2 sold by BOL to Green City Buses $= 150 \times 10^3\ \text{m}^3 \times 0.8 = 120 \times 10^3\ \text{m}^3$.

Total profit for BOL is $120 \times 10^3\ \text{m}^3 \times \$1.9/\text{m}^3 = \$228{,}000$.

Total environmental benefits of Green City buses $= 120 \times 10^3\ \text{m}^3 \times 50\ \frac{\text{kgCO}_2}{\text{m}^3} =$ 1200 ton of CO_2.

Supply Chain Synergies: Involve the participation of a local manufacturer and client (when companies use similar input/produce, they can save on distribution of input/outputs). The local manufacturer could have a dedicated sole supplier of their principal reagents for a core process in the surrounding area. This saves both transportation emissions and costs by avoiding the procurement of materials from another region or state or country or supply intermediary further afield.

Service Synergies: This happens when industries share services and facilities (e.g. transportation, cleaning, security). For example two remotely located neighbouring industries can share the same rail line to deliver materials like ore, or manufactured items at the same time to port or they can share a workshop to repair and service mechanical equipment.

Gaseous Synergy: Gases or pollutants emitted from chimneys or stacks can be captured to produce feedstock or useful materials for neighbouring industries. CO_2 from ammonia production is being cleaned and then sold for beverage production. Secondly, NO_x which is separated from the crude oil during the refining process can be converted to feedstock (KNO_3) through oxidation and absorption processes for fertiliser production instead of releasing them to the atmosphere (Mohammed et al. 2016) (Figure 6.4) as greenhouse gases.

A third example could be an ammonia fertiliser plant supplying carbon dioxide to an alumina refinery to neutralise the bauxite residue. The neutralised residue is known as Red Sand which has been used as a road construction material. The CO_2 actually arrives via pipeline from the nearby ammonia plant, resulting in GHG benefit equal to at least 70,000 tons CO_2/year (Harris, 2007). Box 6.3 shows a hypothetical problem based on this case study developed for problem solving of this gaseous synergy.

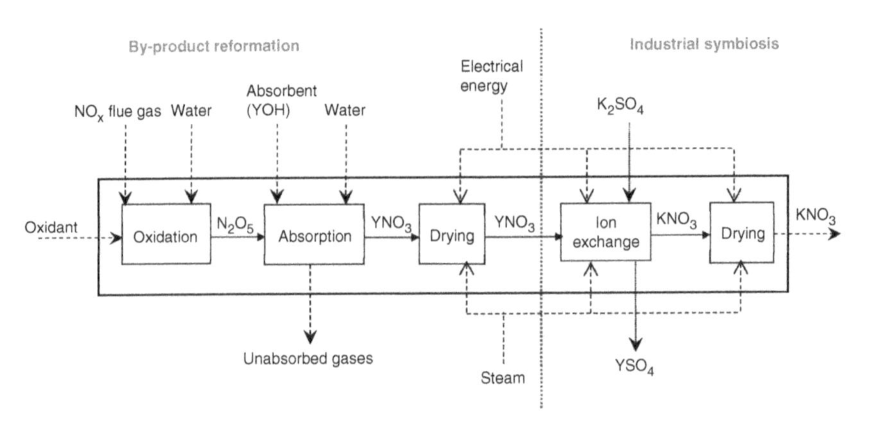

Figure 6.4 Conversion of NO_x emission to feedstock for fertiliser production. Source: Mohammed et al. (2016) / with permission of Elsevier.

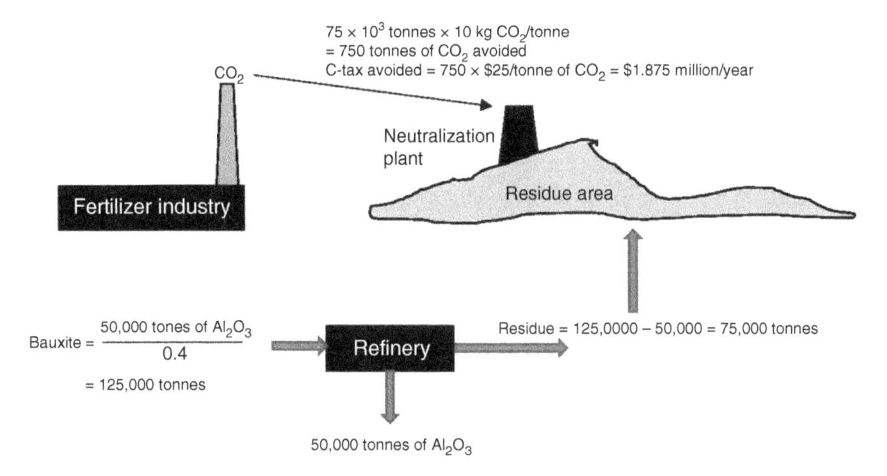

Figure 6.5 Gaseous symbiosis between fertiliser and refinery industries.

Box 6.3 CO_2 gaseous synergy

An alumina refinery is using process carbon dioxide (CO_2) from a nearby fertiliser plant to reduce the alkalinity of its bauxite residue, thus reducing environmental risks and significant ongoing management related to bauxite residue storage areas while also leaving options open for additional processing of the residue into other useful products at a future stage. The alumina refinery uses 10 kg CO_2 to reduce the alkalinity of 1 ton bauxite residue, while the refinery produces 50,000 tons of alumina/year. The bauxite which is refined to alumina is low grade bauxite (40%). The refinery is expected to reduce their carbon taxes by sequestering CO_2 from this neighbouring plant. The Government has just implemented a Carbon tax of $25/ton of CO_2 emitted. Calculate the economic and environmental benefits from this industrial symbiosis.

Solution

Figure 6.5 shows the amount of amount of CO_2 that a fertiliser plant can avoid by using this for turning neighbouring industry's bauxite residue into red sand for road construction purposes. This assists the fertiliser industry to reduce their carbon tax.

6.6 Challenges in By-Product Reuse

There is usually no standard requirements for the reuse of industrial by-products for many applications, e.g. soil conditioning, or for use as a road base or fill material. Sometimes the process of authorisation is very bureaucratic and time-consuming. There appears to be continuous concerns and potential opposition to this reuse, with limited support from government. Often there is a lack of leadership and support from the government when dealing with related community issues and concerns as well as the lack of commercial security, especially where community concerns can effectively end the reuse. As a result, by-products continue to build up in storage facilities. A government initiative is crucial to engage the community, environmental experts and policy makers to make the community aware of the benefits and risks associated with the use of industrial by-products. Some of the risks associated with IS projects include:

Flexibility Risk: Despite the great potential for improving resource efficiency and profitability, participation in exchanges between firms can have a downside because of possible risks, related to the inter-dependency between participants in terms of reliance on the supply, quantity, and quality of particular by-product or material stream.

Financial Risks: There is usually a lack of proven success of the by-product exchange. Furthermore, material exchange agreements can also fail unless they are beneficial for all parties concerned. Typically, the costs of recovered by-product materials need to be less than either their disposal costs or the price of virgin materials.

Regulatory: Regulatory structures often place substantial demands on eco-industrial related activity. For example, red sand which is produced from bauxite residue was found to be a structurally sound road construction material in 2010 in Western Australia, but it is still not used commercially due to the absence of regulatory support and guidelines.

Availability Issue: A neighbouring industry depending on by-products from a mining industry that is closing will be effected and they may have to revise their production process to utilise completely new inputs.

6.7 What Is an Eco Industrial Park?

An Eco Industrial Park (EIP) is the application of IS and is built as eco-industrial networking covering interactions between organisations as well as involving a range of stakeholders in the transportation, human resources, information systems, community connections, marketing, environment, health and safety, and production processes (Valenzuela-Venegas et al. 2017). It is considered as a community business not only co-operating amongst themselves but also with the local community to efficiently share information and resources (information, materials, water, energy, infrastructure, and natural habitat), leading to economic growth, improvement in environmental performance and equitable enhancement of broader social outcomes for the business and local community (Mouzakitis et al. 2003).

The generalised framework that needs to be considered during the planning stage for the design of EIPs consists of five steps (Al-Quradaghi et al. 2020): (i) Create a vision and plan, (ii) Identification of all entities with industrial symbiosis, (iii) Pinpoint the anchor industries, (iv) Determination of industrial symbiosis between at least three entities and two exchange flows, (v) Defining exchange-flow types.

Create a Vision and Plan: This step initially ensures the need for an EIP. The motivation behind having an EIP is to fulfil three objectives of sustainability: economic, environmental, and social. Some companies exchange resources in order to reduce the costs of raw materials and transportation or increase profit (economic pillar). Other companies use industrial symbiosis as a way to reduce greenhouse-gas (GHG) emissions and waste or comply with environmental policy to avoid infringements (environmental pillar). In addition to these two objectives, some companies can create more job opportunities for people in the region through an EIP (social pillar). For example, the creation of recycling, recovering, and remanufacturing businesses involved in collecting and converting one industry's waste to resources for the neighbouring industries.

Identification of All Industries with Industrial Symbiosis: The next step is to identify all possible industries that collaborate with each other in achieving the TBL objectives. Industries in the planned EIP should have potential to exchange flows with other industries from which the EIP benefits. To be highly effective, an eco-industrial park needs companies that provide a diverse range of potential synergies/exchanges to help support collaboration development.

Pinpointing the Anchor Entity: From all industries listed in the previous steps, there should be at least one industry or a few industries that could attract other industries to increase further exchange flows. Their by-products, wastewater, and waste heat can be used by a number of neighbouring industries and the anchor industry should also be capable of using most of the wastes produced by neighbouring industries. The anchor industry is therefore regarded as the largest contributor in the EIP. Figure 6.6 shows the Energy 2 Asnaes station which is an anchor industry in the Kalundborg EIP to collaborate with the neighbouring agricultural, chemical, and utility industries.

This anchor industry needs to market and co-ordinate EIP's operations, is dedicated to the IS network, and earns the network industries' trust. It is important that these anchor companies are firmly committed to the park's operations, as they ensure sufficient financial resources as well as envisage adequate and continuous streams of by-products. The anchor companies should also be able to build trust

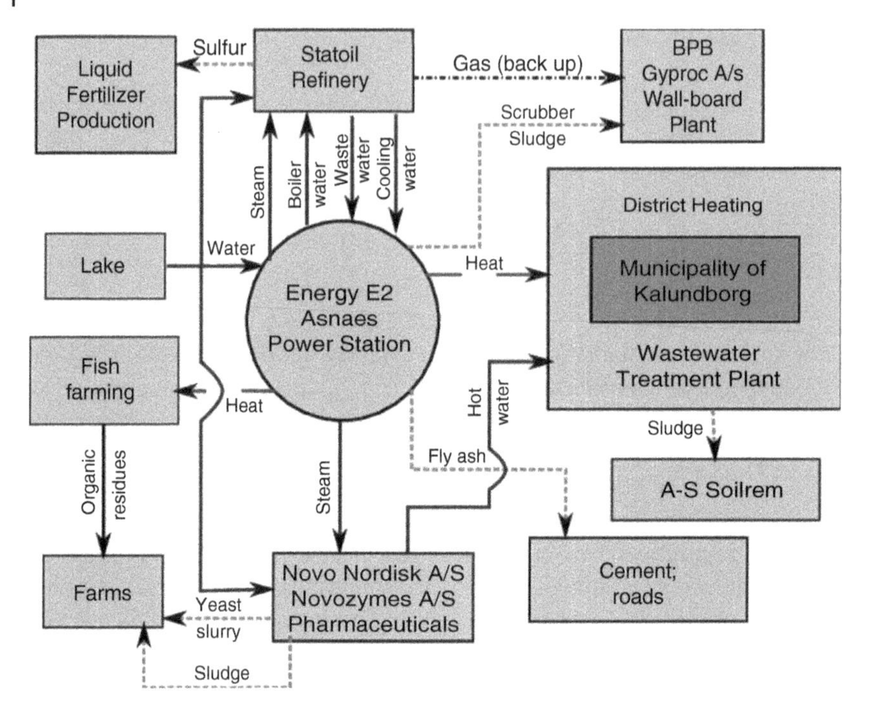

Figure 6.6 EIP in Kalundborg.

between organisations and maintain the park's networking ability. EIP's have been found to function best when it presents a healthy mix of larger and smaller companies operating in the same area across a diverse range of industries.

Industrial Clustering: Exchange flows between the neighbouring industries should be listed and determined. This characterises all possible industrial symbiosis in the EIP. The EIP matrix or network flow chart can summarise all exchange flows and gives detailed information about the network. A good practice is to have an industrial symbiosis between at least three industries and two exchange flows. Figure 6.7 shows that there are seven clusters in KIA with each cluster consisting of two to seven industries.

Defining Exchange-Flow Types: The 'internal exchange' is an industrial symbiosis that exists within the industries (e.g. old equipment from one section in the industry can be used in another section). On the other hand, an 'external exchange' is a symbiotic relationship that goes beyond the boundary of the industry (e.g. old equipment sent to recycling company or a decomposer firm or by-product conversion or waste heat recovery), including collocated, non-co-located, and regional firms. This requirement is very important, as it

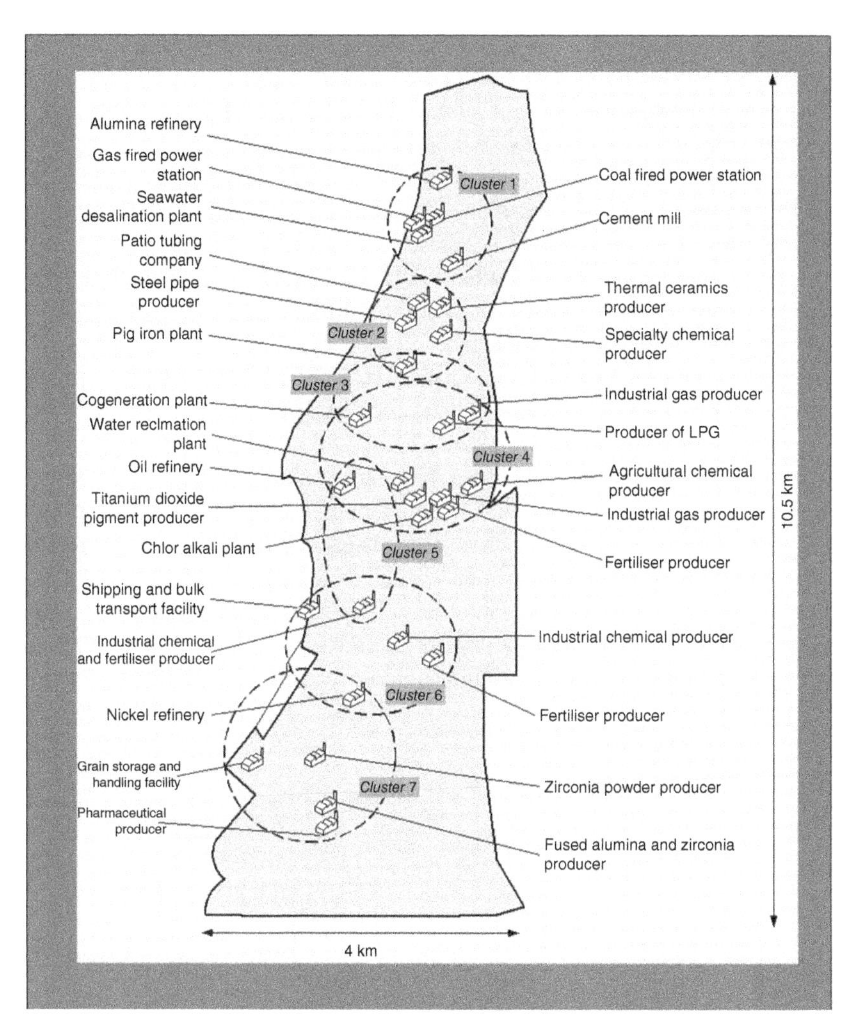

Figure 6.7 Location of companies and defined clusters in the Kwinana Industrial Area. Source: van Beers and Biswas (2008) / with permission of Elsevier.

identifies the distance between industries and the modes of transportation of the exchange flows (distribution networks, e.g. trucks, pipelines, etc.).

Resource Exchange Networks: Each industry must have a cleaner production strategist at their facility supporting resource recovery and reuse. This can be enhanced through 'decomposer' firms whose sole task is to collect one industry's wastes and then convert it a suitable product for use by the neighbouring industries.

The eco-services provider in an EIP is responsible for providing non-toxic chemicals and clean energy to reduce emissions and to reduce the production of toxic wastes. EIP Management is important for initiating collaboration between industries and also for helping neighbouring industries become involved. Finally, it is necessary to help the neighbouring industries to keep going with the industrial symbiotic relationships either creating new decomposer firms or upscaling them.

Environmental management system (EMS) for park management are also important in order to help industries to comply with environmental policy and to implement cleaner production strategies.

Figure 6.8 shows a hypothetical example of a small EIP. The cow dung generated from a dairy farm can be converted to biogas and digested slurry through an anaerobic digestion process. About 50% of the electricity that is generated from biogas meets the total electricity demand of the dairy farm and the remaining 50% is sold to a nearby poultry farm to meet the energy demand for their broiler production. The digested slurry, which is enriched with nitrogen, potassium, and phosphorus, is sold to the nearby citrus farm to meet its demand for fertiliser. This citrus farm is owned by a juice factory producing a large amount of citrus peel, which is used as fodder by the dairy farm. Poultry manure acts as a useful source of protein for a nearby aquaculture farm. This symbiosis has three by-product synergies

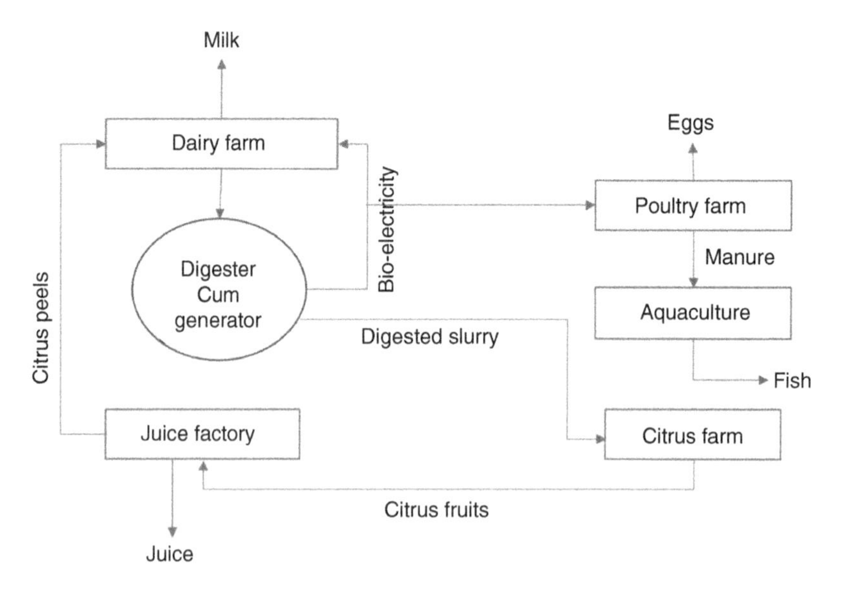

Figure 6.8 An example of an EIP.

and one utility synergy. Digester generation is considered as a decomposer farm in this example as it converts cow dung to electricity and digested slurries for use by neighbouring industries.

6.8 Practice Examples

6.8.1 Development of an EIP

Local Councils are becoming interested in demonstrating their sustainability performance to the local community. The council of which you are employed as an Environmental Consultant, has planned to allocate some portion of land to develop small- and medium-sized enterprises (SMEs) following the concept of EIP. Being a consultant to the Council, you have received a request from your Mayor to develop a plan for an EIP. The EIP will be consisting of SMEs in a way that they are capable of exchanging products and by-products with each other to reduce industrial waste and the importation of raw materials for industrial operation.

You are required to conduct the following tasks in order to develop a plan for your Mayor.

(i) Prepare a list of SMEs so that they can exchange wastes, by-products, and products with each other in a technically feasible way. In order to draw up this list, please conduct a literature review on a material flow analysis of products produced at the small- and medium-sized enterprises chosen. You will need to select at least five SMEs to develop this EIP and provide justification for selecting these industries.

(ii) Draw up a flow diagram of industrial symbiosis showing how these SMEs are linked with each other and how material, energy, wastes, and by-product flow between SMEs. You may also want to include decomposer firms ('scavenger' niche) within these SMEs to convert one SME's waste or by-product to usable product/input for the neighbouring industry.

(iii) Calculate the amount of materials in the form of waste, by-product and product that can be exchanged between these SMEs and then discuss these results briefly from the industrial metabolism perspective. You can gather this quantitative data through literature search or any document published online. The quality of data is not a priority as long as they are reasonably acceptable for use.

(iv) Roughly work out the amount of landfill area and virgin materials that can be conserved due to use of the concept of EIP and then discuss these results from an ecological footprint perspective.

Finally, discuss some potential barriers that may arise in the development of this EIP. Also discuss some 'Park Design and Management' aspects you would like to consider in this development process.

6.8.2 Industrial Symbiosis in an Industrial Area

In a city, a new potential opportunity for the implementation of industrial symbiosis has been identified. There are five industries, which have the potential for industrial symbiosis:

(a) A 1300 MW power plant needs cold water and produces electricity along with steam and hot water
(b) An oil refinery needs hot water and steam and produces gas, cold water, and waste water
(c) Axis Chemical needs electricity and clean water to produce medicine, waste water, and biomass + yeast slurry
(d) A plasterboard producer needs electricity and gas to produce plasterboards and soil slurry
(e) Farms need electricity, nutrient rich soil and water to produce fruits.

In Figure 6.9 below, 11 interactions from industrial symbiosis have been marked up. Students are required to identify each of the 11 numbered interactions. Also they need to indicate which are by-product and which are utility synergies.

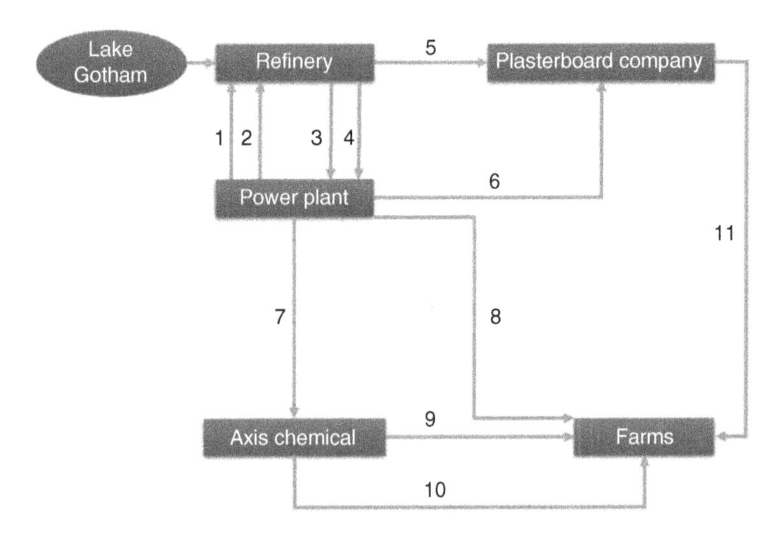

Figure 6.9 Industrial symbiosis in an industrial area.

6.9 Industrial Symbiosis in Kwinana Industrial Area

Kwinana is 40 km south of Perth in Western Australia, is a large industrial area with a coastal strip about 8 km long and 2 km wide. The resource processing industries are alumina, nickel, oil refineries, HIsmelt, etc., and utilities are power, water treatment, co-gen plants and manufacturing plants are cement, chemical, fertiliser plants, etc. KIA provides a best practice example of industrial symbiosis or regional synergy implementation with 47 diverse and matured existing synergies. World leading EIP industry collaboration occurred during the years 1990–2000, with the number of existing interactions increased from 27 to 106, and the total number of industrial symbiosis projects grew to 47 (Figure 6.10). The number of synergies increased from 2 per industry to about 4 per industry. An extensive network of exchanges was developed and specific examples highlight that IS can result in significant sustainability (i.e. social, economic, and environmental) and business benefits.

Existing synergies result in significant economic, environmental, and social benefits. The environmental benefit includes less 'waste' and emissions to environment (all synergies), increased water (e.g. KWRP) and energy efficiency (co-generation facilities). Economic benefits include a reduction in operational costs (gypsum) and an increased company income (lime kiln dust). Social benefits include water and energy security (KWRP + co-gen) and also the generation of new business, jobs, and emission reduction from waste and traffic levels improving the health and environment for the surrounding community.

The favourable IS development features in Kwinana includes Kwinana Industries Council playing a matchmaker role in assisting companies to collaborate with each other. Secondly, the KIA has a diversity of industries, including mining, refinery, chemicals, processing and utilities enabling them to use each other's wastes and by-products. Thirdly, Kwinana has close geographic proximity to many industries, which are non-competitive. This co-location is a significant benefit.

6.9.1 Conclusions

Traditional business practices fundamental to twentieth century economic development have largely overlooked surrounding communities and the natural environment in which they have operated. As we progress into the twenty-first century, a period in which anthropogenic activities are the key drivers of planetary environmental and climate dynamics, it is very important to have more open interface between industries and their surrounding environment. Businesses must alter their view of the firm to one in which they acknowledge their critical

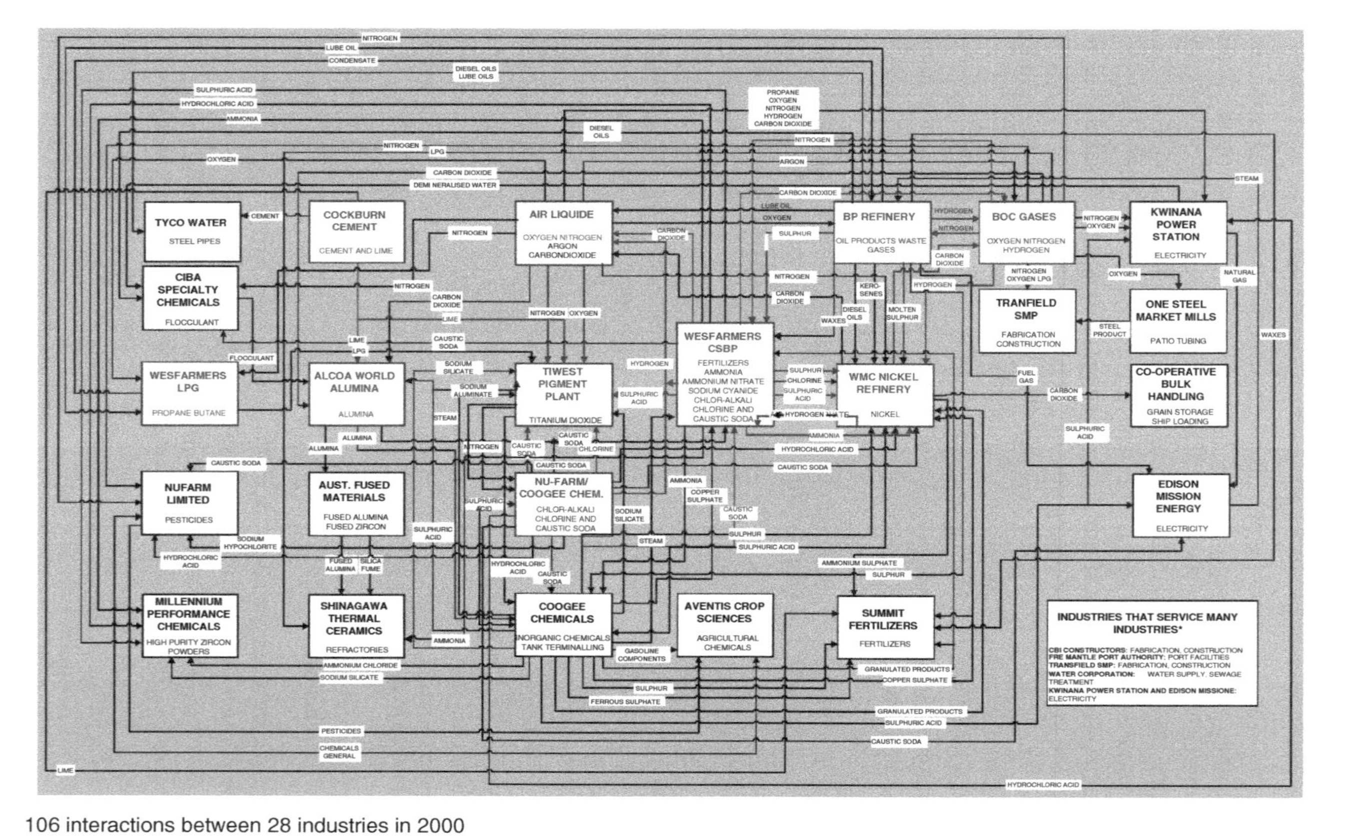

106 interactions between 28 industries in 2000

Figure 6.10 Increase in IS in Kwinana. Source: Modified from van Beers et al. (2007).

role and responsibilities in a larger socio-ecological industrial system (Sullivan et al. 2018). A shift in strategic direction for industrial operations and business is required, one that identifies the competitive advantages associated with socially and environmentally responsible business practices through the concept of industrial ecology, its focus on sustainable industrial eco-systems and its interrelation with new business practices and the community. The contents presented in this chapter contribute to the theories and examples of industrial ecology that prove sustainability practices are strategically beneficial for industries.

References

Al-Quradaghi, S., Zheng, Q.P., and Elkamel, A. (ed.) (2020). Generalized framework for the design of eco-industrial parks: case study of end-of-life vehicles. *Sustainability* 12: 6612. https://doi.org/10.3390/su12166612.

van Beers, D. and Biswas, W.K. (2008). A regional synergy approach to energy recovery: the case of the Kwinana industrial area, Western Australia. *Energy Conservation and Management* 49 (11): 3051–3062.

van Beers, D., Bossilkov, A., Corder, G. et al. (2007). Industrial Symbiosis in the Australian minerals industry. The cases of Kwinana and Gladstone. *Journal of Industrial Ecology* 11 (1): 55–72.

van Berkel, R., van Beers, D., and Bossilkov, A. (2006). Regional resource synergies for sustainable development: the case of Kwinana. *Materials Forum* 30: 176–187.

Chertow, M.R. (2000). Industrial Symbiosis: literature and taxonomy. *Annual Review of Energy and the Environment* 25: 313–337.

Chertow, M.R. (2008). Industrial ecology in a developing context. In: *Sustainable Development and Environmental Management* (ed. C. Clini, I. Musu and M.L. Gullino). Dordrecht: Springer https://doi.org/10.1007/978-1-4020-6598-9_24.

CSBP (2010). CSBP Kwinana: ammonium nitrate production expansion project: phase 2. Public Environmental Review. Prepared by Strategen Environmental Consultants Pty Ltd Level 2, 322 Hay Street Subiaco WA 6008.

Department of Water (2016). Local water supply strategy. https://www.water.wa.gov.au/__data/assets/pdf_file/0004/8887/110703.PDF (accessed 10 September 2021).

Gibbs, D. (2008). Industrial symbiosis and eco-industrial development: an introduction. *Geography Compass* 2 (4): 1138–1154.

Harris, S. (2007). Industrial symbiosis in the Kwinana Industrial Area (Western Australia). *Measurement + Control* 40 (8): 239–244.

ISIE (International Society for Industrial Ecology) (2021). What is industrial ecology? https://is4ie.org/about/what-is-industrial-ecology (accessed 18 October 2021).

Korhonen, J. (2001). Industrial ecology in the strategic sustainable development model: strategic applications of industrial ecology. *Journal of Cleaner Production* 12: 809–823.

Lovins, A.B., Lovins, L.H., and Hawken, P. (1999). A road map for natural capitalism. *Harvard Business Review* 77 (3): 145–158.

Mohammed, F., Biswas, W., Yao, H., and Tade, M. (2016). Identification of an environmentally friendly symbiotic process for the reuse of industrial byproduct – an LCA perspective. *Journal of Cleaner Production* 112: 3376–3387.

Mouzakitis, Y., Adamides, E., and Goutsos, S. (2003). Sustainability and industrial estates: the emergence of eco-industrial parks. *Environmental Research, Engineering and Management* 4 (26): 85–91.

Oughton, C., Anda, M., Kurup, B., and Ho, G. (2021). Water circular economy at the Kwinana Industrial Area, Western Australia—the dimensions and value of industrial symbiosis. *Circular Economy and Sustainability* https://doi.org/10.1007/s43615-021-00076-3.

Sullivan, K., Thomas, S., and Rosano, M. (2018). Using industrial ecology and strategic management concepts to pursue the sustainable development goals. *Journal of Cleaner Production* 174: 237–246.

Valenzuela-Venegas, G., Henríquez-Henríquez, F., Boix, M. et al. (2017). A resilience indicator for eco-industrial parks. *Journal of Cleaner Production* 174: 807–820.

7

Green Engineering

7.1 What Is Green Engineering?

The purpose of green engineering (GE) is to avoid any negative environmental consequences associated with engineering design. This chapter mainly discusses how the principles of GE can be incorporated into engineering design, not only to address environmental sustainability but also social and economic sustainability. GE is considered an engineering-led approach to sustainability. It openly embraces the concept of cleaner production practices and follows a preventative approach. This helps identify the sources of pollution and prevents hazardous substances from being released into the environment prior to recycling, treatment, or disposal. Apart from pollution prevention, these practices also save energy, water, and other natural resources for future generations.

Green engineering is part of an engineer's professional responsibility in designing processes, products, and procedures that are as safe as possible. Traditionally, this has meant identifying hazards, evaluating their severity, and then applying several layers of protection as a means of mitigating the risk of negative impacts. These layers of protection include basic process control systems, alarms and operator interventions, safety instrumented systems, physical protection using relief devices, containing physical contaminants in bunds, fire and gas alarms, and plant and community emergency response mechanisms.

Although these 'layers of protection' have resulted in significant improvements in the safety performance of chemical processes, they have some limitation in their effectiveness. Firstly, these layers are expensive to build and maintain. Secondly, there will still remain some hazards after applying these layers of protection measures given that there is a finite risk that an accident can happen despite the layers of protection.

Alternative to chemical-focused safety, there is inherently the safer design (ISD) approach. ISD is an iterative process that considers the elimination or reduction of a hazard, substitution of a less hazardous material, using less hazardous process

Engineering for Sustainable Development: Theory and Practice, First Edition.
Wahidul K. Biswas and Michele John.

conditions, and designing a process to reduce the potential for, or consequences of, human error, equipment failure, or intentional harm. The application of ISD can therefore help avoid many of these protective layers by identifying the sources of hazardous pollution and then mitigating them. The avoidance of a series of protection layers reduces the costs significantly. Instead of protective measures, safer design follows a preventative approach, which in fact gave birth to the concept of green engineering. In GE, an engineer is challenged to reconsider the design and eliminate or reduce the source of the hazard within the process rather than working with existing hazards in the chemical process by adding layers of protection, making production costly and expensive. Approaches to the design of inherently safer processes can be grouped into four categories.

7.1.1 Minimise

It is important to consider the use of smaller quantities of hazardous substances or to reduce the size of equipment operating under hazardous conditions (Hendershot 2011). There can be three possible strategies to minimise hazardous materials to achieve safer design (Sandia National Laboratories 2012). Firstly, it is done by reducing the inventory which can be either through the use of storage of less hazardous materials; or the use of fewer storage tanks by having just-in-time delivery of these materials; or to generate these hazardous materials on demand (e.g. chlorine, methyl isocyanate (MIC), ammonia, hydrogen); or to receive these hazardous materials by pipeline instead of by truck or rail to avoid large vessels or tanks. For example, nitroglycerine can be made in a continuous pipe reactor with a few kilograms of inventory instead of a large batch reactor with several thousand kilograms of inventory. Secondly, process intensification, which combines a number of chemical processes into one, can also be considered. The third strategy would be to have the process operation closer to ambient conditions. The benefits associated with this include minimising the reduction in consequences of an incident (explosion, fire, toxic material release), reduction in the number of tanks that require maintenance and the improvement of the effectiveness and feasibility of other protective systems (e.g. secondary containment, reactor dump or quench systems). For example, beads stick – NH_4NO_3 is a cheap chemical but it absorbs moisture quickly to form a hard rock which is risky as its exposure to high temperature caused a series of dangerous explosions in a warehouse in Beirut in 2020 (Guglielmi 2020).

7.1.2 Substitute

More hazardous materials, chemistry, and processes have to be replaced with less hazardous ones. This is done by eliminating hazardous raw materials, process intermediates, or by-products as much as possible and by using an alternative process or chemistry (i.e. substitute with non-combustible for flammable solvents). For example, the substitution of solvent-based paints with water-based coatings and paints eliminates fire, toxicity, odour, and environmental hazards for end users and also for the manufacturer (Hendershot 2011). Some strategies for making a process inherently safer by substitution are the use of commercially available alternatives, alternative raw material or intermediates that can be transported and stored more safely and the use of alternative chemistry. An example of alternative chemistry is sulfur burning to generate SO_3 on demand as a replacement for oleum (fuming sulfuric acid, $SO_3 \cdot H_2SO_4$) as a sulfonating agent as large quantities of unreacted sulfuric acid left behind due to the use of oleum can increase the hazardous materials disposal cost (Sandia National Laboratories 2012).

7.1.3 Moderate

This considers the use of less hazardous conditions or facilities which minimise the impacts of a release of a hazardous material or energy. This means that the supply pressure of raw materials needs to be limited to less than the working pressure of the vessels, and the reaction conditions (temperature and pressure) have to be made less severe by using a catalyst, or by using a better catalyst. In addition, hazards can be reduced by dilution (i.e. using aqueous instead of anhydrous form), refrigeration, or process alternatives that operate in less hazardous conditions. Storage of monomethylamine under refrigerated conditions significantly reduces the hazard to the surrounding community by reducing the potential leak to the atmosphere from the storage tank (Hendershot 2011). Another example is where a combustible solid can be used as a pellet instead of a fine powder, to reduce dust explosion hazards.

7.1.4 Simplify

Facilities have to be designed to eliminate unnecessary complexity and make operating errors less likely. For example, confusing control system layouts, equipment controlling switches, and labelling in the plant need to be simplified to reduce the potential for accidents. This simplification happens by considering whether the equipment can be sufficiently designed to totally contain the maximum pressure generated. In addition, a new generation of processes has to be designed that do not generate any wastes instead of minimising environmental impact or the end-of-process treatment of wastes.

7.2 Principles of Green Engineering

The following 12 principles of green engineering can be applied to achieve an inherently safer process (Anastas and Zimmerman 2003) and a more sustainable engineering decision.

7.2.1 Inherent Rather than Circumstantial

Engineers need to make sure that all material and energy inputs and outputs in the production or reaction processes are inherently as nonhazardous as possible. The application of this principle can help in materials recovery as this will prevent recyclers from being exposed to harmful chemicals for example, when recovering rare earth materials from the end-of-life products. Compact fluorescent lamps which contain mercury can contaminate soil and water when disposed of to landfill due to the fact that the lamp cannot be easily disassembled to recover the mercury from it. Each compact fluorescent lamp (CFL) contains between 3 and 5 mg of mercury, which is a neurotoxin and poses a high threat to the environment when failed bulbs are not recycled properly (Martinez 2020). In addition, mercury is also a rare earth material. However, in the case of light emitting diode (LED) manufacturers, they have avoided the use of mercury and have been increasingly reducing mercury-containing lamps in favour of LED light sources for their displays.

In general, manufacturing of electronics is extremely chemically intensive. There are about 60 elements from the periodic table which are used in our smartphones or laptop computers (Gordon 2017). Typically, the manufacturing process uses large amounts of chemical that end up in the body of the manufactured electronic product or component.

7.2.2 Prevention Rather than Treatment

It is better to prevent waste than to treat or clean up waste after it is formed. The use of clean fuel or renewable energy to supply electricity during production will automatically reduce emissions through a chimney or stack. The 3D manufacturing process can avoid the generation of wastes if what is inputted into the printer (i.e. filament) is mostly converted to product and is wasted less. Modular types of design can also help in end-of-life management where products can be dismantled and remanufactured or recovered and made into the same product.

7.2.3 Design for Separation

Separation and purification operations should be designed to minimise energy consumption and material use. In order to meet water scarcity pressures from the

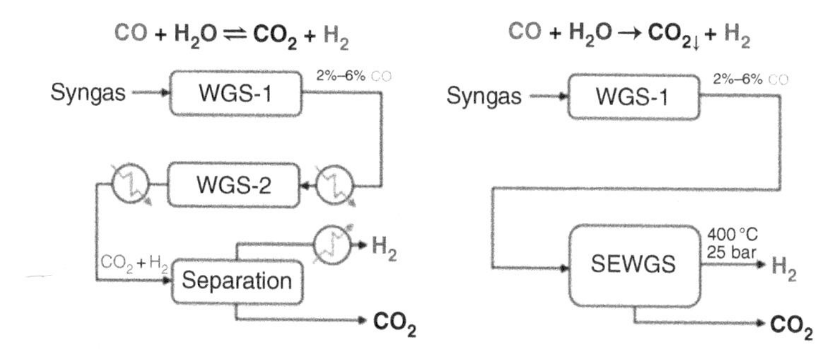

Figure 7.1 Process intensification: SEWGS case. Source: Modified from van Dijk et al. (2018).

ground and surface water sources, reverse osmosis systems have been developed to separate salts from seawater. However, this reverse osmosis (RO) process is very energy intensive and requires significant pressure to push the seawater through the infinitesimal filter pores to separate the salts from seawater. If the electricity required to run the RO is sourced from renewable energy technologies instead of conventional fossil fuel, then it will not only reduce the demand for fossil fuel for water treatment but will also reduce the associated GHG emissions.

7.2.4 Maximise Mass, Energy, Space, and Time Efficiency

Products, processes, and systems should be designed to maximise mass, energy, space, and time efficiency through process intensification. Several unit operations can be combined to reduce specific costs ($ per litre of chemical). Sorption enhanced water gas shift (SEWGS) processes could be a relevant example as they combine water gas shift (WGS) reaction with downstream separation into a single unit operation (Figure 7.1). This means that the shift reaction and separation are happening in one reactor/vessel (i.e. SEWGS). Firstly, CO is reacted with steam at increased temperatures and pressures, typically 350–500 °C and 10–50 bar to produce H_2 and CO_2 (van Dijk et al. 2018). This temperature and pressure are high enough to convert all CO to CO_2 which is then separated from H_2 in a single unit operation.

This process intensification helps reduce the number of units, energy, and materials which reduced extra units or reaction vessels, and finally maximised the conversion of CO to CO_2 for better sequestration.

7.2.5 Output-Pulled vs Input-Pushed

Products, processes, and systems should be 'output-pulled' rather than 'input-pushed' through the specific selection of energy and materials. This actually minimises the amount of resources consumed to transform inputs into desired outputs. This follows an approach requiring you to cut your cloth to suit your coat. Radio-frequency identification (RFID) is a technology wherein digital data encoded in RFID tags or smart labels (defined below) are captured by a reader via radio waves and assist producers to produce goods or items according to the demand or to avoid overproduction or oversupply of raw materials or feedstock (Nordic ID Group 2017). An RFID-based system transmits information embedded in a tag to a reader via radio waves and from there to an online database. This makes the flow of information faster and more reliable and enhances green supply chain management (GSCM) practices by reducing waste, emissions, and energy use in the supply chain. Secondly, it can optimise storage space, and attain customers' satisfaction by providing faster delivery and invoicing.

7.2.6 Conserve Complexity

Embedded entropy and complexity must be seriously considered when making end-of-life design choices on recycle, reuse, or beneficial disposal. The product should be designed for disassembly. More focus on product disassembly should be given to manufacture less complicated products for easier recycling. If a product is complex in design, engineers should make sure that it should be designed for reuse. For example, IBM PCs used to be made with 15 different types of screws (unnecessary complexity) (Anthony 2017). Later they replaced them with one single type of screw, making the personal computer easier to disassemble and recycle. Fuji Xerox machines as discussed in Chapter 5 are modular in design, making them more than 90% reusable. This principle, therefore, helps avoid the use of virgin materials and associated upstream processes, landfill, waste and emissions in mining, mineral processing and manufacturing, and achieve cost savings benefits.

7.2.7 Durability Rather than Immortality

Targeted durability, not immortality, should be a green engineering design goal. It is therefore necessary to design products with a targeted lifetime to avoid the immortality of undesirable materials in the environment. However, this strategy must be balanced with the design of products that are durable enough to withstand anticipated operating conditions.

Use of fly ash as a partial replacement for cement in concrete for marine infrastructure (e.g. seaport buildings, desalination plants) has not only been found to be structurally sound but also offers significant environmental benefits in terms of increased durability, reductions in carbon intensive cement production, and the conversion of waste to resources (Nath et al. 2018). The other indirect environmental benefits are reductions in land use changes and loss of biodiversity by avoiding land use both in cement production as well as the fly ash storage area at the coal fired power plants.

7.2.8 Meet Need, Minimise Excess

Design for unnecessary capacity or capability is considered a design flaw. It is important not only to over design processes but also to keep contingency factors low. Extra size means wasted material and energy. More recently, genetic engineering has been used to produce microorganisms that contain all the steps for a particular series of reactions in one cell, thereby turning it into a miniature chemical factory.

Gary Chan, an architect in Hong Kong, was able to effortlessly transform his tiny living space into no fewer than 24 different configurations (i.e. everything from a bathroom, to a fully-equipped kitchen, to a wet bar and/or 'gaming room') through a series of sliding panels and walls, known as the 'Domestic Transformer'. In Hong Kong, where space is in short supply, and where population density is very high, Gary engineered a 344 ft^2 apartment into what most people could only dream of with 10 times the usable area (Huffpost 2010).

7.2.9 Minimise Material Diversity

Material diversity in multi-component products should be minimised to promote disassembly and value retention. For example, circuit boards are manufactured of composites and are difficult to recycle. As a result, e-waste levels are incredibly high and highly toxic.

7.2.10 Integration and Interconnectivity

Design of products, processes, and systems must include integration and interconnectivity with available energy and material flows. For example, combined-cycle power generation technology uses both a gas turbine and a steam turbine to generate electricity by burning natural gas, a fuel with a lower environmental impact than many other fossil fuels. The exhaust heat from the gas turbine is used to convert water to steam to generate additional electricity. For the same input (i.e. gas or diesel to run the gas turbine) more electricity is generated by combining the gas turbine with the steam turbine. As a result, the thermal efficiency of the combined cycle power plant (60%) is higher than the gas turbine (40%) (Mitsubishi Heavy Industries 2021). Since more output or electricity is produced for less inputs for gas or diesel, less emissions (i.e. CO_2, NO_x, SO_2) are emitted in a combined cycle than the gas turbine. Also low volumes of hot wastewater released into the ocean or water bodies as waste heat can be used for steam generation to produce additional electricity and of course results in lower fuel costs, and lower resource use. This is discussed in more detailed in the energy recovery section of this chapter.

Another example of the role and value of integration and interconnectivity is the industrial symbiosis process discussed in Chapter 6; e.g. Kalundborg Industrial Park, Kwinana Industrial Area.

7.2.11 Material and Energy Inputs Should Be Renewable Rather than Depleting

Bamboo is not just strong but also a durable, attractive, and sustainable resource. It is stronger than steel in terms of tensile strength and is a cheaper and more sustainable resource. It also takes less electricity to make a bamboo bike than a metal one. The frame is completely recyclable. Cycling is well known as being a low-carbon form of transport. Bambo bikes reduce emissions associated with the mining, processing and manufacturing of steel for metal bike frames (World Economic Forum 2020).

7.2.12 Products, Processes, and Systems Should Be Designed for Performance in a Commercial 'After Life'

Disassembly of equipment for the reuse of components can enhance resource conservation and avoid environmental impact. For example, modular buildings, which include prefabricated buildings, are constructed almost entirely in off-site factories, as opposed to conventional buildings, which are constructed on site. Modular construction involves fabricating a number of smaller units, known as modules, which are later transported to a construction site and assembled. The modular building offers environmental, social, and economic benefits in terms of worker safety, site efficiency construction costs, and resource conservation. The benefits of prefabricated construction begin in the factory, continue on to the building site as no landscaping is required and much less waste is generated particularly as the building generates no demolition waste. Modular construction projects are also completed twice as fast, reducing energy usage at the building site. Using modular construction reduces emissions by cutting deliveries to the construction site by 90% (BigRentz, Inc 2020). Modular designs are also easily disassembled for use in other projects or for recycling, reducing debris from demolition and the consequential landfill waste.

7.3 Application of Green Engineering

Green Engineering is a general concept that applies to many industries and sectors. The four main areas of GE are as follows:

- Chemical
- Material
- Thermal energy recovery and
- Renewable energy generation (to be discussed in the next chapter)

7.3.1 Chemical

Green chemistry designs chemical products and processes that reduce or eliminate the use or generation of toxic and hazardous substances and applies these principles across the life cycle of a chemical product, including its design, manufacture, use, and ultimate disposal. Green chemistry is aimed at reducing the dispersion and accumulation of synthetic chemical products into the environment and is aimed at less-hazardous materials, processes, and products. The objectives that can be achieved by applying green chemistry principles are as follows:

- Reduction in costs
- Avoiding/minimising health hazards and risks
- Reduction in energy consumption
- Reduction in material use
- Minimization of waste generation

There are 12 principles of green chemistry which are as follows:

7.3.1.1 Prevent Waste

Chemical synthesis needs to be designed to prevent waste or to leave no waste to treat or clean up. In other words, it is better to prevent than to treat or clean up waste after it is generated. The cost of treatment and disposal of chemical substances has grown considerably over the last few decades. The more hazardous the substance, the higher the cost to deal with it. A widely used measure of waste is the E-factor, which relates the weight of a waste coproduced to the weight of the desired product (Sheldon 2007). Figure 7.2 and Eq. (7.1) further describe this E-factor.

$$E = \frac{\text{kg of waste}}{\text{kg of product}} \tag{7.1}$$

Pharmaceutical industries experience huge waste problems. More recently, the American Chemical Society Green Chemistry Institute Pharmaceutical Roundtable (ACS GCIPR) estimated the ratio of the weights of all materials (water, organic solvents, raw materials, reagents, process aids) used to the weight of the active drug ingredient (API) produced. For example, a large amount of waste is coproduced during drug manufacturing – more than 100 kilos per kilo of API in many cases (Anastas and Warner 1998). Following Eq. (7.1), the E-factor of these drugs is more than 100.

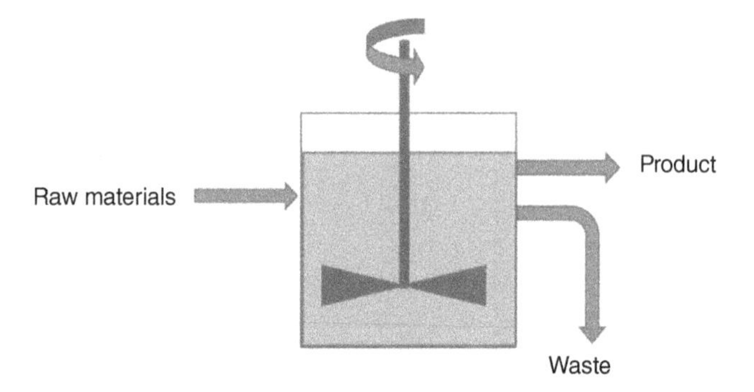

Figure 7.2 The E-factor. Source: Based on Sheldon (2007).

7.3.1.2 Maximise Atom Economy

Atom efficiency is defined as the ratio between the mass of a desired product to the total mass of reactants, expressed as a percentage. In the following Eq. (7.2), A and B react to give product C in high yield and high purity, but with the formation of by-products (or waste) D and E. In engineering or manufacturing, it can be common to produce more waste than product. The atom efficiency increases by reducing waste and by increasing the mass of the desired product.

$$A + B \rightarrow C + D + E \tag{7.2}$$

Maximising atom economy requires designing reactions with minimal waste with the reactions designed to be atom efficient, with as many of the reacting atoms as possible ending up as useful products. This is both environmentally and economically beneficial. The formula for atom efficiency (η_{atom}) is as follows:

$$\eta_{atom} = \frac{\text{Molecular weight (MWt) of desired product}}{\sum \text{Molecular weight (MWt) of all input materials}} \tag{7.3}$$

Sodium benzene sulfonate reacts with sodium hydroxide to make sodium phenoxide (Figure 7.3). Using Eq. (7.3), the atom efficiency of this chemical reaction is as follows:

$$\acute{\eta}_{atom} = 116/260 \text{ (or } 44.6\%)$$

For example, the method for ibuprofen production uses six steps with an overall atom economy of just 40.1%. In the 1990s, the Hoechst Celanese Corporation developed a fewer stage (three-stage) process with an atom economy of 77.4%, which is a classic example of improving the production route for a

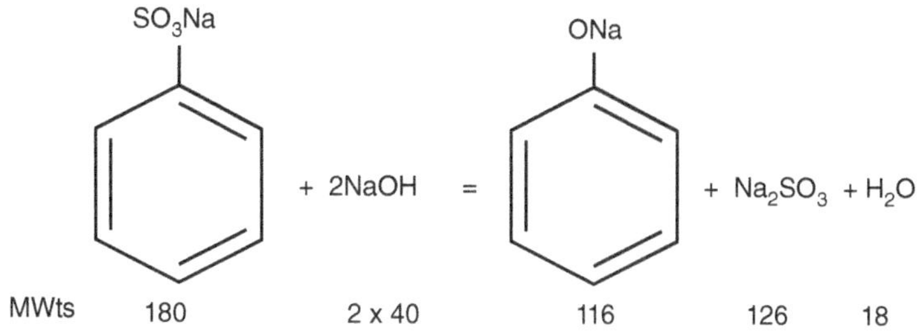

$C_6H_5SO_3Na + 2NaOH = C_6H_5ONa + Na_2SO_3 + H_2O$

MWts 180 2 x 40 116 126 18

Figure 7.3 Atom efficiency.

commercial product. The fewer the chemical stage processes, the less waste is produced.

7.3.1.3 Design Safer Chemicals and Products

Minimising toxicity, while maintaining function and efficacy at the same time, is found to be one of the most challenging aspects of designing safer chemical processes. Achieving this goal requires an understanding of not only chemistry but also of the relationship between chemicals and the surrounding environment. Highly reactive chemicals are not only quite valuable at affecting molecular transformations but they are also more likely to react with unintended biological targets, human and ecological, resulting in unwanted adverse effects. For example, dichlorodiphenyltrichloroethane (DDT), is one of the most effective insecticides, but this chemical kills other unintended microorganisms and insects other than the target ones. It should be noted that the presence of microorganisms in soil is crucially important for the production of food for plants by converting leaves and twigs to inorganic nutrients (nitrogen, potassium, phosphorus). This nutrient cycling is essential in agricultural production.

Per- and polyfluoroalkyl substances (PFASs) are extremely persistent chemicals linked to cancer, liver toxicity, and other negative health effects (Safer Chemicals 2021). Firefighters have used and trained with PFAS-based foam at commercial and military airports for many years. The PFAS in foam has contaminated drinking water for millions of people nationwide whilst also putting firefighters at a greater risk of toxic chemical exposure that can cause cancer and other chronic diseases.

7.3.1.4 Use Safer Solvents and Reaction Conditions

The chemical processes need to be designed in a way that will help avoid using solvents or separation agents, or other auxiliary chemicals. If these chemicals are necessary, innocuous chemicals should be used. For example, in the case of dry cleaning, perchloroethylene (PERC) is an excellent solvent for organic materials but was found to be carcinogenic and its disposal can contaminate ground water and cause eco-toxicity. Instead, liquid CO_2 can be used as a safer solvent to dissolve grease. This method is now being used commercially by some dry cleaners.

7.3.1.5 Use Renewable Feedstocks

The chemical processes have to be designed to use raw materials and feedstocks that are renewable rather than non-renewable or depleting feedstocks. Renewable feedstocks could either be made from agricultural products or from the wastes generated in other processes; depleting feedstocks typically are made from fossil fuels (petroleum, natural gas, or coal) or are mined.

For example, Cargill Dow LLC in the United States developed polylactic acid (PLA), a biopolymer that not only involves the use of bioprocesses that are energy and materials efficient but also utilises a renewable agricultural feedstock (e.g. corn) (Gruber 2004). PLA is not only recyclable but also biodegradable, and can be composted. It can functionally replace plastics such as nylon, polyethylene tetrachloride (PET), polyester, and polystyrene. Life cycle assessment analysis show that the replacement of PLA with PET can reduce the fossil fuel consumption by 20–50% and reduce manufacturing costs. The avoidance of the use of fossil fuel not only enhances the security of energy supply but also reduces ecological damage and solid waste management problems. Currently, developing countries have banned the import of plastic, wastes from developed nations which is why the reprocessing capacity of plastic waste recyclers has been extended. It is therefore important to find substitutes for PET to address this waste management challenge.

Another example is biofine, which is produced from relatively low-grade waste cellulosic by-products coming from paper mills and other sources (Hayes 2005). They can be converted into levulinic acid ($C_5H_8O_3$), a versatile chemical that is an intermediary to several other products. Levulinic acid can be used to develop biodegradable herbicides or as an economic fuel additive that makes gasoline burn more efficiently. The technology also has the potential to revolutionise the handling of cellulosic wastes. In addition to the by-products of paper mills, the process could reduce the accumulation of sewage sludge, waste wood, agricultural residues, and municipal solid waste.

Recent research from Nanyang Technological University, Singapore, is tackling food waste by turning discarded durian husks into antibacterial gel bandages (World Economic Forum 2021). The process extracts cellulose powder from the fruit's husks after they are sliced and freeze-dried, then mixed with glycerol. This mixture becomes a soft hydrogel, which is then cut into bandage strips.

7.3.1.6 Avoid Chemical Derivatives

There are some reactants that require protection from other reactants. The use of protecting agents like catalysts, polymers, and pharmaceuticals to protect the reactants increases the number of steps in a chemical reaction and derivatives (Janes 2017). Minimising the use of derivatives in a chemical synthesis can be achieved by avoiding the use of protecting groups, which will result in the increase of atom efficiency of the reaction. Decreasing the number of chemical reaction steps saves time, energy, and wastes.

A good example is the production of antibiotics based on penicillin, where several chemical steps can be replaced by an enzymatic reaction. As a result, reactions requiring low temperatures (−60 °C), organic solvents, and completely unsuitable conditions that increased and complicated the production were avoided.

7.3.1.7 Use Catalysts

Catalytic reactions minimise waste by maximising the conversion of reactants to products. They can be used in small amounts and be recycled indefinitely, and they do not generate any waste. They play an important role in the transition from a chemical industry based on non-renewable fossil resources to a more sustainable bio-based economy utilising renewable biomass as a raw material (Thangaraj et al. 2019). For example, methanol is a better reactant or catalyst with vegetable oils to produce biodiesel. This catalyst can be used over and over again.

7.3.1.8 Increase Energy Efficiency

Chemist often follow protocols to get a reaction to go to completion and to separate the desired product at as high a yield as possible without considering the amount of energy used in the reaction. Energy consumption, from their perspective, may be considered irrelevant and for all intent and purposes, not needing to be managed. From a sustainability perspective, energy use does need to be managed.

7.3.1.9 Design Less Hazardous Chemical Syntheses

Design syntheses to use and generate substances with little or no toxicity to humans and the environment. For example, chlorofluorocarbons are a non-flammable, non-toxic, inexpensive and effective refrigerants due to their other important properties such as volatility, and having boiling points close to 0 °C. These physical properties make them ideal for use as refrigerant gases in air conditioners, freezers, and refrigerators. However, chlorofluorocarbons (CFCs) are stable meaning that they are long-lived, and when they are released, they can enter the upper atmosphere causing ozone depletion, which then lead to the entrance of harmful radiation into the earth's atmosphere (Anthony 2017). These highly reactive UV radiations significantly affect the life on earth.

7.3.1.10 Design Chemicals and Products to Degrade After Use

Design chemical products to break down to innocuous substances after use so that they do not accumulate in the environment. They have to be biodegradable and recyclable. When batteries end up in landfills, they can leak dangerous substances like lithium, cadmium, and mercury. We should always dispose of batteries responsibly. A team of researchers from the University of Wollongong, Australia, have created a biodegradable electrode (University of Wollongong 2015). The thin, flexible battery is made of silk, and can dissolve in water completely in just 45 days.

Hemp fibre has long been used, instead of fibreglass, to make surfboards as it is lighter, more flexible and water resistant, and gives very good grip and buoyancy (Gibson and Warren 2017). The use of hemp not only entails ecological benefit, it also can improve the quality and performance of the product containing it.

Meanwhile, plastic waste, including 'microplastics', results in a great deal of air, sea, and land pollution (Sensi Seeds 2020). A team of scientists have recently demonstrated that these microplastics are dispersed through the air and reach many remote natural areas, with concentration levels similar to those found in large cities. Hemp is proving to be a clean alternative to the highly polluting plastics that are damaging our environment.

7.3.1.11 Analyse in Real Time to Prevent Pollution

In-process, real-time monitoring and control during syntheses help to minimise or eliminate the formation of unwanted by-products.

An example is an exothermic reaction, in which energy is released as heat. At the bench scale (grams), one can use a simple ice bath to cool down an exothermic reaction (Waked 2018). Even if the solution's temperature does end up rising, this usually does not pose a great risk due to the small scale of the reaction.

7.3.1.12 Minimise the Potential for Accidents

We need to design chemicals and their physical forms (solid, liquid, or gas) to minimise the potential for chemical accidents including explosions, fires, and releases to the environment. For example, the Bhopal disaster, resulted from corrosion of pipes due to poor maintenance (Mandavilli 2018). Water then accidentally entered into an MIC tank through leaks from the corroded pipes. In the presence of water, an exothermic reaction took place which resulted in an explosion that caused thousands of deaths in surrounding areas.

Box 7.1 Which pathway is greener?

There are two pathways to convert nitrobenzene (feedstock) to aniline (product). Using the principles of green chemistry, we can analyse these pathways.

Traditional approach/pathway

$$4\,C_6H_5NO_2 + 9Fe + 4H_2O \xrightarrow[HCl]{FeCl_2} 4\,C_6H_5NH_2 + 3Fe_3O_4$$

Alternative approach/pathway

$$C_6H_5NO_2 + 3H_2 \xrightarrow[300\,°C,\ 5\ psi]{Ni\ (catalyst)} C_6H_5NH_2 + 2H_2O$$

The gram molecular weights (GWt) of chemical in these reactions are given below.

Chemicals	GWt
Nitrobenzene	123
Aniline	93.13
Fe	55.84
O	16
H	1

Solution

Traditional pathway

The summation of GWt of reactants on the left hand side of the equation

$$= 4 \times 123 + 9 \times 55.84 + 4(2 \times 1 + 16) = 1067$$

The GMW of the product or aniline $= 4 \times 93.13 = 373$.

Using Eq. 7.3, atom efficiency $= (373/1067) \times 100\% = 35\%$.

The summation of GWt of reactants on the right hand side (RHS) of the equation

$$= 4 \times 93.13 + 4(3 \times 55.84 + 4 \times 16) = 1.067$$

Therefore, LHS = RHS.

Alternative pathway
The summation of GWt of reactants on the left hand side (LHS) of the equation

$$= 123 + 3 \times 2 \times 1 = 129$$

The GMW of the product or aniline = 93.13.
Using Eq. 7.3, atom efficiency = (93.13/129) × 100% = 72%.

Analysis The atom efficiency of the alternative pathway is higher than the traditional one, which means that the former produces less wastes than the latter. Secondly, the waste produced by the reaction in the alternative pathway is water, which is non-toxic. On the other hand, iron oxide which is produced as a waste in the traditional approach is hazardous as exposure to its fumes can cause metal fume fever (NJDHSS 2007). Also, the reactants used in the traditional pathway are relatively energy intensive (iron) and resource scarce (water). While alternative approach appears to perform better than the traditional one from the green engineering perspective, it has some drawbacks. The reaction occurs at a higher temperature (300 °C) and pressure (5 psi) and the approach utilises a nickel metal as a catalyst which is energy intensive to produce.

7.3.2 Sustainable Materials

Sustainable materials are materials that have less embodied energy, are durable, locally available, recyclable, can be produced without depleting non-renewable resources and without disrupting the ecosystem. Such materials vary enormously and can include bio-based polymers derived from polysaccharides, or industrial by-products, wastes, or highly recyclable materials such as glass and steel that can be reprocessed an indefinite number of times without requiring additional material/mineral resources.

Construction industries are material intensive-industries. The construction industry's appetite for raw materials is vast; it is the world's-single largest consumer of materials and accounts for 25–40% of global carbon emissions (World Economic Forum 2016). Construction is the largest steel-consuming sector accounting for approximately 50% of global steel consumption (Lee and Dai 2016). However, steel consumes a significant amount of energy in its manufacture. Concrete is widely used as a structural element and in building foundations. However, cement manufacture produces significant quantities of GHG emissions. Currently, composite cementitious materials are being developed as environmentally friendly cement alternatives to address the aforementioned issues.

Composite materials have exceptional properties such as being lightweight, high strength-to-weight ratio, low thermal conductivity, and corrosion resistance. For example, carbon fibre composite materials are the prime alternative material for fuel pipes for automobiles.

A composite is made by physically combining two or more materials (components) to produce a combination (blend) of structural properties not present in any individual component. The filler/reinforcement typically consists of carbon, glass, or Kevlar strands. The matrix is typically a resin or epoxy. Examples of composite materials include carbon fibre reinforced polymer, glass fibre reinforced polymer, and Aramid/Kevlar reinforced polymer. Composite materials are designed to offer the following benefits over conventional materials (Raco 2019):

Low Weight: Because of this property, it can be easily transported by truck at the end-of-life stage from one place to another. Lightweight fibre reinforced plastic (FRP) requires smaller crane systems for loading and offloading and weighs approximately 1400 kg/m^3, while concrete is a heavier material weighing approximately 2400 kg/m^3.

Installation: Fibreglass tanks can be transported on a single truck and are delivered to the site as a finished product, making installation easier and faster. In the case of a concrete tank, its step-by-step installation is on site, as concrete structures have to be poured and set. The transportation distances of concrete ingredients increase the ton-kilometres travelled and, by association with the environmental emissions, time, and overall expenses.

Stiffness and Strength: This property allows FRP to be used in places where concrete, steel, and other energy intensive construction materials are used. Concrete being used in high strength applications (e.g. column, beam) is enriched with energy-intensive cement, steel, and aggregates, which can be completely replaced with FRP providing the same level of structural performance. A car body panel made of carbon fibre plastic makes the car lighter to help achieve more fuel economy.

Low Coefficient of Expansion: Fibreglass has a low coefficient of expansion under heat and weather conditions. This property can potentially avoid the expansion of construction materials, which reduces the risk of cracking and enhances durability. Concrete expands and contracts regularly with temperature which causes cracks to appear with increased risk of corrosion due to the existence of reinforcement steel.

Bacterial Resistance: FRP products are manufactured with smooth interior resistant layers and exterior waterproof layers making it a good environment to combat bacteria and algae accumulation. Therefore, FRP tanks or pipes can be more suitable than concrete in moisture-laden conditions.

Resistance Against Fatigue: This property also increases the service life of products, which lowers the dependency on virgin materials. The durability or service life of products is usually the main criteria as prolonged service life of materials slows down the extraction of virgin materials and reduces the product's ecological footprint on the built environment. As a result, the life span of buildings using FRP increases, often providing a 75 year guaranteed lifetime, up to a maximum of 150 years. On the other hand, concrete buildings have between 15 and 50 years of expected lifespan.

Ease in Manufacturing Complex Shapes: FRP is generally made using an injection moulding process. The shape of FRP is determined by the mould cavity, but the thing that shapes the mould is the pattern. This pattern can be a complicated shape used for making the mould cavity. Melted plastic and reinforcement materials are poured into the mould to make a desired shape for a product. In a conventional construction process, some wastes are generated as all materials like sand and bricks are not fully utilised. FRP systems can be made as a whole, monolithic system, and usually coming pre-assembled.

Simple Repair of Damaged Structures: This property of FRP offers preventative maintenance by reducing repairing, plastering, and the use of chemicals.

An FRP tank is easier to clean, which extends the associated life of pumps and filtering equipment. It is more difficult to clean and maintain a concrete tank, requiring higher frequency of care that increases the cost over the life of the system.

Resistance to Corrosion: FRP does not use metal and therefore it is not as susceptible to corrosion, which increases the durability of FRP products. Concrete often uses steel rebars for reinforcement and the rebar is susceptible to rust through water permeating the surface. Also FRP has resistance to most chemical corrosion. Hydrogen sulfide creates sulfuric acid when in underground water systems, and this is a common cause of deterioration of concrete, but does not affect FRP concrete.

FRP systems are also watertight and have an external resin rich watertight penetration barrier, whereas concrete is a porous material allowing water to penetrate the surface. Even with sealants, concrete can leak to the external environment.

Environmental Comparison: FRP can replace concrete, which is made of resource intensive materials like cement and aggregate. In addition, the processes to produce cement generate significant CO_2 emissions, whereas composite materials can utilise lower GHG emission materials in its manufacture.

Recyclability: Steel is infinitely recyclable, while concrete is difficult to separate and recycle, whilst composite materials have a modular structure, which allows reuse/recycling.

7.3.2.1 Applications of Composite Materials

Composite materials are mainly applied in the building and construction industry. They need less maintenance, due to their low density and ability to dissipate energy in seismic activities. It has high strength and mechanical performance. The physical properties of these materials allow their assembling and disassembling. Therefore, they are very suitable in earthquake regions.

Composite materials (CM) can potentially be used in the automotive industry as they offer more efficient construction processes and also more durable performance with better dissipation of energy allowing more deformation. CM are highly flexible, requiring no welds and have been recognised in international technical design codes: American Society of Civil Engineers (ASCE), European Committee for Standardization (CEN), and International Standard Organization (ISO) (EuCIA 2021). CM reduce (lifetime) construction costs as less material and fewer working hours are required. The weight of the vehicle is reduced due to the use of composite materials. A comparative analysis has been performed among steel 4340, aluminium 6061 alloy, and Carbon FRP (CFRP) fuel pipes. It was observed that the use of aluminium 6061 and CFRP fuel pipes can lead to about 65–83% weight reduction in the fuel pipe (Sinha et al. 2018). For every 10% reduction in vehicle weight, fuel consumption is reduced by about 4–7%, offering both economic and environmental benefits (Sinha et al. 2018).

7.3.2.2 The Positives and Negatives of Composite Materials

Although composite materials offer lower energy requirements and GHG emissions, greater durability with less maintenance, lower life-cycle costs, high strength to weight ratio, (i.e. high strength and lightweight), corrosion resistance in marine/acidic environments, and provide fair heat and noise insulation, they are still more expensive than conventional materials due to the economies of scale associated with cement production and also it is still a new material compared with concrete.

7.3.2.3 Bio-Bricks

Both concrete and clay can be used to make bricks (Figure 7.4). In the case of clay bricks, clay is not an energy intensive material but the manufacturing can include energy intensive processes to convert clay into bricks. In the case of concrete bricks, there is no need for energy in manufacturing as the cement which is a binder material binds all ingredients together to form high strength bricks. Concrete constituents like binder and aggregates are energy intensive. Binder or ordinary Portland cement is a very energy and carbon intensive material as energy is consumed in raw material extraction, transportation, and fuel sources for heating kilns. These conventional bricks are used in over 80% of global construction and so 1.23 trillion bricks are created annually worldwide. It is estimated the fabrication of bricks emit over 800,000,000 tons of CO_2 each year (Allen 2016).

Therefore, another alternative brick type has been designed that avoids the use of carbon intensive materials and also manufacturing energy to reduces the overall

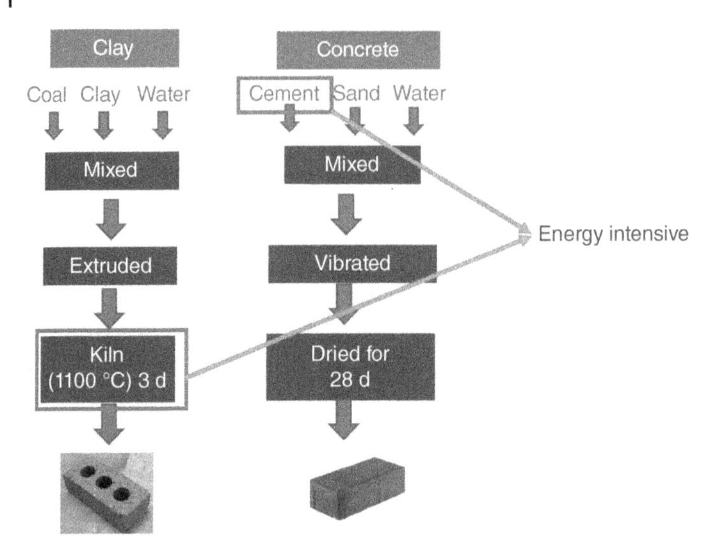

Figure 7.4 Concrete bricks vs clay bricks.

energy consumption and GHG emissions. Bio-bricks reduces the energy demand to produce bricks as they use biologically controlled structural cement and remove the toxic materials which are often used in brick making (glazes, artificial pigments). A bio-brick is naturally occurring and biodegradable.

For example, the bioMASON process begins with sand. It is placed into moulds and inoculated with *Sporosarcina pasteurii* bacteria, which are then fed with calcium ions suspended in water (Designing Buildings 2021). The ions are attracted to the bacterial cell walls, creating a calcium carbonate shell, which causes particles to stick to each other. Anything made by a biological process may appear to take longer than a physical process. This is not the case here as a single bacterial brick takes 2–5 days to grow, compared with 3–5 days to make a kiln-fired version. There can be emissions associated with nurturing bacteria used for manufacturing bricks, but this is not expected to increase the overall GHG emissions significantly.

7.3.3 Heat Recovery

Globally, heat accounts for nearly half of all energy consumption and 40% of energy-related carbon dioxide emissions (BBC 2020). In this regard, the use of waste heat recovery (WHR) systems in industrial processes has been one of the major areas for research to reduce fuel consumption, lower harmful emissions, and improve production efficiency.

Industrial waste heat is the energy that is generated in industrial processes which is not put into any practical use and is wasted or released to the environment. Sources of waste heat mostly include heat loss transferred through conduction. The conductive heat losses happen through building walls, floors, ceiling, glass, or convection and radiation from industrial products, piping, equipment, and processes. Also a significant amount of heat is discharged though chimneys or stacks with flue gases resulting from the combustion processes to generate heat or electricity. WHR can be conducted through various WHR technologies to provide valuable energy sources and reduce the overall energy consumption. The recovery strategies and the application of recovered waste heat depend on the temperature of the heat more than the content of heat. Equation (7.4) represents the recovered heat energy.

$$Q_r = mc\Delta T \tag{7.4}$$

where

m = mass flowrate (kg/s) of the fluid that has extracted or recovered energy from the waste heat sources,

c = specific heat of the fluid (kJ/kg °C) that recovered the waste heat. This is a thermodynamic property specific to the fluid used to transfer heat,

ΔT (°C) = the difference in temperature rise (°C) of the fluid used to recover the waste heat

It is not possible to recover all the heat from the waste heat sources Q_s which is greater than Q_r due to heat losses in the heat exchangers through the heat exchanging process. For example, plate heat exchangers offer a maximum efficiency of 80% with normal variations between 55% and 65% (The Renewable Energy Hub UK 2018).

The value of Q_r can be the same for different heat exchanging processes with different amounts of mass and temperature difference, but their thermal applications may vary. For example, in two different heat exchanging processes for the same fluid of specific heat 'c' (0.8 kJ/kg °C), the heat recovered is same (i.e. 800 kJ), but one process has a mass flow of 50 kg/s and a temperature difference of 20 °C, while another process has a mass flow of 20 kg/s and a temperature difference of 50 °C. In this case, the former has a limited number of thermal applications than the latter due to having a lower temperature heat.

Heat loss can be classified into high temperature, medium temperature, and low temperature grades.

7.3.3.1 Temperature Classification

High temperature WHR consists of recovering waste heat at temperatures greater than 650 °C, the medium temperature range is 230–650 °C, and the low temperature range is for temperatures less than 230 °C (US Department of

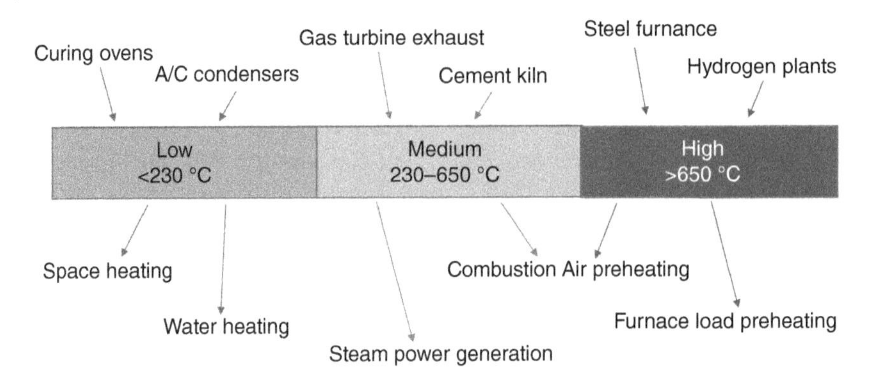

Figure 7.5 Classification of waste heat in terms of temperature. Source: Based on US Department of Energy (2008).

Energy 2008). Figure 7.5 shows the classification of heat in terms of temperature and applications.

Usually most of the waste heat in the high temperature range comes from direct combustion processes, in the medium range from the exhaust from combustion units, and in the low temperature range from parts, products, and the equipment process units. The waste heat temperature is a key factor in determining the feasibility of WHR.

High Temperature Waste Heat Examples of high temperature waste heat sources include a nickel refining furnace, steel electric arc furnace, basic oxygen furnace, aluminium reverberatory furnace, copper refining furnace, steel heating furnace, copper reverberatory furnace, hydrogen plants, fume incinerator, glass melting furnace, coke oven, and iron cupolas.

The advantages are high-quality energy available for a diverse range of end uses with varying temperature requirements, high efficiency power generation, and high heat transfer rate per unit area. However, high temperature could create increased thermal stresses on heat exchange materials, which increases chemical activity and corrosion.

Typical recovery methods/technologies could be an air pre-heater, and heat exchangers to extract heat to run a steam generator for process heating or for mechanical/electrical work, furnace load pre-heating, and also for transferring to medium low temperature processes.

Medium Temperature Waste Heat Medium temperature waste heat (MTWH) sources are steam boiler exhaust, gas turbine exhaust, reciprocating engine exhaust, heat treating furnace, drying and baking ovens and cement kiln. MTWH has an advantage over high temperature WHR processes in that it is more compatible with heat exchanger materials, and the recovered waste heat is practical for power generation, preheating air for combustion/power generation, furnace load preheating (i.e. introduction of significant heat into furnace charge material prior its to placement into the furnace), and feedwater preheating.

Low Temperature Waste Heat Low temperature waste heat sources are exhaust gases exiting recovery devices (gas-fired boilers, ethylene furnaces, etc.), process steam condensate, cooling water from furnace doors, annealing furnaces, air compressors, internal combustion engines, air conditioning and refrigeration condensers, drying, baking, and curing ovens, and hot processed liquids/solids.

The advantage is that a large quantity of low temperature heat is contained in numerous product streams. However, there are more disadvantages than advantages such as there are few end uses for low temperature heat resulting in low-efficiency power generation. Secondly, for combustion exhaust, low-temperature heat recovery is impractical due to acidic condensation and heat exchanger corrosion.

Typical recovery methods/technologies are space heating, and domestic water heating upgrading via a heat pump to increase temperature for end use.

7.3.3.2 Heat Recovery Technologies

Various technologies exist and/or are undergoing development to make use of recovered thermal energy to do 'work'.

- Heat exchangers
- Low temperature heat recovery
- Power generation

Heat Exchangers These devices are most commonly used to transfer heat from combustion exhaust gases to combustion air entering the furnace. Since preheated combustion air enters the furnace at a higher temperature, less energy must be supplied by the fuel.

Economisers or finned tube heat exchangers that recover low–medium waste heat are mainly used for heating liquids. The system consists of tubes that are covered by metallic fins to maximise the surface area of heat absorption and the heat transfer rate (low temperature energy recovery).

Low Temperature Heat Recovery There are various applications where low-grade waste heat has been cost-effectively recovered for use in industrial facilities.

Figure 7.6 displays an example of the heat pump that transfers heat from lower temperature to higher temperature. It uses a closed compression cycle to recover heat from cooling water leaving a sterilizer in a dairy plant at 55 °C. The heat

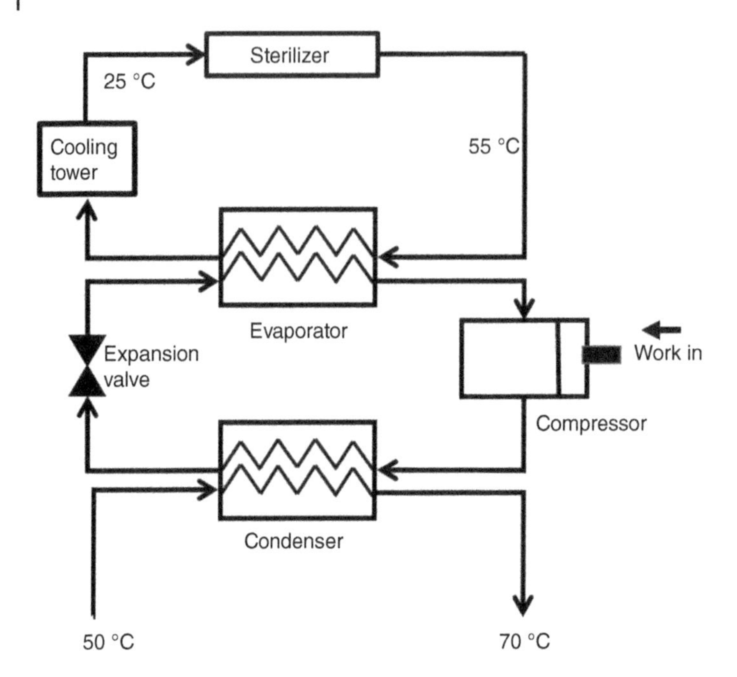

Figure 7.6 Dairy farm utilising a heat pump to transfer heat for use elsewhere.

pump lowers the temperature of the cooling water, which uses the extracted heat to increase the temperature of process water that can be used in other processes in the plant. The heat pump consists of these components: evaporator, compressor, condenser, and expansion valve. In the evaporator, the temperature of the waste heat is reduced from 55 to 25 °C by transferring heat energy to the refrigerant. Then the refrigerant enters the compressor, where its temperature is increased further. Superheated refrigerant then enters the condenser and transfers heat to the heat sink or water. The temperature of the water increases from 50 to 70 °C, and this hot water can be used in other applications in the dairy farm. Finally, refrigerant leaving the condenser is throttled in an expansion valve, where its temperature is further reduced before returning to the evaporator.

Power Generation Generating power from waste heat typically involves using the waste heat to convert working fluids to gaseous products in the boiler to create thermal energy which is then converted to kinetic or mechanical energy to drive an electric generator.

The efficiency of power generation is heavily dependent on the temperature of the waste heat source. There are two types of boilers: high temperature waste heat boilers and low temperature waste heat boilers.

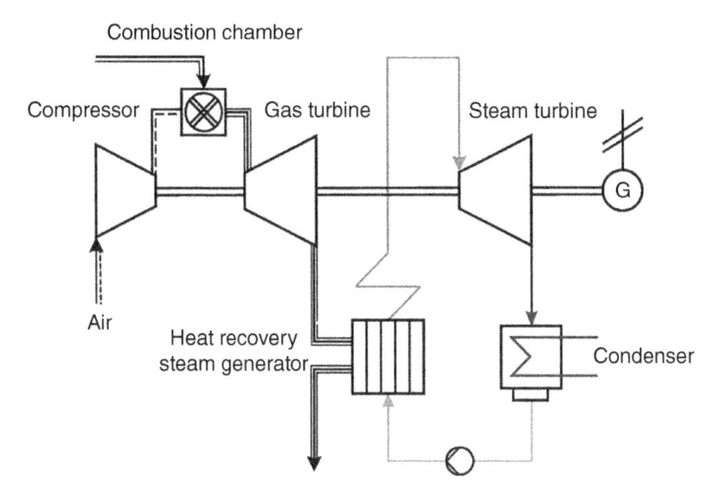

Figure 7.7 Combined cycle power plant.

High Temperature Waste Heat Boiler Waste heat boilers consist of a number of water tubes that are positioned in parallel to each other and in the direction of the waste heat leaving the system. The system is suitable for recovering heat from medium – high temperature exhaust gases, which would have otherwise been discharged into the atmosphere through the chimney or stack. The recovered heat is used to generate steam as an output. It needs high heat or more than 250 °C to convert water to superheated steam to produce sufficient thermal energy to produce kinetic energy in the turbine to generate electricity. The steam can then be used for power generation or directed back to the system for energy recovery.

Figure 7.7 shows that the exhaust leaving a gas turbine after generating electricity flows through a heat exchanger, in which the heat from the exhaust gas is transferred to water to convert it to steam for running the turbine to generate additional electricity. The gas turbine is known as a topping cycle or Brayton cycle. On the other hand, the steam turbine is known as a bottoming cycle or Clausius-Rankine cycle or steam cycle. Using this additional arrangement or a steam turbine, the same amount of energy input like gas or diesel used in running a gas turbine can generate an additional amount of energy. The capital cost of combine cycle gas turbine (US$ 1000/kW) was found to be higher than the simple gas turbine (US$ 350/kW), but the life cycle cost of electricity of the former (US$ 167/MWh) is lower than the latter (US$ 215/MWh) (Aji et al. 2018). This is because the additional amount of electricity generated recovers the additional capital cost spent on the steam power plant.

In summary, a typical Rankine cycle consists of a pump, a condenser, an evaporator, and a generator. Fuel is burnt in the evaporator and the water as the working

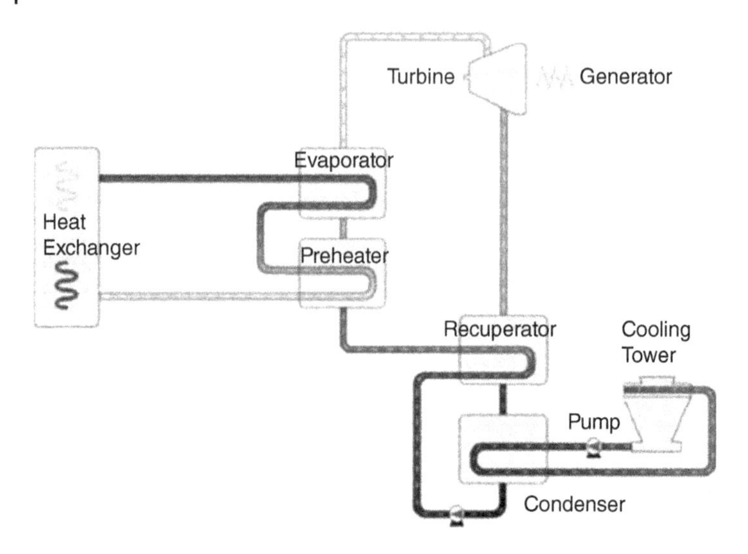

Figure 7.8 Organic Rankine cycle. Source: Jouhara (2018) / with permission of Elsevier.

fluid is heated to generate superheated steam. This is then directed to the turbine to generate power and then passed through the condenser, losing heat and turning back into its liquid state. The liquid water is then pumped into the evaporator and the cycle is repeated.

Low Temperature Waste Heat Recovery The organic Rankine cycle works on the principle of the Rankine cycle; however, the system uses organic substances as a working fluid with lower boiling points and high vapour pressures to generate power instead of water as a working fluid (Figure 7.8). It has been shown that the use of an organic fluid as the working fluid makes the system suitable for utilising low temperature waste heat for power generation.

A variant of the organic Rankine cycle, the Kalina cycle (KC), uses the working fluid in a closed cycle to generate electricity and demonstrates 40% more thermal performance than that of organic Rankine cycle (ORC) at moderate pressures (Elsayed et al. 2013). KC uses a binary working fluid of ammonia and water, whereby ammonia permits efficient use of waste heat streams allowing boiling to start at lower temperatures. In addition to the turbine condenser, evaporator, pump and throttling valves, KC uses an absorber, separator, and regenerator to run the thermodynamic cycle (Figure 7.9). In the evaporator, the ammonia–water mixture is heated up by the low-temperature heat source, which then flows through the separator as not all the ammonia-water mixture is converted to vapour. In the separator, the saturated vapour part of the mixture is separated from the liquid. The rich ammonia saturated vapour mixture then expands through the turbine, producing electricity, which then passes through the absorber. Here, the weak solution and the strong solution streams are mixed at low temperatures.

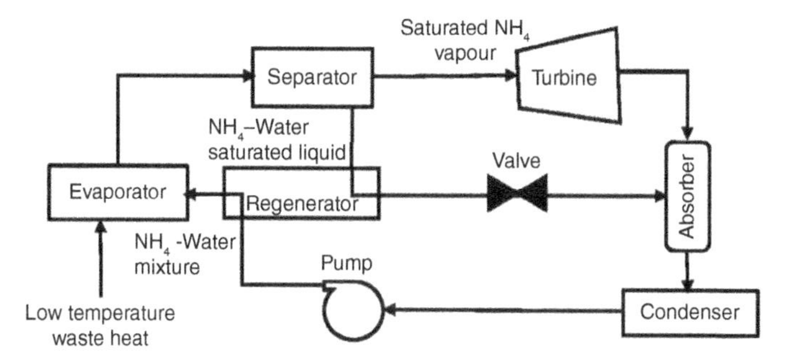

Figure 7.9 Kalina cycle.

This solution then undergoes condensation in the condenser, after which it is pumped to increase its pressure to that corresponding to the evaporator. The hot weak ammonia–water saturated liquid mixture that was separated earlier (from the saturated ammonia vapour in the separator) is directed to the regenerator to be cooled and then to a throttling valve to lower its pressure before mixing with the rich ammonia solution in the absorber.

Although KC is found to be more efficient than ORC, this power generation process needs special safety procedures that prevent the leakage of ammonia to avoid eco-toxicity effects.

7.3.3.3 The Positives and Negatives of Waste Heat Recovery

Green engineering principles can increase the overall material and energy efficiency of engineering processes. They can reduce the operating costs of the process and provide additional income streams (industrial symbiosis, heat recovery). The negative aspects are that additional costs are involved, and also the use of heat exchangers in the system increases the complexity of the entire system due to the increase in the number of heat pipes to recover heat from the waste gas, which creates further dependencies. Pinch analysis a way to find the optimum heat recovery without entailing excessive costs. This analysis is done by calculating thermodynamically feasible energy targets (or minimum energy consumption) and achieving them by optimising heat recovery systems, energy supply methods, and process operating conditions.

References

Aji, S.S., Kim, Y.S., Ahn, K.Y., and Lee, Y.D. (2018). Life-cycle cost minimization of gas turbine power cycles for distributed power generation using sequential quadratic programming method. *Energies* 11: 3511. https://doi.org/10.3390/en11123511.

Allen, D. (2016). Going green: bricks and microbes. Independent Santa Barbara. https://www.independent.com/2016/07/20/going-green-bricks-and-microbes (accessed 2 November 2021).

Anastas, P.T. and Warner, J.C. (1998). *Green Chemistry: Theory and Practice*, 30. New York: Oxford University Press. By permission of Oxford University Press.

Anastas, P.T. and Zimmerman, J.B. (2003). Design through the twelve principles of green engineering. *Environmental Science and Technology* 37 (5): 94A–101A.

Anthony, J.L. (2017). Green engineering: principles and practice. Department of Chemical Engineering Kansas State University. https://sckool.org/green-engineering-principles-and-practice.html (accessed 26 October 2021).

BBC (2020). How to cut carbon out of your heating? Smart Guide to Climate Change. https://www.bbc.com/future/article/20201116-climate-change-how-to-cut-the-carbon-emissions-from-heating (accessed 25 October 2021).

BigRentz, Inc (2020). How modular construction benefits the environment? https://www.bigrentz.com/blog/modular-construction-environmental-benefits (accessed 26 September 2021).

Designing Buildings (2021). Biocement. www.designingbuildings.co.uk/wiki/Biocement (accessed 1 November 2021).

van Dijk, H.A.J., Cobden, P.D., Lukashuk, L. et al. (2018). STEPWISE project: sorption-enhanced water-gas shift technology to reduce carbon footprint in the iron and steel industry. *Johnson Matthey Technology Review* 62 (4): 395–402.

Elsayed, A., Embaye, M., AL-Dadah, R. et al. (2013). Thermodynamic performance of Kalina cycle system 11 (KCS11): feasibility of using alternative zeotropic mixtures. *International Journal of Low-Carbon Technologies* 8: i69–i78.

EuCIA (European Commission Industry Association) (2021). Prospect for new guidance in the design of FRP structures. https://eucia.eu/userfiles/files/Eucia_Prospect%20for%20New%20Guidance%20in%20the%20Design%20of%20FRP%20Structures_web2.pdf (accessed 30 October 2021).

Gibson, C. and Warren, A. (2017). Surfboard making and environmental sustainability: new materials and regulations, subcultural norms and economic constraints. In: *Sustainable Surfing* (ed. G. Borne and J. Ponting), 87–103. Abingdon: Routledge.

Gordon, S. (2017). Four materials illustrate hazards of electronics manufacturing. https://www.wiscontext.org/four-pollutants-illustrate-hazards-of-electronics-manufacturing (accessed 23 October 2021).

Gruber, P.R. (2004). Cargill Dow LLC. *Journal of Industrial Ecology* 7 (3–4): 209–213.

Guglielmi, G. (2020). Why Beirut's ammonium nitrate blast was so devastating? *Nature* https://doi.org/10.1038/d41586-020-02361-x.

Hayes, D.J. (2005). The biofine process: production of levulinic acid, furfural and formic acid from lignocellulosic feedstocks. Chapter 7. In: *Biorefinery (8b)* (ed. B. Kamm, P.R. Gruber and M. Kamm), 1–21. Wiley https://doi.org/10.1002/9783527619849.ch7.

Hendershot, D. (2011). Inherently safer design – an overview of key elements. *Professional Safety*, February: 48–55. https://www.assp.org

Huffpost (2010) The transformer (video): Hong Kong architect's tiny apartment has 24 room options. https://www.huffpost.com/entry/the-transformer-video-hon_n_558639 (accessed 10 November 2021).

Janes, T. (2017). Green chemistry principle #8: reduce derivatives. https://greenchemuoft.wordpress.com/2017/03/20/greenchemprinciple8 (1 November 2022).

Jouhara, H. (2018). Waste heat recovery technologies and applications. *Thermal Science and Engineering Progress* 6 (2018): 268–289.

Lee, E. Y. and Dai, A. (2016). Oversupply in the global steel sector: challenges and opportunities. DBS Group Research.

Mandavilli, A. (2018). The world's worst industrial disaster is still unfolding. *The Atlantic, November 30*. https://www.theatlantic.com/science/archive/2018/07/the-worlds-worst-industrial-disaster-is-still-unfolding/560726 (accessed 15 October 2021).

Martinez, C. (2020). Products with added mercury and risks for the environment and health. Chile: Ministry of Environment.

Mitsubishi Heavy Industries (2021). Combined-cycle power plants. https://power.mhi.com/group/msc/business/products/combined (accessed 31 October 2021).

Nath, P., Sarker, P.K., and Biswas, W.K. (2018). Effect of fly ash on the service life, carbon footprint and embodied energy of high strength concrete in the marine environment. *Energy and Buildings* 158: 1694–1702.

NJDHSS (New Jersey Department of Health and Social Services) (2007). Hazardous substances fact sheet. https://nj.gov/health/eoh/rtkweb/documents/fs/1036.pdf (accessed 10 November 2021).

Nordic ID Group (2017). How RFID technology can make your business more sustainable? https://www.nordicid.com/resources/blog/how-rfid-can-make-your-business-more-sustainable (accessed 25 October 2021).

Raco, N.D. (2019). FRP (fibre reinforced polymers) Vs concrete. *Protector*. https://protector.com.au/frp-fibre-reinforced-polymers-vs-concrete (accessed 5 November 2021).

Safer Chemicals (2021). FAA must end the use of polluting PFAS firefighting foam. https://saferchemicals.org/2021/10/05/faa-must-end-the-use-of-polluting-pfas-firefighting-foam (accessed 1 November 2021).

Sandia National Laboratories (2012). Inherently safer design. Prepared for the U.S. Department of Energy's National Nuclear Security Administration, Paper No. SAND2012-3406c.

Sensi Seeds (2020). Hemp Plastic: What Is It and How is it Made?. https://sensiseeds.com/en/blog/hemp-plastic-what-is-it-and-how-is-it-made (accessed 30 November 2021).

Sheldon, R.A. (2007). The E factor: fifteen years on. *Green Chemistry* 9: 1273. https://doi.org/10.1039/B713736M.

Sinha, M., Tyagi, R.K., and Bajpai, P.K. (2018). Weight reduction of structural members for ground vehicles by the introduction of FRP composite and its implications. In: *Proceedings of the International Conference on Modern Research in Aerospace Engineering. Lecture Notes in Mechanical Engineering* (ed. S. Singh, P. Raj and S. Tambe), 277–290. Singapore: Springer https://doi.org/10.1007/978-981-10-5849-3_28.

Thangaraj, B., Solomon, P.R., Muniyandi, B. et al. (2019). Catalysis in biodiesel production—a review. *Clean Energy* 3 (1): 2–23. https://doi.org/10.1093/ce/zky020.

The Renewable Energy Hub UK (2018). How efficient are heat recovery systems? www.renewableenergyhub.co.uk/main/heat-recovery-systems-information/heat-recovery-system-efficiencies (accessed 5 November 2021).

University of Wollongong (2015). Battery to power implant, then disappear. www.uow.edu.au/media/2015/battery-to-power-implant-then-disappear.php (accessed 29 October 2021).

US Department of Energy (2008). Waste heat recovery: technology and opportunities in U.S. industry. USA. https://www1.eere.energy.gov/manufacturing/intensiveprocesses/pdfs/waste_heat_recovery.pdf

Waked, A. (2018). Green chemistry principle #11: real-time analysis for pollution prevention. https://greenchemuoft.wordpress.com/2018/09/13/green-chemistry-principle-11-real-time-analysis-for-pollution-prevention (accessed 1 November 2022).

World Economic Forum (2016). Can the circular economy transform the world's number one consumer of raw materials? https://www.weforum.org/agenda/2016/05/can-the-circular-economy-transform-the-world-s-number-one-consumer-of-raw-materials (accessed 20 October 2021).

World Economic Forum (2020). Meet the woman making cycling even more sustainable in Ghana. https://www.weforum.org/agenda/2020/07/ghana-bamboo-bike-cycling-sustainability (accessed 3 November 2021).

World Economic Forum (2021). These scientists are making antibacterial bandages out of fruit waste. https://www.weforum.org/agenda/2021/09/scientists-singapore-antibacterial-bandages-food-waste-innovation-nanyang-technological-university (accessed 27 September 2021).

8

Design for the Environment

8.1 Introduction

Careful considerations at the design stage can prevent unanticipated environmental consequences during the production and service life cycle stages of a product. The main theme of this chapter is aligned with the philosophical statement, '*look before you leap*'. In order to materialise this philosophy in engineering design, there needs to be a gradual change of focus from the production process to the product itself. Figure 8.1 presents a product system, where feedstock or raw materials are converted to products, waste is generated and the pollutants are emitted from the production process to the atmosphere. The design for the environment (DfE) concept requires us to choose raw materials during the design process in a way that they are recyclable, bio-degradable, have less embodied energy, and create no or little environmental damage in the production system. Secondly, manufacturing processes have to be designed in a way that less waste is generated, and any wastes or by-products are non-toxic and can potentially be used as resources either by the industry itself or by other industries. Thirdly, the pollutants emitted from the production processes can be captured and used in suitable applications. Fourthly, the product can then be disassembled to enable remanufacturing, recovery, or recycling at the end of its life. Also the product has to be less energy and material intensive and reduce impacts during the use stage by avoiding the use of consumable items for maintenance purposes. All these main considerations need to be taken into account in the design process to decrease the ecological footprint and to enhance 'ecosystem services' (e.g. food, water, soil, primary energy).

8.2 Design for the Environment

DfE can be defined as the systematic application of environmental and health considerations at the design stage of the product. It aims to reduce significant

Engineering for Sustainable Development: Theory and Practice, First Edition.
Wahidul K. Biswas and Michele John.

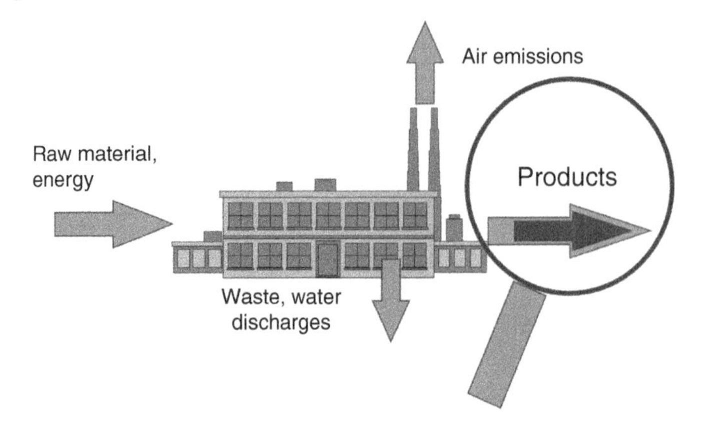

Figure 8.1 Production system.

environmental impacts and increase resource efficiency at all stages of a product's life cycle, including the extraction and processing of raw materials, production/manufacturing, packaging and distribution, product use, and end-of-life. DfE thus goes beyond the commonly considered environmental management strategies (e.g. waste treatment, recycling materials or proper packaging or disposal). Importantly, deep thinking, innovative and holistic approaches at the initial design of the product are important as they can make the end of life products more suitable for recovery strategies (i.e. reuse, recycling, and remanufacturing). A modern engineer of the twenty-first century should not only design technologies but also think about avoiding unanticipated consequences that may result from the design process.

The DfE approach broadly deals with the environmental impacts across five stages of the life cycle of a product (GRDC 2021), and enables industries to be more environmentally friendly in their production.

- *Material Extraction Stage*: This considers the sourcing of those materials which are non-toxic, less energy intensive, readily available and renewable resources to reduce hazardous pollution and the pressure on scarce non-renewable resources. Secondly, product design should enable the use of less materials as well, as reduce the number of parts for doing the same job so that the extraction/mining of materials can be reduced. The inclusion of multifunctional devices (MFDs) (e.g. smart phones) also helps dematerialisation.
- *Production Stage*: The product needs to be designed in a way that the energy use during the production stage can be optimiseds for example, by using heat exchangers by minimising the losses and maximising the use of insulation in walls, pipes, and ceilings, and by avoiding leaks and the oversizing of components/vessels. Chemical reactions need to be designed in a way that most of the reactants are converted to products, resulting in higher atom efficiency. The

design should also enable reduction in the amount of consumables, like grease and lubricants, as much as possible, and they should be renewable (e.g. using compressed air as coolant) and biodegradable materials.

- *Transportation, Distribution, and Packaging*: Transportation by ship and train is preferable to the transportation by lorries and air crafts to avoid air pollution. Secondly, the packaging of the product needs to be designed in a way that it accommodates the maximum number of items without exceeding the carrier's capacity in order to reduce the number of trips and associated air emissions. Thirdly, the packaging materials should be made of biodegradable and renewable materials and the industries should maximise the use of refillable or reusable containers, where appropriate. Transportation fuels for distribution purposes need to be clean, such as hydrogen, so that the emissions per tonne of kilometre travel are reduced.
- *Use*: The products need to be designed in a way that they are compatible with using power from renewable resources during the use of product. The design should enable the upgrade of the product during its use rather than designing it to become obsolete. It should be designed for easier maintenance and repairing to enhance the life span of the product. The product should minimise the use of periodic consumables such as batteries, cartridges, and containers, liquid materials for maintenance such as cooling liquid or lubricants, and any consumables containing toxic or otherwise hazardous materials.
- *End of Life*: The product needs to be designed in a modular way, where components can be disassembled for performing 6R strategies, including, reuse, recycling, recovery, remanufacturing, refurbishment, and retrofitting. The inclusion of non-toxic additives during the design process will reduce the health risk of recovery or recycling of materials at the end of its life.

8.3 Benefits of Design for the Environment

Not only do environmental benefits result from the application of the concept of DfE, but there are also economic, operational, and marketing benefits, which are as follows:

8.3.1 Economic Benefits

Maintaining Competitiveness: The application of DfE reduces energy and material consumption and waste generation during the production of a product. It enables the use of recyclable, reusable, and renewable materials, efficient and clean technologies, which also reduces operational cost. There can be some additional costs involved due to product modification (e.g. research and development, infrastructure changes in the production processes), but typically there are long-term benefits as the operational savings can offset these additional costs over a period of time. The reduction in operational cost could also reduce the price of the product and this is how industries can gain a competitive advantage by selling their products cheaper than their competitors.

Eco-efficient Product Design: As discussed in Chapter 5, eco-efficient products can do more with less. DfE considers how products can be modified to enhance eco-efficiency throughout the product lifecycle using a life cycle assessment (LCA) approach as it reviews economic and environmental hotspots or the inputs/processes with the highest economic and environmental impacts to enable manufacturers, engineers, and designers to innovate/explore and design suitable solutions to produce more environmentally friendly products without entailing excessive extra cost.

Improving Company Value: Environmental issues are gradually gaining more public attention. Industries producing products in an environmentally unfriendly way are often criticised by the public and may receive media attention, which affects company reputation. For example, Volkswagen were fined $125 million for misleading customers about their car emissions (ACCC 2021). Shell sank the Spar platform in the North Sea as its disposal cost ($19 million) was 3.5 times cheaper than dismantling it on land. Green peace campaigned against Shell's action which caused an unprecedented consumer boycott of Shell products and services (Jefferson 1995). There was a 20% drop in Shell sales in Germany alone. The improvement of the environmental performance of company products can maintain or enhance the company corporate image.

The stakeholders in the downstream process therefore tend to source more environmentally sensitive products to reduce their environmental impact. For example, a canola-based biodiesel producer will source canola seeds with a lower carbon footprint in order to reduce the overall carbon footprint from their biodiesel production.

Identifying New Business Opportunities: Will create market for recycled, remanufactured, recovered, and reused products. Secondly, the market for clean energy will enable manufacturers to reduce the impacts of products during use. Thirdly, the expansion of the end of life (EoL) waste (main bodies of the used EoL products) collection area and training to increase workers' skills can also provide technically and environmentally feasible remanufacturing solutions. The market niche for green products will facilitate the DfE activities. The manufacturer can gain both corporate image and competitive advantage as a supplier of the green products.

8.3.2 Operational Benefits

Improving Relationships with Regulators: Applying the concept of DfE will assist manufacturers in avoiding emissions and waste. This will prevent infringement notices and enable companies to renew their licence to operate and to improve risk management.

Building Competencies: DfE will help manufacturers to produce durable products with a reduced level of environmental performance in a cost-competitive manner.

Improved Staff Morale: The participation of all staff with different capacities in an industry is required to sustain environmentally friendly manufacturing activities. Staff are trained to operate clean technologies, and work collectively to prevent activities generating waste and emissions, and practice energy and material conservation measures to produce green products.

8.3.3 Marketing Benefits

Addressing Customer Needs: Manufacturers or businesses look for environmentally friendly raw materials and technologies to reduce energy consumption and enhance their environmental performance or to simply comply with environmental laws and regulations. The niche market for green chemicals, technologies, and feedstocks can also address the needs of customers in the downstream supply chains (e.g. supply of recycled concrete aggregates to builders/pavers).

Improved Public Relations: Industries producing products in an environmentally friendly manner are well accepted by the surrounding communities, environmentalists, pressure groups and regulatory bodies for complying with environmental regulations and for selling cheaper and often more durable products.

8.4 Challenges Associated with Design for the Environment

The challenges associated with the application of the concept of DfE are as follows:

Shortening of the Life Span of Products: This design strategy discourages resource conservation as it does not allow the product to be used for a long period of time. The hardware and software of many electronic items are upgraded so quickly that it shortens the life of the product and compels customers to buy new ones. For example, a mobile phone might be just 6 months old and suddenly there is a software update which makes the phone very slow. An increased number of throw-away products are being developed, which by their very nature, has led to increased material and energy cycles. The pace of replacement is so quick that the borderline between durable goods and throw-away products is disappearing so fast. Secondly, electronic goods can be equipped with components with inferior quality, requiring constant repairing (Behrendt et al. 1997). The repairing may be possible, but it is cheaper to buy a new one than to repair it. The company may choose materials and manufacturing techniques that create a lower quality product. The product can be made in a way that it is difficult or cost-prohibitive to repair. Thirdly, the financial implications of shortened product life may appear daunting to a majority of individuals as they struggle financially to fulfil the myriad of consumption expectations.

Shortening of Innovation Cycles: Due to the shortening of product life cycles, the need for a rapid design process has become an emerging issue. An important cause of shorter life cycles is that the innovation cycle is becoming shorter. As a result, products quickly become obsolete, despite their technical life span not being reached. Shorter lifespans have usually been defended on the grounds of promoting technological innovation, business growth, and healthy economics (De Beukelaer and Spence 2018); however, these occurrences have also been linked to negative environmental consequences like resource depletion, pollution, and greenhouse gas emissions.

Fashion: This contributes to increasing consumption of products. Due to changes in fashion, product completely changes, which do not allow the upgrading of the existing product. As a result, new products come to the market and the existing products become obsolete very quickly.

Inbuilt Obsolescence: As discussed earlier, companies tend to plan obsolescence by designing a product with a limited useful life. This requires the consumer to buy a replacement more often. It creates a style that will have limited choice and contemporary appeal. The manufacturers change the way people interact with the product so that previous versions become unusable or unfashionable.

Insufficient Collection, Return, and Reuse of Logistics: One drawback of the application of 6R strategies lies in the organisation of the registration and collection of old appliances. The customer needs to be encouraged to register on-line or via e-mail, to show their intention to return a product. In this way, the firm acquires information in advance on which products are to be returned. It could be very useful if this information is registered or accessed. The software developed by the German recycler covertronic can read the configuration of a computer and compute costs and revenues of subsequent recovery (de Brito et al. 2002). Based on this, an appropriate bonus is offered to the final user when the computer is returned. Reverse logistics in this case are very important as they involve the movement of products or materials in the opposite direction for the purpose of creating or recapturing recovery value, or for proper disposal.

Low Use of Renewable Resources: Non-renewable resources are still being considered as the dominant source of materials for product design. The modern economy results in the excessive use of non-renewable resources. Resources are finite in nature and it is important to either slow down the use of non-renewable resources using R strategies (e.g. reuse, recycling, remanufacturing, recovery) or to maximise the use of renewable resources. There are technical difficulties in maximising the use of renewable resources in appliances, machinery, and infrastructure and therefore, products need to be designed in a way that maximises the use of recycled, reused, or recovered non-renewable resources. This is because these non-renewable energy resources are depleting rapidly with the increased dependence on technology. Of the 77 elements studied, human activities dominate 54 elements meaning that they have already been excessively exploited to run the modern economy, and nature dominates only 23 elements (Klee and Graedel 2004). The World model shows that recycling will be the dominant source of iron, nickel, and aluminium after 2080, 2035, and 2020, respectively (Ragnarsdóttir and Sverdrup 2015). Because of the relatively efficient recycling of these metals, annual supply is much larger than the actual mining rate.

Burn-off time gives an order of magnitude indication for how long the reserves will last and is defined as estimated extractable reserves divided by the present net yearly extraction rate or the ratio between reserves estimated to be extractable, and present consumption (expressed in years). Out of 25 widely used materials in the table below, about 11 materials have a burnout time of less than 50 years and 17 materials are expected to become scarce after 2050 (Ragnarsdóttir and Sverdrup 2015) (Table 8.1). Therefore, it is important to take into account in design process that these materials are easily recovered in a technically and economically feasible way.

Energy return on investment (EROI) of oil is sharply increasing due to the excessive use of oil in the power and transport sectors. For the same investment, the supply of oil decreased from 100 years to around 40 years in the United States in 2005, and is below 10 years for deep oil and 2 years for Shell oil (Ragnarsdóttir and Sverdrup 2015).

End of Life Management/Disposal: The use of some chemicals like flame retardants and chlorinated plastics can potentially cause eco-toxicity once they are disposed of along with the product into landfill.

A manufacturing process involving both complex and composite material construction, utilising a variety of materials, and harmful substances such as additives, makes it difficult to disassemble for reuse, remanufacturing, recovering, and recycling. Especially when a product predominantly consists of electronic items, it has to be thrown away as electronic devices are made of composite materials. Due to the existence of harmful additives, there are toxic emissions affecting the health of recyclers.

Table 8.1 Material scarcity in near future (Ragnarsdóttir and Sverdrup 2015).

			Scarcity risk			
Material	**Burn-off, years**	**Peak supply**	**2025**	**2050**	**2100**	**2200**
Iron	214	2080	No	No	No	Yes
Aluminium	478	2070	No	No	No	No
Copper	31	2115	No	Yes	Yes	Yes
Chromium	225	2060	No	Yes	Yes	Yes
Manganese	29	2060	No	Yes	Yes	Yes
Nickel	42	2065	Yes	Yes	Yes	Yes
Germanium	100	—	No	Yes	Yes	Yes
Gallium	500	2040	No	Yes	Yes	Yes
Indium	19	2060	No	Yes	Yes	Yes
Zinc	20	2095	No	Yes	Yes	Yes
Lead	500	2095	Yes	Yes	Yes	Yes
Antimony	25	2050	No	Yes	Yes	Yes
Lithium	25	2150	No	Yes	Yes	Yes
Rare earths	660	2440	No	No	No	Yes
Gold	37	2025	Yes	Yes	Yes	Yes
Helium	9	—	Yes	Yes	Yes	Yes
Silver	14	2090	No	Yes	Yes	Yes
Platinum	73	2060	Yes	Yes	Yes	Yes
Palladium	61	2060	Yes	Yes	Yes	Yes
Oil	44	2040	Yes	Yes	Yes	Yes
Coal	78	2065	No	No	Yes	Yes
Natural gas	64	2015–2020	Yes	Yes	Yes	Yes
Uranium	144	2220	No	No	No	Yes
Thorium	187	2370	No	No	No	Yes
Phosphorus	161	2160	No	No	Yes	Yes

8.5 Life Cycle Design Guidelines

Life cycle design guidelines help achieve DfE with the continuous application of environmental improvement strategies to the design of products, considering production, distribution, consumption, and disposal systems, with a view to reducing the net environmental burden caused during the life cycle stages of the product.

Continuous improvement of the design of products, services, and systems can be attained through the application of sound industrial design principles and practice, with simultaneous consideration of different, but partially overlapping, strategies and the information on the environmental impacts of the product or service over its product life cycle. The life cycle design strategies are discussed as follows.

Achieving Environmental Efficiency: This is related to the concept of eco-efficiency discussed in Chapter 5. The main aim is to produce more goods or services with reduced levels of environmental degradation. An example could be hydrogen fuel which is produced by a renewable-energy-powered electrolyser and the water that is supplied to the electrolyser is produced from a renewable energy powered desalination plant. This is because seawater desalination is an energy-intensive process, where the use of renewable and clean electricity to power the system avoids the emissions from electricity generated from fossil fuels. Secondly, seawater is abundant in nature and its use to produce water for hydrogen production could avoid the sourcing of water from groundwater and fresh surface water sources. Thirdly, the combustion of hydrogen does not produce any environmental emissions.

Saving Resources: Reduction, reuse, recycling, recovery, remanufacturing, and redesign strategies can help avoid the use of virgin materials, which avoids the need for extraction, processing, and manufacturing of raw materials. Secondly, efficiency improvements in technology can reduce the use and sourcing of non-renewable resources to generate energy or electricity.

Use of Renewable and Sufficiently Available Resources: Wind, solar, wave, and biomass energy can now be used to produce electricity cost effectively to replace non-renewable energy sources like coal, gas, oil, and nuclear energy. These renewable energy technologies are so advanced in terms of efficiency and usability that they can potentially be able to meet a significant portion of energy demand for automotive, refinery, transportation, building, and mining industries. Besides, renewable resources like bamboo and wood, can replace energy intensive construction materials while maintaining the required level of structural performance. However, this depends on the availability of renewable resources which vary across regions.

Increasing Product Durability: As discussed in Chapter 7, the use of more durable products can slow down the rate of material flow through economic sectors. If the service life of a building is 70 years instead of 50 years, quarrying, processing, and the manufacturing of brand new/virgin construction materials can be delayed by 20 years. This also slows down the degradation of land and the exhaustion of virgin materials and associated energy and material consumption during the upstream processes of the product supply chain.

Design for Product Reuse: Glass milk bottles are classic examples of a product that can be reused after sterilisation. Printer cartridges can be found to be reused and refilled. Some products have filters that can be washed rather than disposing of them to landfill. Designers need to consider repairing strategies with few replacements of parts to give EoL products a new life.

Design for Material Recycling: The largest single end-use of steel for reinforcement purposes in concrete increased from 210 Mte in 2008 (Cooper and Allwood 2012) to 947 Mte in 2019 (Kwon et al. 2021) (approximately one-fifth of all steel). This reinforcing steel is not only energy intensive but it is also a non-renewable resource. Currently, conventional concrete design does not allow the reuse of these rebars as they cannot be separated from the aggregates, cement, and sand. Iron ore, which is a feedstock for rebar production, is expected to run out by 2070. At present, it is difficult to recover steel bars undamaged from concrete, but by developing modular precast designs that are easier to disassemble and reuse, greater reuse may occur in future.

Design for Disassembly: As discussed in Chapters 5 and 7, ideally products should not be made of composite materials as they are very complicated to separate. Design for disassembly can help the remanufacturing of end of life products. The remanufacturing or rebuilding process often includes upgrading that returns a used product to at least the original equipment manufacturer (OEM) performance specification, which gives the resultant product a warranty that is often at least equal to that of a newly manufactured equivalent (Smith and Keoleian 2004). This differs from the recycling of part materials in that the value added items during the original fabrication, including labour, energy, and equipment expenditures, are conserved or avoided. Therefore, remanufacturing can be regarded as a sustainable manufacturing process as it focuses on enhancing use-productivity in the total life cycle, where the product can be used over and over again (Seliger et al. 2008). For example, a compressor can be completely disassembled, which makes it easier for remanufacturing. In one study (Biswas and Rosano 2011), 96.5% of the total parts were reused and the rest were replaced by new parts (3.5%). The remanufacturing consisted of five stages: disassembling, cleaning and washing (C&W), machining, reassembling, and testing. Biswas and Rosano (2011) found that the remanufactured compressors of 15 horsepower produce about 89–93% less greenhouse gas (GHG) emissions than those associated with a new (OEM) compressor and are also 50% cheaper than a new compressor.

Minimising Harmful Substances: Products need to be made of environmentally benign materials. Australians began to use asbestos cement (fibro) as a house-building material at the beginning of the twentieth century to meet the demand of the post-war housing boom. Fibro was commonly sold as fireproof, durable, easily transportable, and suitable materials for Australia's temperate climates. Above all, it was relatively cheaper than other cementitious materials. However, workers were found to suffer from asbestosis, lung cancer, and mesothelioma about 20–30 years after their first exposure to asbestos. As a result, asbestos has been completely banned in Australia since December 2003.

Environmentally Friendly Production: Dry machining, for example, is obviously the most ecological form of metal cutting as there are no environmental impacts from coolant use or disposal to consider. Traditional coolants that used to keep the cutting tool cool during the machining operation have been replaced with minimum quantity liquid (MQL). A small to medium sized enterprise (SME) in Western Australia has reduced the GHG emissions and eco-toxicity associated with the disposal of the contaminated coolant liquids, while increasing the performance of the metal cutting operation (Ginting et al. 2015). Another example can be the use of cleaner consumables, where synthetic paints have been found to be replaced with vegetable based paints, using linseed oil as a non-toxic solvent, and sourced within 30 km of a factory to avoid transportation emissions from feedstock. Manufacturing involving chemical reactions should be designed to increase the atom efficiency by reducing the number of steps/chemical reactions using green chemistry principles such as process intensification, catalysts, and bio-technology.

Minimising Environmental Impacts of Products in Use: This involves a product being designed in a way that it consumes less energy during the use stage and requires no or very minimum amounts of chemicals/consumables for maintenance. Designing energy efficient products mitigates emissions during the use stage by reducing fossil fuel generated electricity consumption.

Using Environmentally Friendly Packaging: Packaging needs to be designed in a way that is easy to recycle, is non-hazardous and is made of recycled materials. The materials and manufacturing practices for packaging should also have minimal impact on energy consumption and natural resources. It is also important to consider biodegradability, insulating quality, and the weight and volume of the packaging. It needs to be reusable, and has to be designed in a way that it can contain more items than conventional packaging so that it can reduce the number of trips for transporting goods. These strategies help lessen the amount of product packaging, promote the use of renewable/reusable materials, cut back on packaging-related expenses, eliminate the use of toxic materials in the production of packaging, and provide options to recycle packaging more easily.

Environmental Disposal of Non-recyclable Materials: Energy recovery from waste is a common practice to handle non-recyclable wastes. These wastes are converted to heat, electricity, or fuel through a variety of processes, including combustion, gasification, pyrolisation, anaerobic digestion, and landfill gas recovery. This process is often called waste to energy (W2E). Once non-recyclable wastes are burnt to produce electricity, emissions and ash (or incinerated bottom ash/IBA) are produced. These emissions are treated through a number of air pollution control technologies before they are released to the environment (NBCI 2000, NAP 2000). Firstly, NO_x is reduced by injecting ammonia or urea into the furnace via jets. This treatment is done immediately after the combustion because it requires temperatures of 1600–1800 °F, which are attained immediately after the combustion of wastes. Then dioxins and furans are removed the exhaust gas along with mercury by injection of activated carbon powder. Then acid gas (HCl, SO_2) with single stage dry injection of sodium bicarbonate techniques and scrubbers and particulate are successively used to achieve the highest degree of reduction of mercury, dioxins and furans, particulates and acid gases. Incinerated bottom ash which is taken out using the conveyor belt from the W2E plant can potentially be used in construction materials.

Optimisation of Distribution System: The product and distribution system need to be designed in a way that the transportation of products to the retailer and user occurs in an environmentally sound manner by considering reusable packaging and energy efficient transport modes, logistics, and distribution systems. The inputs need to be sourced locally. It requires the implementation of environmentally friendly logistics by using clean fuels (hydrogen fuels), sourcing local inputs such as chemicals and feedstock and applying reverse logistics to move used goods from customers back to the sellers or manufacturers.

New Concept Development: Develop products based on new ways to meet specific customer needs by focusing on product function, product ownership and performance, considering dematerialisation, the shared use of product, and integration of functions and functional optimisation of the product. For example, a lecture theatre equipped with a video conferencing system could help dematerialisation by avoiding participants travelling to the conference. This saves transportation fuels and the emissions from the combustion of transportation fuels. More than 20 years ago, an office secretary used to have a printer, a fax, a scanner, and a photocopier to perform official activities. Now all these devices are in one machine which has not only achieved dematerialisation by reducing the amount of materials but has also made the operation more efficient than previous individual devices.

Optimisation of Initial Lifetime: A product needs to be developed with extended technical and prolong aesthetic lifetimes. A longer technical lifetime is preferable if technology is static or stable and a shorter technical lifetime preferable if new innovation will lead to improved efficiency by improving reliability and durability. Some other factors needing consideration are easier maintenance and repairing, modular product structure, user-friendliness, and product adaptability (is less complicated to operate or does not require specialised skills).

Optimisation of End-of-Life System: The product needs to be designed in a way that valuable product components are reused and others are safely disposed of. The EoL system includes collecting, sorting, recycling, and disposal of non-recyclable or unusable wastes. The design should enable the product to be easily separated and disassembled. Secondly, the manufacturer should not use chemical products that have unknown effects with end-of-life disposal. The producer can accept the product at its end-of-life for reuse using extended producer responsibility approaches discussed at the end of this chapter. There should be government policy to encourage remanufacturing of products so that more re-manufacturers enter into the market. Utilise down-cycling or up-cycling strategies. An example of down-cycling is the conversion of used car tyres to recycled aggregates for concrete production. The car tyre is a higher value product than the recycled aggregate and this results in the upgrading of the end of life tyre waste. In the case of upcycling, a suitable example would be the use of lime kiln dust (e.g. residue generated from cement production) in road construction as a high value product.

8.6 Practice Examples

8.6.1 Design for Disassembly

The Purchasing Manager of van der Laan Proprietary Limited is about to order some office equipment for their new office in Fantastic Island. They want to be a leading sustainability focused company in the region. The Purchasing Manager is in consultation with the Environmental Manager to help purchase environmentally friendly office equipment. They are now investigating whether or not to purchase a photocopier, printer, scanner, and fax equipment separately or to purchase a MFD combining all the functions. You are the only engineer on-site and have been assigned to advise as to which option is more environmentally friendly in terms of carbon saving benefits. The following information has been made available from the office equipment suppliers to enable you to conduct an environmental analysis (Table 8.2).

The weight of the MFD is 300 kg and van der Laan Proprietary Limited usually uses the equipment for 6 years before it is sent to either recyclers or remanufacturers. Therefore, you need to conduct your comparative assessment over 6 years of use of the equipment.

The emission factors for remanufactured cartridges, new cartridges, raw materials, tonne-kilometres travelled (tkm), and Fantastic Island's electricity supply are

Table 8.2 Information on office equipment.

	MFD	Photocopier	Printer	Scanner	Fax
Consumable (cartridges)	100 Remanufactured cartridges/year	40 New cartridges/year	40 New cartridges/year	15 New cartridges/year	10 New cartridges/year
Raw material	10% of the weight	50 kg	25 kg	10 kg	15 kg
Device production	20 kg CO_2 e-	100 kg CO_2 e-	75 kg CO_2 e-	50 kg CO_2 e-	50 kg CO_2 e-
Distribution	60 km	20 km	20 km	20 km	50 km
Use	0.15 MWh/year	0.1 MWh/year	0.1 MWh/year	75 kWh/year	0.05 MWh/year

2 kg CO_2 e-/remanufactured cartridge, 5 kg CO_2 e-/new cartridge, 15 kg CO_2 e-/kg of raw material, 1 kg CO_2 e-/tkm, and 0.6 kg CO_2 e-/kWh, respectively.

a) Calculate the carbon saving benefits/losses due to the replacement of one option with another.
b) How is this case study linked to design for the environment?
   Solution of (a)

Calculation of GHG emissions of a photocopier

New cartridge	40 × 5 kg CO_2/cartridge/year × 6 years = 1200 kg CO_2
Raw materials	50 kg × 15 kg CO_2/kg raw material = 750 kg CO_2
Manufacturing	100 kg CO_2
Distribution	50 kg × 1 t/1000 kg × 20 km × 1 kg CO_2/tkm = 1 kg CO_2
Energy use	0.1 MWh/year × 6 years × 0.6 kg CO_2/kWh × 1000 kWh/MWh = 360 kg CO_2
Total	2411 kg CO_2

Calculation of GHG emissions of a printer

New cartridge	40 × 5 kg CO_2/cartridge/year × 6 years = 1200 kg CO_2
Raw materials	25 kg × 15 kg CO_2/kg raw material = 375 kg CO_2
Manufacturing	75 kg CO_2
Distribution	25 kg × 1 t/1000 kg × 20 km × 1 kg CO_2/tkm = 0.5 kg CO_2
Energy use	0.1 MWh/year × 6 years × 0.6 kg CO_2/kWh × 1000 kWh/MWh = 360 kg CO_2
Total	2011 kg CO_2

Calculation of GHG emissions of a scanner

New cartridge	15×5 kg CO_2/cartridge/year $\times$ 6 years = 450 kg CO_2
Raw materials	10 kg $\times$ 15 kg CO_2/kg raw material = 150 kg CO_2
Manufacturing	50 kg CO_2
Distribution	10 kg $\times$ 1 t/1000 kg $\times$ 20 km $\times$ 1 kg CO_2/tkm = 0.2 kg CO_2
Energy use	0.075 MWh/year $\times$ 6 years $\times$ 0.6 kg CO_2/kWh $\times$ 1000 kWh/MWh = 270 kg CO_2
Total	920 kg CO_2

Calculation of GHG emissions of a fax

Remanufactured cartridge	10×5 kg CO_2/cartridge/year $\times$ 6 years = 300 kg CO_2
Raw materials	15 kg $\times$ 15 kg CO_2/kg raw material = 225 kg CO_2
Manufacturing	50 kg CO_2
Distribution	15 kg $\times$ 1 t/1000 kg $\times$ 50 km $\times$ 1 kg CO_2/tkm = 0.75 kg CO_2
Energy use	0.05 MWh/year $\times$ 6 years $\times$ 0.6 kg CO_2/kWh $\times$ 1000 kWh/MWh = 180 kg CO_2
Total	775 kg CO_2

Calculation of GHG emissions of MFD

New cartridge	100×2 kg CO_2/cartridge/year $\times$ 6 years = 1200 kg CO_2
Raw materials	0.1 $\times$ 300 kg $\times$ 15 kg CO_2/kg raw material = 450 kg CO_2
Manufacturing	20 kg CO_2
Distribution	300 kg $\times$ 1 t/1000 kg $\times$ 60 km $\times$ 1 kg CO_2/tkm = 18 kg CO_2
Energy use	0.15 MWh/year $\times$ 6 years x 0.6 kg CO_2/kWh $\times$ 1000 kWh/MWh = 540 kg CO_2
Total	2228 kg

GHG saving = 2411 + 2011 + 920 + 775 − 2228 = 3889 kg CO_2

b) This case study is an application of DfE in terms of:

New Concept Development: The MFD was developed based on new ways to meet specific customer needs. It is multifunctional as it performs multiple operations including scanning, fax, photocopying, and printing. It is also helping dematerialisation in two ways. Firstly, it is done by integrating four devices into one machine. Secondly, 90% of the old parts are reused to remanufacture this MFD.

Use of Low Impact Materials: Only 10% of the total weight of the MFD contains virgin materials, meaning that it increased the use of lower energy content recycled materials. It can do the same job as four individual devices while reducing material consumption.

Material Intensity: The weight of the virgin materials in MFD (i.e. 0.1 s 300 = 30 kg) is less than the total weight of four individual devices made of virgin materials. The integration of four devices into the MFD reduced the size of office area needed for performing official operations.

Optimisation of Production Techniques: The manufacturing of the MFD produces far less emissions than those emitted during the production of the four individual devices.

Optimisation of Distribution System: Instead of procuring four devices from four different places, the MFD can be brought from one place, thereby reducing transportation distance or tonne kilometres travelled and the emissions associated with the transportation of office equipment.

Reduction of Impact During Use: Recycled cartridges are used in the MFD. There is energy saving due to the replacement of all individual devices.

Optimisation of End-of-Life System: The MFD can be remanufactured as it is a modular design; some of its components are designed for separation (e.g. labels, plates, tray handle, cartridge plate) while some are designed for disassembling (e.g. gear train, magnet catch, rubber roll, housing).

8.6.2 The Life Cycle Benefits of Remanufacturing Strategies

Here we are assessing the environmental and economic implications of the replacement of a new 15 kW compressor with a remanufactured compressor of the same capacity with the same durability (15 years) and a weight of 167 kg.

In the case of a new compressor, you need to consider '*cradle to gate*' stages, including mining and material production, foundry processes, assembling, and transportation of a complete compressor to its destination in Malaga, Western Australia, and assess all life cycle environmental impacts. The foundry process consumes 4618 MJ of energy to manufacture a 15 kW compressor. It was manufactured in Queensland, which is 3550 km away from Kwinana, Western Australia. Once transported by ship to Kwinana, it is taken to a local supermarket to run the refrigeration unit. The distance between the port and the supermarket is 20 km.

A large number of impacts are created due to the mining, metal processing, manufacturing, and transportation of this 15 kW compressor to the supermarket. You are only estimating the carbon footprint or global warming potential (GWP) in terms of kg CO_2 equivalent (kg CO_2 e-) in this practice example.

The emission factors of inputs that are required during the cradle to gate stages of this new compressor are given in Table 8.3.

Table 8.3 Emission factors for inputs for a new compressor production.

Inputs	Values	Units
Mining operation	0.29	kg CO_2 e-/kg of compressor
Metal processing	4.0	kg CO_2 e-/kg of compressor
Foundry energy	0.19	kg CO_2 e-/MJ
Transport		
Sea	0.02	kg CO_2 e-/tkm
Road	0.06	kg CO_2 e-/tkm
Assembly	0.53	kg CO_2 e-/kW

tkm = tonne kilometre travelled.

Now the supermarket is thinking of buying a remanufactured compressor, which has the same durability, capacity, and weight, as a new compressor, which will not only save money but also avoid the carbon tax that is currently being imposed by the State Government of Western Australia (i.e. \$50/t of CO_2 e- for both goods and services). A local remanufacturing company, Sustainable Energy Solution Pty Ltd. (SES Pty Ltd.), which is the sole supplier of remanufactured compressors in Western Australia, has provided the life cycle inventory for a remanufactured compressor (Figure 8.2) in order to estimate the global warming impacts or carbon footprint of a remanufactured compressor. As you can see in Figure 8.2, materials equivalent to 99% of the total weight of the 'End of Life' (EoL) compressor are reused through the remanufacturing and re-winding process. Only components, such as the bearings, accounting for 1% of the total weight of the compressor have to be replaced with new products as these are critical components necessary for the long-term performance of the compressor.

Here you are considering a 'grave to gate' approach for estimating the environmental impacts. SES Pty Ltd. collects the EoL compressors from different industries and supermarkets, which avoids their unnecessary disposal to landfill.

Table 8.4 shows the emission factors of the inputs to produce a remanufactured compressor. Since SES Pty Ltd. is located next to the supermarket, the emissions associated with transportation of the remanufactured compressor to the supermarket are negligible and can be ignored.

a) Calculate the GWP potential in terms of kg CO_2 e- of transporting one 15 kW new compressor to the supermarket in Western Australia.

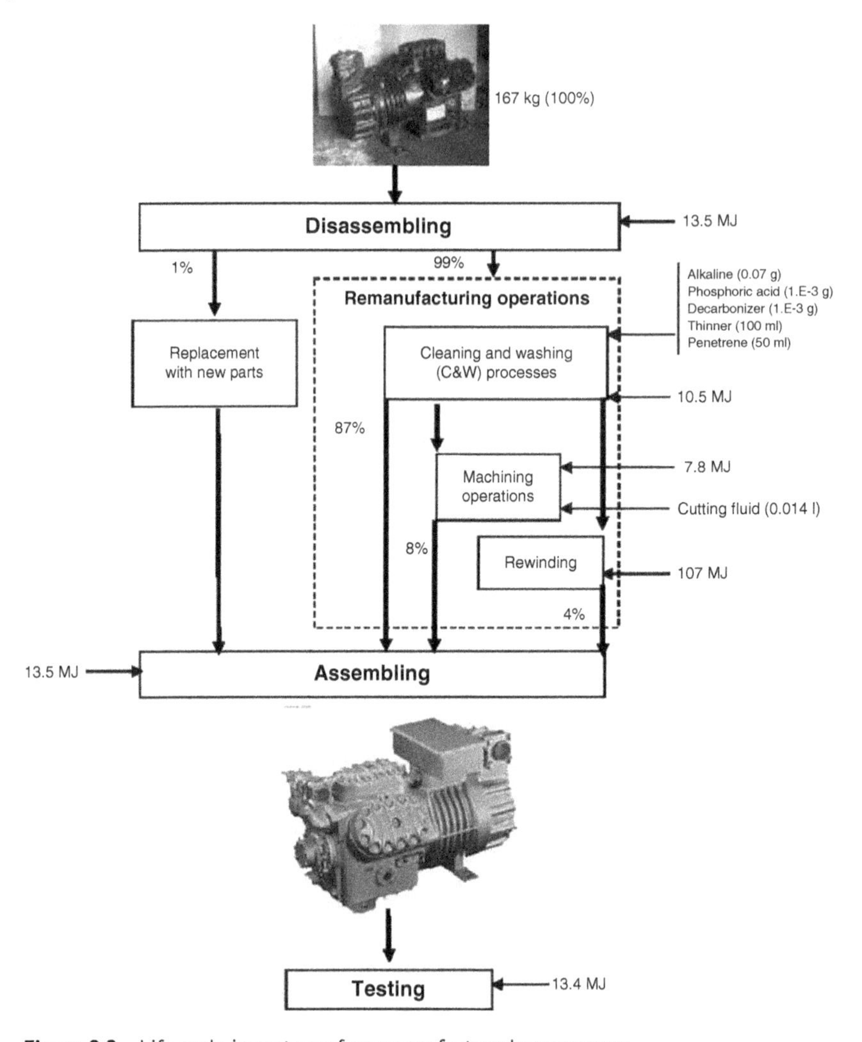

Figure 8.2 Life cycle inventory of a remanufactured compressor.

Table 8.4 Emission factors for inputs for remanufactured compressor production.

Inputs	Values	Units
Energy	0.24	kg CO_2 e-/MJ
Copper	7.6	kg CO_2 e-/kg of copper
Cutting fluid	167	kg CO_2 e-/litre of cutting fluid/lubricating oil
Chemicals	50,000	kg CO_2 e-/kg of chemical
Bearings	6.3	kg CO_2 e-/kg

Inputs	Emission factors (EF)		Inputs (I)		EF × I
	Values	Units	Values	Units	GHG
Mining operation	0.29	kg CO_2 e-/kg of compressor	167	kg	48.4
Metal processing	4.0	kg CO_2 e-/kg of compressor	167	kg	668.0
Foundry energy	0.19	kg CO_2 e-/MJ	4618	MJ	877.4
Transport					
Sea	0.02	kg CO_2 e-/tkm	592.85	tkm	11.9
Road	0.06	kg CO_2 e-/tkm	3.34	tkm	0.2
Assembly	0.53	kg CO_2 e-/kW	15	kW	8.0
			Total GHG emissions (kg CO_2 e-)		1613.9

b) Calculate the GWP potential of the remanufactured compressor in terms of kg CO_2 e-.

Inputs	Values	Units	Inputs obtained from Figure 8.2	GHG
Energy	0.24	kg CO_2 e-/MJ	Total = 165.7 MJ	39.8
Copper	7.6	kg CO_2 e-/kg	7.8 kg	59.3
Cutting fluid	167	kg CO_2 e-/l	0.014 l	2.3
Chemicals	50,000	kg CO_2 e-/kg	5.00E-05 kg	2.5
Bearings	6.3	kg CO_2 e-/kg	1.67 kg	10.5
			Total GHG emissions (kg CO_2 e-)	114.4

c) Calculate the carbon saving benefit associated with the replacement of a new compressor with a remanufactured compressor.

$$1614 - 114 = 1500 \text{ kg of } CO_2 \text{ e-}$$

d) The cost of a new compressor that has been imported from Fantastic Island, is $15,000. The cost of this new important items is 3 times more than the cost of a locally remanufactured compressor. Calculate the cost saving benefits associated with the replacement of a new compressor with a remanufactured compressor.

$$\$15,000 - \$15,000/3 = \$10,000$$

8.7 Zero Waste

DfE can potentially help achieve zero waste (ZW) by minimising waste and by converting wastes to resources. One of the ZW main aims is to achieve a design for disassembling that converts an end of life product to a usable product, thereby preventing the disposal of the product waste to landfill. The conversion of EoL to a recovered/remanufactured product not only reduces the size of the landfill but also reduces the amount of land used for extracting materials from which virgin materials are sourced. Therefore, the generation of wastes during upstream stages (i.e. mining, processing, manufacturing) and downstream stages (use and landfilling) associated with the production of virgin materials can also be avoided.

Waste generation is currently a major global environmental issue resulting from increasing population, booming economies, rapid urbanisation, and a significant rise in community living standards. The continuous depletion of natural finite resources will undoubtedly lead to an uncertain future. With our current industrial revolution of mass production, civilization has actually become synonymous with increasing wasteful habits. At present, the most-practiced energy and mass consumption options are possibly the most inefficient that mankind has ever used/employed. There are some possibilities to expand our production and consumption on the basis of net zero waste principles either at the input of any process or at the output level of any process using the concept of DfE.

There are a number of ways to measure the amount of waste reduced or diverted from landfill. Waste diversion rates have currently been used as an important indicator by local governments, researchers, waste authorities, and city corporations to measure the performance of waste management. A higher diversion rate from landfill is considered a benchmark of success.

8.7.1 Waste Diversion Rate

The diversion rate can be defined as the percentage of the total waste that is directed to reduction, reuse, recycling, and composting programs instead of being disposed of to permitted landfills and transformation facilities such as waste to energy plants. The diversion rate can be measured by either knowing the information on the amount of waste diverted or from a disposal-based measurement system.

The waste diversion rate can be expressed as follows (Zaman and Lehmann 2013):

$$\text{Diversion rate} = \frac{\text{Weight of recyclables}}{\text{Weight of garbage} + \text{Weight of recyclables}} \times 100\% \quad (8.1)$$

Recyclables = Waste that is reused, recycled, recovered, composted or digested
Garbage = Waste that is landfilled or incinerated

For example, out of 1000 t of wastes generated in a town per year, only 400 t of construction and demolition waste are recyclable as they can potentially be used in pavement as recycled aggregates. Therefore, the diversion rate is 40%.

Box 8.1 Waste diversion rate

In a small manufacturing plant, waste is generated on a daily basis. Of this waste, about 100 lb was recycled, 100 lb was composted, 100 lb was collected by another company for reuse, and 25 lb was trashed. Confirm as to whether this plant is achieving a zero waste target. Note that a zero waste strategy must have a waste diversion rate of 90% or more.

Solution

The amount of recyclables is 100 + 100 + 100 = 300 lb.

Using Eq. (8.1), the waste diversion rate has been calculated as follows:

$$\text{Diversion rate} = \frac{300}{300 + 25} = 92.3\%$$

This confirms that this plant has achieved a zero waste target.

The diversion rate as per Eq. (8.1) does not consider waste avoidance through industrial design. The diversion rate of waste management strategies may be the same, but they have different levels of implications on the product supply chain, and waste management performance. For the same rate, the diversion of waste in one area can replace more useful virgin materials than in another area.

8.7.2 Zero Waste Index

The zero waste index (ZWI) is considered a tool to measure the potential of replacing/substituting virgin materials by using ZW management strategies such as recycling, remanufacturing, composting, etc. One of the important goals of the ZW concept is the zero depletion of natural resources by avoiding upstream processes like unnecessary mining, quarrying and deforestation, as well as landfill wastes themselves. Therefore, ZW performance is the measurement of the resources that are extracted, consumed, wasted, recycled, recovered, and finally substituted for virgin materials. The ZWI is also a useful tool to compare different waste management systems The zero waste index is expressed as follows (Zaman and Lehmann 2013):

$$\text{ZWI} = \frac{\sum_{i=1}^{n} \text{WMS}_i \times \text{SF}_i}{\sum_{i=1}^{n} \text{WMS}_i} \tag{8.2}$$

WMS_i	=	Amount of waste managed by a strategy i (i.e. $i = 1, 2, 3\ldots n$ = type of waste management strategy, e.g. recycling, composting, incinerated, landfill)
SF_i	=	Substitution factor/efficiency for different waste management strategies based on their virgin material replacement efficiency

Box 8.2 A hypothetical example of ZWI

In an industrial area A, 100 t of waste is generated per year. Of this waste, about 50 tonnes of energy intensive materials were recycled, 25 t of organic wastes were composted to produce fertiliser, and the remaining waste was disposed to landfill. The substitution efficiencies of recycling and composting of these waste materials generated in this industrial area are 95% and 70%, respectively. Determine the zero waste index (ZWI) of this waste management system.

In another industrial area B, 200 t of waste is generated per year. Of this waste, about 100 t of organic wastes were composted to produce fertiliser, 25 t of energy intensive materials were recycled, 25 t are recovered and the remaining waste was disposed to landfill. The substitution efficiencies of recycling, recovery, and composting of these waste materials generated in this industrial area are 90%, 90%, and 65%, respectively. Compare the waste management performance of industrial areas A and B.

Solution

Using Eq. 8.2, the ZWI of industrial area A can be determined as follows:

$$\text{ZWI} = \frac{50 \times 0.95 + 25 \times 0.7}{100} = 0.65$$

The diversion rates for industrial areas A (75%) and B (75%) are same, but their ZWI are not the same. The ZWI of an industrial area B has been calculated as follows:

$$\text{ZWI} = \frac{100 \times 0.65 + 25 \times 0.9 + 25 \times 0.9}{200} = 0.55$$

It appears that the waste management systems (WMSs) in these two industrial areas contribute the same to the reduction of landfill size. However, the WMS in industrial area A could offer more land saving in the upstream processes by substituting more virgin materials than that in industrial area B. The ZWI of two industrial areas could be the same, but they could have different levels of environmental impacts as it also depends on the type of materials replaced and where these materials are replaced. For example, on average 700 acres-years of life time land required to produce 1 t of gold as opposed to 0.03 and 0.004 acres-year for copper and aluminium, respectively (Spitzley and Tolle 2004). Avoiding more land by converting waste to resources can avoid deforestation, loss of biodiversity, and the loss of vegetation coverage for CO_2 sequestration. Most importantly, a higher ZWI of a WMS means that it increases the circulation of materials throughout the economy and reduces the dependence on virgin materials.

8.8 Circular Economy

The introduction of DfE in engineering design can help keep materials 'circulating' in the techno-sphere for longer periods of time and is fundamental to enhance resource efficiency and to reduce the adverse impacts of our day-to-day economic activities on the environment. The actions (e.g. reuse, recycling, remanufacturing, refurbishment, cascading use, dematerialisation) of varying nature, as discussed before, need to be considered in the chain of consumption.

Our economy is still dominated by a linear system as only 9% of our world has circular economy production (PACE 2019). The Circular Gap is not closing, as most of the materials we use end up either in landfill or water bodies. This is an enormous challenge for engineers in coming up with resource conservation technologies to close this gap.

The application of cleaner production strategies (CPS), including input substitution, production technology modification, good housekeeping, and on-site recycling, as discussed in Chapter 5, can help small and medium sized enterprises to reduce energy and material consumption and waste generation (Figure 8.3). As discussed in Chapter 6, the exchange of wastes between industries through

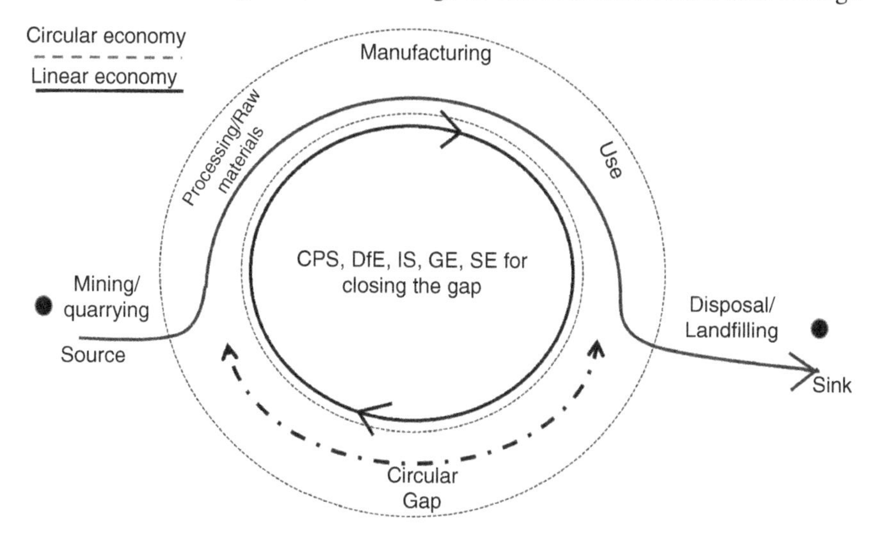

Figure 8.3 Engineering strategies for reducing circular gap.

industrial symbiotic (IS) relationships can reduce residue storage areas and avoid the use of virgin materials. The application of green engineering principles (as discussed in Chapter 7) such as process intensification, atom efficiency, energy recovery strategies, and the use of composites, and durable materials can also improve resource efficiency and reduce waste generation. Products that are designed for disassembly would enable the conversion of end of life product to a usable product, thereby reducing the pressure on virgin materials. The use of renewable energy not only reduces the energy generation from non-renewable resources but also avoids the extraction of coal and natural gas, which avoids land use in upstream processes. Using these engineering strategies, the circular gap between source and sink can be reduced to a large extent.

A circular economy aims to make better use of materials, components, and products by minimising the amount of resources taken from the natural environment, maximising the prevention of waste, and optimising their economic, social, technical, and environmental values throughout consecutive lifecycles.

Box 8.3 A hypothetical problem of circular economy

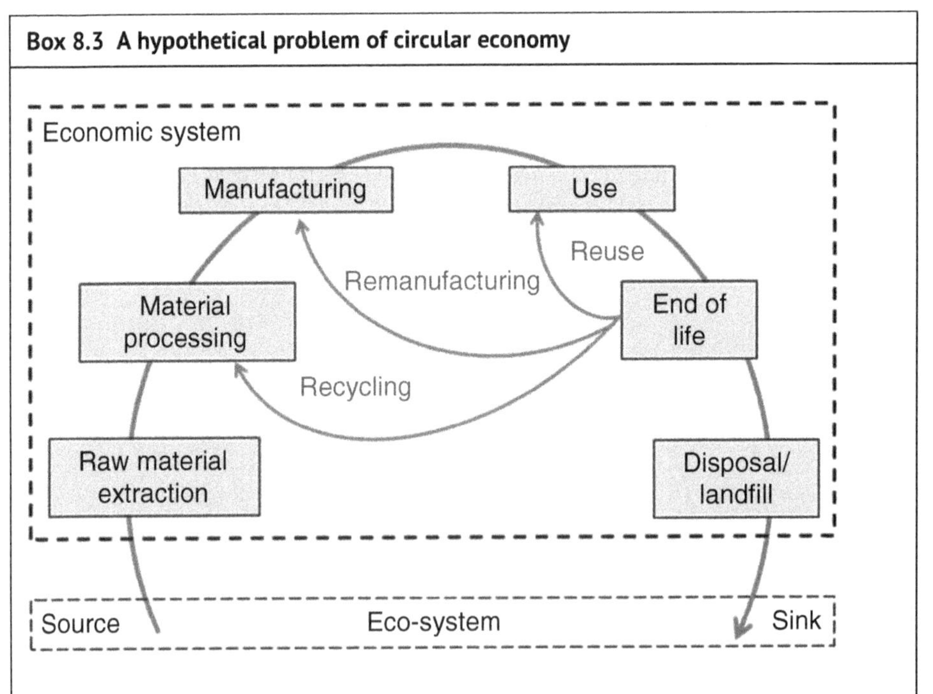

The amount of land required for raw material acquisition, material processing, and manufacturing and the assembling of an electronic device are 2 hectares (ha), 0.5 ha, and 0.1 ha, respectively. Calculate the amount of land use that can be avoided with three waste management strategies as mentioned in the above figure. Reuse, remanufacturing, and recycling of this electronic device require 0.05, 0.1, and 0.01 ha of land, respectively. The amount of landfill required is 0.01 ha if this device is disposed after the end of life.

Solution

All waste management options avoid end of life products going to landfill and so the amount required for disposing at EoL is avoided through this option.

Land conservation due to the reuse of a WMS avoids upstream activities, including raw material extraction, material processing, and manufacturing.

$$\begin{aligned}\text{Land avoided} &= 2\text{ ha (raw material extraction)} + 0.5\text{ ha (material processing)}\\ &\quad + 0.1\text{ ha (manufacturing)} + 0.01\text{ ha(landfill)} - 0.05\,Ha\ (reuse)\\ &= 2.56\text{ ha}\end{aligned}$$

In the case of a remanufacturing strategy, land use can be avoided for raw material extraction and material processing activities, while a recycling strategy can only avoid the use of land in raw material extraction. Similarly land saving for remanufacturing and recycling activities have been calculated as 2.41 ha (2 ha + 0.5 ha + 0.01 ha − 0.1 ha) and 2 ha (2 ha + 0.01 ha − 0.01 ha), respectively.

This means that different strategies can contribute differently to the conservation of land and other natural resources in upstream processes and therefore, products need to be designed accordingly to maximise resource efficiency, reduce land use changes and other associated environmental consequences like the loss of biodiversity, and vegetation coverage to sequester CO_2 emissions.

8.8.1 Material Flow Analysis

To assess circular economy (CE), theories we use material flow analysis (MFA) and LCA. LCA has already been discussed in detail in Chapters 3 and 4. MFA is a tool that helps us to understand material flow and is considered a starting point of thc environmental assessment. Furthermore, MFA reveals opportunities for improving and monitoring recycling targets.

MFA includes the comprehensive measurement of the flow of inputs and outputs in terms of kilograms or tonne for a system at a specific period of time. The input flows involve all raw resources extracted from nature. Hidden flows are products not evident from economic records which are needed to produce the resource used at the end, for example, earth-removed products (e.g. bauxite residue resulting from the production of alumina) and overburden mining materials or wood harvest reductions in forestry. Both extracted materials and hidden flows are associated with the production of a product and are called an environmental backpack or 'Rucksack', which is measured by means of a Material Intensity per Unit of Service (MIPS) following the Wuppertal Institute (Ritthof et al. 2002).

The MIPS gives the amount of materials (including energy in terms of the material required to produce it; e.g. tonnes of coal/MWh of electricity transmitted) needed for a specific benefit in mass units (kilogram or tonne). It mainly focuses on pressures (such as resource consumption, wastes, emissions) rather than impacts (such as acidification, global warming etc.) calculated following an LCA approach. The MIPS methodology was mainly developed to determine the opportunities for improving the resource efficiency of products. It also allows for the assessment of the rate of resource flow and the environmental changes associated with the extraction of resources from their natural ecosystems and the important determination of carrying capacity and ecological footprint.

MIPS can help compare the life cycle material consumption of two products doing the same job and is a useful environmental indicator. MIPS can provide the baseline for the development of different sustainability strategies across Product-Service-Systems. MIPS takes into account the movement of mass to produce a good regardless of the quality of the mass flows, and solely focusses on material movement as a pressure factor. It is strictly an input-oriented analysis quantifying the total life cycle material resources used (Material Input, MI), which are necessary to provide a specific service (S). A number of products are used to deliver a service (e.g. motor, generator, transformer, power cable and switches are used to deliver electricity at home). It is described by the following equations

$$\mathrm{MI} = \sum_i P_i MI_i \ (\mathrm{kg}) \tag{8.3}$$

$$\mathrm{MIPS} = \frac{\mathrm{MI}}{S} \left(\frac{\mathrm{kg}}{\mathrm{unit}}\right) \tag{8.4}$$

$\mathrm{MI_i}$ = Material intensity of product part (kg/kg) of product i

$\mathrm{P_i}$ = Mass of product part (kg) of product i

S = Unit services per product

In the MIPS concept, the material inputs are divided into five different input categories.

a. Abiotic material
   - Mineral raw materials (e.g. ores, sand, gravel, slate, granite)
   - Fossil energy carriers (e.g. coal, petroleum oil, petroleum gas)
b. Unused extraction (overburden, gangue etc.)
   - Soil excavation (e.g. excavation of earth or sediment)

c. Biotic material
   - Plant biomass from cultivation
   - Biomass from uncultivated areas (e.g. plants, animals)

d. Water
   - Surface water
   - Ground water
   - Deep ground water (subterranean)

e. Air

The categories 'water' and 'air' are examined separately and are not be aggregated with others. for example, air used to aid combustion in an internal combustion engine, and chemical reactions during the production process and water used in processing and cooling purposes.

Another parameter is the ecological rucksack (ER), which is defined as the total quantity (in kg) of materials removed (TMR) from nature to create a product or service, minus the actual weight (W_p) of the product. TMR considers the entire production process, from the cradle to the point when the product is ready for use.

This is mathematically presented as

$$\mathrm{ER} = \mathrm{MI}\,(\mathrm{TMR}) - W_\mathrm{p} \tag{8.5}$$

8.8.2 Practice Example

A simple problem for providing better understanding of MIPS is given below. MIPS follows the same approach as LCA except that it calculates the amount of material consumed instead of the environmental impacts made during the life cycle of a product or service.

Table 8.5 shows the LCI consisting of inputs used during the life cycle of carpet sweeper and carpet cleaner. The raw data used in this example were sourced from the Wuppertal Institute report (Ritthof et al. 2002). While both products provide the same service, which is to clean carpet, it is important to determine which one is more resource efficient in terms of delivering the actual cleaning service.

Table 8.6 shows the material factor (MF) values of the inputs in the life cycle. MI are subdivided into five different categories. It is not always the case that a material will consist of all these categories. In the case of LCA discussed in Chapter 3, emission factors are multiplied by the corresponding inputs to determine environmental impacts (i.e. kg CO_2 e- or kg SO_2 e- per kg of gold). For MIPS, material factors (i.e. kg/kg, kg/kWh) are multiplied by the corresponding inputs to determine the material intensity.

Using Eq. (8.3), the material factors are multiplied by the corresponding material/energy values in the life cycle inventory to determine the material intensity (MI) of these cleaners (Table 8.7).

Table 8.5 Material factors.

Options	Carpet sweeper	Carpet cleaner
Service life (years)	30	10
Weight	3.27	4.42
Life cycle inventory		
Materials (kg/appliance)		
Steel	3.15	1.85
Plastic	0.04	2.1
Aluminium		0.247
Copper (primary)		0.06
Copper (secondary)		0.06
Oil		0.002
Tin		0.001
Renewable (cotton)	0.08	0.1
Use (per year per appliance)		
Brush (renewable resource (cotton)) kg	0.08	
Cleaning agent (renewable resource) kg	0.1	
12 Dust bags		12[a)]
Electricity (kWh/year)		108
Recycling/disposal (km)		
Distance from the recycling station	50	50

a) Each dust bag weighs 0.1 kg and is made of paper.

These appliances are used three times a week for 48 weeks a year. The MIPS values for water and air for one time use of the sweeper and carpet cleaner are calculated using Eq. (8.4)

$$\text{MIPS}_{\text{sweeper}} \frac{79\text{ kg}}{\frac{3\text{ times}}{\text{week}} \times \frac{48\text{ weeks}}{\text{year}} \times 30\text{ years}} = 0.02$$

$$\text{MIPS}_{\text{cleaner}} \frac{5235\text{ kg}}{\frac{3\text{ times}}{\text{week}} \times \frac{48\text{ weeks}}{\text{year}} \times 10\text{ years}} = 3.63$$

Table 8.6 Material factors of LCI inputs.

	Abiotic	Biotic	Erosion	Water	Air
Material (kg/kg)					
Steel	6.97			44.6	1.3
Plastic	5.4			64.9	2.1
Aluminium	85.38			1378.6	9.78
Copper (primary)	500			260	2
Copper (secondary)	9.66			105.6	0.72
Oil	1.5			11.5	0.03
Tin	6800				
Renewable (cotton)	8.6	2.9	5.01	6814	2.74
Renewable (cleaning agent)	6			98	0.7
Renewable (paper)	1.2	5		14.7	0.24
Electricity					
kg/kWh	4.7			83.1	0.6
Transport (kg/tkm)					
Road goods transport without infrastructure (here: truck-trailer >8 t)	0.107			0.927	0.1
Infrastructure	0.749			5.16	0.017

tkm = tonne kilometres travelled.

Ecological rucksacks for carpet sweeper and carpet cleaner are calculated using Eq. (8.5)

$$ER_{\text{carpet sweeper}} = 79.81 - 3.37 = 76.54\ \text{kg}$$

$$ER_{\text{carpet cleaner}} = 5234.78 - 4.42 = 5230.36\ \text{kg}$$

Since water and air are calculated separately, the calculation of their MIPS for both appliances (Table 8.8) is based on the information in Tables 8.5 and 8.6.

Analysis: The MIPS and ecological rucksack of a carpet cleaner is higher than that of a carpet sweeper, meaning that the former creates a greater environmental burden in terms of land use change than the latter during its life time. Also the service life of a carpet cleaner is three times lower than the carpet sweeper, which indicates that we will need three carpet cleaners during the life time of a carpet

Table 8.7 Material intensity per appliance.

	Carpet sweeper	Carpet cleaner
Materials		
Steel	21.96	12.89
Plastic	0.22	11.34
Aluminium	0.00	21.09
Copper (primary)	0.00	30.00
Copper (secondary)	0.00	0.58
Oil	0.00	0.00
Tin	0.00	6.80
Renewable (cotton)	1.32	1.65
Sub-total	22.17	84.36
Use		
Brush renew	39.62	
Cleaning agent renew	18.00	
12 dust bags		74.40
Electricity		5076.00
Sub-total	57.62	5150.40
km recycling/disposal		
Road goods transport without infrastructure	0.02	0.02
Infrastructure	0.12	0.17
Sub-total	0.14	0.19
MI (kg/appliance)	**81.13**	**5234.78**

sweeper unless the former has been designed for disassembly for remanufacturing. The breakdown of MI shows that electricity consumption accounts for the largest portion of the total material consumption, which significantly increased the MIPS of the carpet cleaner. Firstly, electricity is not a once off use item as it is used throughout the life of the carpet cleaner. Secondly, this electricity is produced from fossil fuel materials (i.e. coal, gas) which needs to be mined by removing soils or by creating overburden. All these non-renewable resources need to be refined prior to use for electricity generation. One way of avoiding the 'ecological rucksack' could be either to reduce the use of fossil fuel significantly or to use 100% renewable energy like solar or wind. It should be noted that the same carpet cleaner can have a different MIPS or ecological rucksack in different areas due to variations

Table 8.8 MIPS of water and air of carpet cleaners.

Inputs	Carpet sweeper		Carpet cleaner	
	Water (kg)	Air (kg)	Water (kg)	Air (kg)
Materials				
Steel	140.49	4.10	82.51	2.41
Plastic	2.60	0.08	136.29	4.41
Aluminium			340.51	2.42
Copper (primary)			15.60	0.12
Copper (Secondary)			6.34	0.04
Oil			2.3E−02	6.0E−05
Tin				
Renewable (cotton)	545.12	0.22	681.40	0.27
Use (over service life)				
Brush renew	16,353.60	6.58		
Cleaning agent renew	294.00	2.10		
12 dust bags			8176.80	3.29
Electricity			8974.80	64.80
End of life				
Road goods transport without infrastructure	0.15	0.02	0,2	2.2E−02
Infrastructure	0.84	2.78E-03	1.14	3.8E-03
MI (kg/appliance)	17,336.80	13.09	10256.46	74.78
MIPS (kg/cleaning)	4.01	3.03E−03	2.37	0.02

in energy mixes in different regions. For example, renewable energy accounts for around 40% of total electricity in Denmark, while fossil fuel accounts for 85% of the electricity mix in Australia (European Commission 2020; Australian Government 2021). Another key message that is derived from this example is that technological sophistication creates more pressure on the environment and will put more pressure on future generations. Land use changes associated with the consumption of materials causes loss of biodiversity and the potential extinction of species, and can affect the provision of ecosystem services like food, water, and clean air.

8.9 Extended Producer Responsibilities

Extended producer responsibility (EPR) facilitates the implementation of DfE goals in a way that the producer's responsibility for a product is extended to the post-consumer stage of the product's life cycle. This means that the producer does not finish the relationship with the customer after selling a product but keeps the relationship by potentially maintaining and operating the product. An EPR policy is thus characterised by the shifting of responsibility (physically and/or economically, fully or partially) upstream towards the producer or to a service business model that promises benefit not just to the participating businesses but also to the entire economy (Dalhammar et al. 2021). Elevator giant Schindler prefers leasing vertical transportation services than selling elevators' because leasing lets it capture the savings from its elevators lower energy and maintenance costs (Gunningham 2009). These elevators are taken back after the leasing period for remanufacturing, thereby reducing operating cost and environmental impact.

EPR helps customers to reduce their need for capital goods like carpets or elevators, and rewards suppliers by extending and maximising their asset values. The adoption of this service model benefits the customer in that it will reduce unnecessary turnover of capital goods and improves the cost efficiency of the service.

Considering 'dematerialisation' initiatives (i.e. avoiding virgin materials, turning EoL products to usable products), several multinational companies successfully de-couple turnover and profits from resource consumption and manufacturing volumes. Under EPR, the products used to perform the service remain the property of the service company. The product is taken back after EoL use and cleaned or remanufactured prior to its reuse. As mentioned in Chapter 5, Fuji Xerox is offering custom-made copy reproduction services instead of just selling photocopiers.

EPR policy has become a popular tool that holds producers liable for their EoL products. It requires OEMs to take responsibility for the entire life cycle of a product, including collection, dismantling, and recycling at the end of the product life. EPR aims to put the responsibility on the producers to manage the EoL waste. It also sets out to improve product design to enable the reuse of EoL product to reduce costs. This creates a provision of incentives for producers to take environmental considerations into account in designing their products. In doing this, EPR also forces producers to consider environmental impacts from their operations and promotes waste management by incentivising producers to consider environmental protection and implement effective recycling processes. EPR has been widely adopted in Europe and other developed countries, resulting in improved resource recovery from waste material.

Switzerland is the first country within the European Union to implement an EPR policy to manage electrical and electronic equipment waste (widely known as WEEE) effectively (Wang et al. 2017). In 1998, the Swiss government enacted the Ordinance on the Return, the Take-Back, and the Disposal of Electrical and Electronic Equipment by autonomously forming Producer Responsibility Organizations (PROs).

Switzerland uses a fund operating model in which the consumer ultimately bears the responsibility for paying the fund. PRO charge a visible advance recycling fee (ARF) to a producer, and specific information about this fee is clearly annotated in the product invoice and subsequently, this fee accompanies the product invoice through various levels of resellers successively and is ultimately paid by the consumer. The ARF collection fee is the difference between the expenses from all reverse logistics and disassembly and the value of the disassembled products. PROs, which are well supported by the Swiss government, are also responsible for improving transparency in fund collection/distribution and reduce the degree of monopoly in PROs.

Under this EPR program, consumers can trade in generated WEEE through retailers, recovery network nodes in public places, and recovery facilities free of charge. Retailers are responsible for clearly noting the specific amounts of ARF in the invoices they give to consumers. Recovery network nodes are responsible to ensure that the recovered items are not stolen or illegally transferred across borders. The Swiss government is responsible for supervising the operation of the EPR system.

As can be seen in Figure 8.4, the ARF that is collected by the retailer is spent on collectors, carriers, and recyclers to deliver EoL products in a useful form (e.g. core or the used products/parts, are essential resources for remanufacturing) to the manufacturer or producer. Also some of this fund is used by the manufacturers to

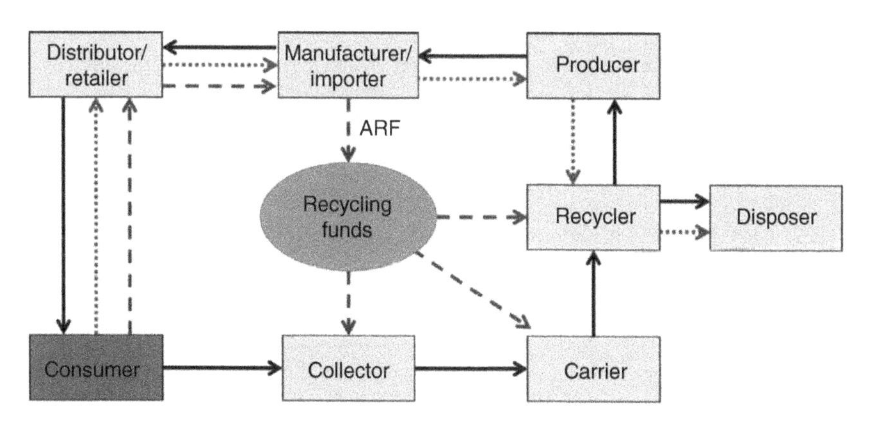

Figure 8.4 Swiss EPR model.

make final products. Although the ARF appears to be an additional cost to the customer, the overall cost of the product is not increased because there are operational savings with the avoidance of expensive virgin materials and other EoL waste charges.

References

ACCC (Australian Competition and Consumer Commission) (2021). High Court denies Volkswagen leave to appeal $125 million penalty. www.accc.gov.au/media-release/high-court-denies-volkswagen-leave-to-appeal-125-million-penalty (accessed 15 December 2021).

Australian Government (2021). Energy supply. Department of Industry, Science, Energy and Resources. www.energy.gov.au/government-priorities/energy-supply (accessed 22 December 2021).

Behrendt, S., Jasch, C., Peneda, M.C., and van Weenen, H. (1997). *Life Cycle Design: A Manual for Small and Medium-Sized Enterprises*. Springer.

Biswas, W. and Rosano, M. (2011). A life cycle greenhouse gas assessment of remanufactured refrigeration and air conditioning compressors. *International Journal of Sustainable Manufacturing* 2 (2/3): 222–236.

de Brito, M.P., Flapper S.D., and Dekker, R. (2002). Reverse logistics: a review of case studies. Econometric Institute Report EI 2002–21. Erasmus University Rotterdam.

Cooper, D.R. and Allwood, J.M. (2012). Reusing steel and aluminum components at end of product life. *Environmental Science & Technology* 46: 10334–10340.

Dalhammar, C., Wihlborg, E., Milios, L. et al. (2021). Enabling reuse in extended producer responsibility schemes for white goods: legal and organisational conditions for connecting resource flows and actors. *Circ. Econ. Sust.* 1: 671–695. https://doi.org/10.1007/s43615-021-00053-w.

De Beukelaer, C. and Spence, K. (2018). *Global Cultural Economy*. Routledge.

European Commission (2020). Renewable Energy Progress Report. Report from the commission to the European parliament, the council, the European economic and social committee and the committee of the region. Brussels.

Ginting, Y., Boswell, B., Biswas, W., and Islam, N. (2015). Advancing environmentally conscious machining. *Procedia CIRP* 26: 391–396.

GRDC (The Global Development Research Centre). (2021) Urban Environmental Management. https://www.gdrc.org/uem/lca/lca-for-cities.html (accessed 18 December 2021)

Gunningham, N. (2009). *Corporate Environmental Responsibility*. London: Routledge.

Jefferson, T. (1995). The ocean, like the air, is the common birthright of mankind. *Environmental Health Perspectives* 103 (9): 786–787.

Klee, R.J. and Graedel, T.E. (2004). ELEMENTAL CYCLES: a status report on human or natural dominance. *Annual Review of Environment and Resources* 29: 69–107. https://doi.org/10.1146/annurev.energy.29.042203.104034.

Kwon, K., Kim, D., and Kim, S. (2021). Cutting waste minimization of rebar for sustainable structural work: a systematic literature review. *Sustainability* 13: 5929. https://doi.org/10.3390/su13115929.

NAP (The National Academies Press) (2000). Incineration Processes and Environmental Releases. https://www.nap.edu/read/5803/chapter/5 (accessed 10 December 2021)

NBCI (National Center for Biotechnology Information) (2000). Incineration Processes and Environmental Releases. National Center for Biotechnology Information, U.S. National Library of Medicine.

PACE (The Platform for Accelerating the Circular Economy) (2019). The Circularity Gap Report. https://www.legacy.circularity-gap.world (accessed 12 December 2021).

Ragnarsdóttir, K.V. and Sverdrup, H.V. (2015). Limits to growth revisited. *The Geological Society* www.geolsoc.org.uk/Geoscientist/Archive/October-2015/Limits-to-growth-revisitedh.

Ritthof, M., Rohn, H.A., and Liedtke, C. (2002). *Calculating MIPS Resource Productivity of Products and Services.* Environment and Energy at the Science Centre North Rhine-Westphalia: Wuppertal Institut for Climate.

Seliger, G., Kim, H.-J., Kernbaum, S., and Zettl, M. (2008). Approaches to sustainable manufacturing. *International Journal of Sustainable Manufacturing* 1 (1–2): 58–77.

Smith, V.M. and Keoleian, G.A. (2004). The value of remanufactured engines life-cycle environmental and economic perspectives. *Journal of Industrial Ecology* 8 (1–2): 193–221.

Spitzley, D.V. and Tolle, D.A. (2004). Evaluating land-use impacts selection of surface area metrics for life-cycle assessment of mining. *Journal of Industrial Ecology* 8 (1–2): 11–21.

Wang, H., Gu, Y., Li, L. et al. (2017). Operating models and development trends in the extended producer responsibility system for waste electrical and electronic equipment. *Resources, Conservation and Recycling* 127: 159–167.

Zaman, A.U. and Lehmann, S. (2013). The zero waste index: a performance measurement tool for waste management systems in a 'zero waste city'. *Journal of Cleaner Production* 50: 123–132.

9

Sustainable Energy

9.1 Introduction

Energy is an indispensable commodity of human life. We are currently meeting this basic need with very significant carbon emissions. Still, nonrenewable or finite resources, which take thousands of years to form and are the main cause of climate change, account for a significant proportion of the global energy mix. Increased population growth and the rapid increase in number of different types of energy technologies have increased the rate of energy consumption.

The growth of energy consumption is now faster than the growth of the human population. An addition of one person to the world population adds energy consumption across a number of sectors, including mining, processing, commercial, electricity, industrial, agricultural, transport, and residential sectors to meet demand (housing, food, clothes, health). During 2000–2015, the global population increased by 25%, while electricity consumption increased by 57% (EIA 2020). Electricity is the most convenient form of energy used widely in buildings for lighting and other end-use appliances, in industrial processes for producing goods, and in transportation for powering rail and light-duty vehicles. However, energy is lost during the conversion of primary energy (e.g. coal) to electricity (i.e. secondary energy), and emissions are generated during this conversion process. Energy consumption is quite steady in developed nations, but it is rising very rapidly in developing economies. For example, in the Guangzhou region in China, for a 20% increase in population size, energy consumption increased by 43.6% (Huang et al. 2018). If this population size is decreased by 20%, the energy demand could be decreased by 44.6%. Energy consumption is very sensitive to population growth.

While a significant portion of human-induced greenhouse gases come from the burning of fossil fuels, they are extremely attractive energy sources as they are highly concentrated and relatively easy to distribute: compare the energy intensity of solar (1 kWh/m^2) to oil (10 kWh/l). In 2019, almost two-thirds (63.3%) of global

Engineering for Sustainable Development: Theory and Practice, First Edition.
Wahidul K. Biswas and Michele John.

electricity came from fossil fuels, resulting in significant land use changes and the loss of bio-diversity (y Leòn and Lindberg 2022). A shortage of nonrenewable resources sets limits to the planet's human carrying capacity, meaning that there is not enough finite energy resources to keep the global economy growing without more alternative and sustainable energy sources. Substantial rises in the price of fossil fuels have previously caused world-wide economic disruption. During this time, industries cut jobs in order to remain competitive in the market and increasingly became dependent on automation to replace labour costs that accounted for a significant portion of total production costs.

Current patterns of energy consumption are unsustainable. The term 'Sustainable Energy' takes into account the social, economic, and environmental consequences of energy production and use. There is an urgent need for alternative energy sources and efficient technologies for powering the future, while fulfilling the TBL objectives of sustainability. This is where sustainable energy production is critical.

9.2 Energy, Environment, Economy, and Society

There is a clear link between energy, environment, economy, and society. Both the extraction and use or combustion of fossil fuels cause environmental impacts. However, without energy, we cannot produce goods and services and create jobs and meet societal needs. Like water, energy is needed to meet the basic needs of life including water treatment and sanitation, education, shelter, food production, textile, medicine, and transportation.

9.2.1 Energy and the Economy

Energy and the economy go hand in hand. The size of the economy has a strong relationship with per capita energy consumption. The correlation factor between these two variables is $R = 0.76$ (Brown et al. 2014). Economic growth is represented as gross domestic product (GDP), which is the monetary value of all finished goods and services made in a country during a specific period. Developed countries produce more GDP than developing nations by producing more goods and services than the latter and so their energy consumption is higher (Figure 9.1). For example, the per capita GDP and energy consumption of India are US$ 1750 and 638 kg of oil equivalent (KOE), respectively, while for Norway, per capita GDP is US$ 67,390 and per capita energy consumption is 5818 kg of oil equivalent (The World Bank 2022a,b).

GDP cannot be a measure of sustainable development as it measures only economic growth without considering the finite limit of world resources. A growing

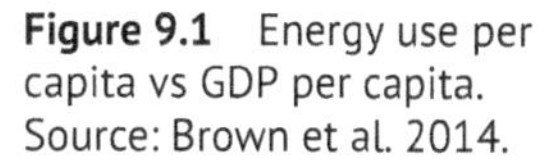

Figure 9.1 Energy use per capita vs GDP per capita. Source: Brown et al. 2014.

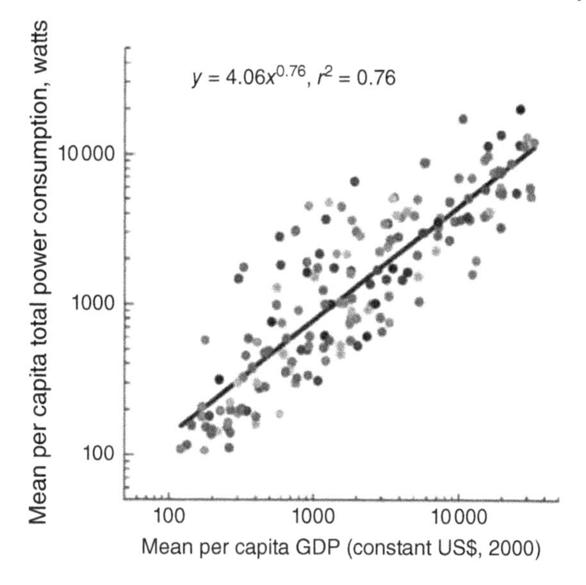

world population requires more energy, but there are differences in perspectives in meeting the global energy demand. According to "conventional economists", energy scarcity is seen as a relative scarcity, addressed by price increases. Their suggestions are to move or switch to gas if there is a high price of oil (due to scarcity) without taking into account the biophysical limits of these nonrenewable gas and oil resources. Ecological economists see scarcity as absolute scarcity, considering the biophysical limits of nonrenewable resources and encouraging the use of renewable sources of energy to strengthen inter-generational and intra-generational equity.

Growth of energy reflects the pace of economic development. The world population can be classified on the basis of energy consumption. The *First group* accounts for most of the consumption of energy worldwide. This group is wealthy enough to afford to protect the environment as they can afford to bear the cost of abatement or clean technologies, such as a solar home power system, renewable-powered desalination plants, electric vehicles, etc. For example, 100% renewable energy is used to desalinate seawater to produce drinking water in Western Australia as it is a rich state and has the ability to invest in capital-intensive renewable technologies. The *Second group* represents nations making the transition to wealth. There exists great disparities between individuals and national abundance. This group does not spend much on the environment but keeps the economy flourishing without taking environmental considerations into account. In the case of the *Third group*, the population is largely outside the cash economy. They meet most of their energy needs via biomass, and animal-based energy. They often depend on free energy sources like leaves and twigs to meet their cooking energy demands, without considering the fact that the absence of leaves and twigs deprives soils of the valuable nutrients needed to sustain an ecological balance. This group does not have any alternative to these free fuels to meet their energy demands.

The majority of energy consumption is occurring in the first group countries (80%), but with slow growth most of the remainder is occurring in the second group of countries (15%), and the balance occurring in the last group (Tester et al. 2005).

Sustainable energy scenarios should include the provision of creating new jobs and local businesses and industries. The decentralised electricity business model as discussed above was found to encourage rural development through local investment and employment, as opposed to the centralised model which concentrates financial returns and employment benefits in urban centres, and in other states and countries. For example, Solar Home Systems (SHS) using micro-finance has led to the roll-out of around 3 million SHS since 2003 and the creation of 114,000 jobs in the renewable energy sector, including the manufacture of all the componentry for the SHS within Bangladesh (Centre for Public Impact 2017).

A decentralised system is also increasing resource security by removing our dependence on the generation, transmission and distribution of finite depleting energy resources. The use of coal to provide electrification to densely populated developing nations will require high levels of investment and finance, and incur mortality and health costs from pollution with limited employment benefit for the local communities. The roll out of small, decentralised renewable energy systems to villages in rural areas could relieve energy poverty without a requirement for industrial load. Initially, it will entail skills and financing challenges but in the longer term, it is likely to provide greater benefit to the villagers through the potential for development of local skills, business, and employment from within the electrification program.

Decentralising energy generation both in developed and developing countries can create a sustainable society by increasing reliable electricity supply and energy services to rural and remote communities on the edges of the grid network (Outhred 2000). Further, decentralised energy generation reduces transmission losses by supplying power closer to points of demand (Thompson 2008). A centralised renewable energy approach can also reduce the load management issues associated with centralised power production, including large-scale renewable energy, and increase the voltage stability of the grid network (Walker 2008).

9.2.2 Energy and the Environment

Reducing or eliminating environmental impacts while generating enough energy for our society should be one of the key objectives of the energy technology design.

The United Kingdom has announced plans to ban the sale of all petrol and diesel cars from the year 2040 as a part of the government's 'Clean Air Plan'. The move puts a heavy emphasis on local authorities implementing their own strategies for reducing NO_2 levels and improving air quality in these areas. This initiative was not actually driven by environmental concerns but for social reasons. This will tackle rising public health risks from air pollution which is linked to 40,000 premature deaths a year and deemed the 'biggest environmental risk to public health in the UK' (BBC 2016).

Apart from the transportation of fuels, harmful emissions are released to the atmosphere during the life cycle of electricity generation from fossil fuels. Table 9.1 shows that the centralised electricity network using coal and gas generating the bulk amount of electricity is responsible for causing not only global warming but also other associated environmental impacts (Atilgan and Azapagic 2015). Table 9.1 also shows the stages responsible for different environmental impacts from fossil fuel power plants. The commonly used fossil fuels for power generation include hard coal, lignite, and natural gas, which are responsible for a large quantity of different emissions, including CO_2, N_2O, CH_4, NO_2, PO_4, and heavy metals, resulting in atmospheric environmental impacts like acid rain, eutrophication, global warming, ozone layer depletion, and photochemical smog. All these impacts cause damage to ecosystems, affect food production, and contaminate air, water, and soil. Not only does the combustion of these fuels to generate electricity causes emissions, but also other activities during the pre-combustion stages of their life cycles, including fuel extraction, fuel processing, and transportation cause further emissions. There are also energy losses in ,the transmission and distribution of electricity. The aforementioned environmental impacts can easily be overcome through the implementation of decentralised renewable energy power systems, by avoiding, pre-combustion activities and the need for transportation and distribution fuels.

9.3 Sustainable Energy

The definition of sustainable energy is based on the definition of sustainable development, which looks at meeting the needs of the present generation without compromising the ability of future generations to meet their own needs (Brundtland 1987). According to Tester et al. (2005), sustainable energy is a dynamic harmony between the equitable availability of energy intensive goods and services to all people and the preservation of Earth for future generations. For further clarification, the definition has been synthesised as follows:

- *Dynamic Harmony*: This deals with coping with the change in the energy mix, the need for appropriate technology and an energy supply–demand balance.

Table 9.1 Environmental impacts of electricity generation from fossil fuel power plants using lignite, hard coal and natural gas.

	Pressure (emissions/ resource utilization)	Type of impacts	Description	Characterisation	Hotspot
Lignite	Fossil fuels such as coal, natural gas	Abiotic depletion potential (fossil)	Depletion of scarce and nonrenewable or finite energy from the earth	MJ/kWh	Fuel extraction (≅ 90%)
Hard coal					
Natural gas					
	Minerals used such as coal for combustion to generate electricity, oil for transportation of fuels, and other inputs	ADP (elements)	This provides an evaluation of the availability of natural elements	kg of Sb e-/kWh	Mining, transportation (25–80%)
	SO_2, NO_x emissions	Acidification	SO_2 and NO_x (i.e. NO, NO_2) react with the water in the cloud and then precipitate as rain	SO_2 e-/kWh	Exploration and combustion of fuels in power plants (60–80%)
	NO_x, phosphates (PO_4)	Eutrophication	The emissions of phosphates, NO_x to fresh water increase nutrients and the productivity of phytoplankton or algae which decreases the oxygen content in water and affects marine life	kg PO_4 e-/kWh	Mining of fuels and the combustion of fuels in power plants (≅60%)
	Emissions of nickel, beryllium, cobalt, vanadium, copper, and barium	Fresh water aquatic ecotoxicity potential (FAETP)	Emissions of metals to fresh water during mining	kg dichlorobenzene (DCB) e-/kWh	These metals are released to water during exploration/mining stages (e.g. coal tailings) (40–80%)

CO_2, N_2O, CH_4	Global warming potential (GWP)	GWP is the heat absorbed by any GHG gases (CO_2, N_2O, CH_4) in the atmosphere	kg CO_2 e-/kWh	Combustion of fuels (74–97%)
Emissions of heavy metals to air, including chromium, arsenic, and nickel	Human toxicity potential (HTP)	This reflects the potential harm of a unit of chemical released into the environment, making people disable or sick to perform normal activities.	g DCP (dichloropropane) e-/kWh	Power plant construction and mining (45–65%)
Same as above	Marine aquatic ecotoxicity potential (MAETP)	The emissions of heavy metals to water. Marine ET refers to the impacts of toxic substances on marine aquatic ecosystems	Same as above	Same as above
The emissions of halons 1211 and 1301 used as fire suppressants and coolants in gas pipeline distribution systems	Ozone layer depletion potential (ODP)	The ODP of a chemical compound is the relative amount of degradation to the ozone layer. It can cause skin cancer (Tasmania, Perth and Melbourne already affected), leaf damage (reducing agricultural production), and affect the growth of phytoplankton, affecting marine life, and accelerating global warming	µg R11e-/kWh	Mining and transportation (60%)

(Continued)

Table 9.1 (Continued)

	Pressure (emissions/ resource utilization)	Type of impacts	Description	Characterisation	Hotspot
	Emissions of SO_2, NO_x, and CO from coal combustion	Photochemical oxidant creation potential (POCP)	Photochemical smog NO_x + Organic HC → Brown smog Causing painful irritation of the respiratory system, reduced lung function and difficulty breathing	g C_2H_4e-/kWh	Combustion (60%)
	Emissions to air and soil of mercury, chromium, vanadium, and arsenic	Terrestrial ecotoxicity potential (TETP)	Have the potential to cause ecotoxic impacts on aquatic and terrestrial ecosystems, leading to damage to ecosystem quality.	g DCP (dichloropropane) e-/kWh	Combustion

- *Availability*: Both rich and poor people have accessibility to and affordability for clean energy sources.
- *Less Energy Intensive Goods and Service*: An example could be the use of recycled steel as a replacement for virgin steel for fabrication, construction, manufacturing, and transportation.
- *All People*: This considers intra-generational factors providing access to electricity to meet the basic needs of all income groups of people.
- *Future Generations*: This considers inter-generation factors by increasing resource efficiency.
- *Preservation of Earth*: This deals with combating climate change, deforestation, and loss of biodiversity.

9.4 Pathways Forward

The pathways that are needed to achieve sustainable energy scenarios are described below. The pros and cons of these pathways were analysed to discuss their sustainability implications and the actions that need to be considered to implement these pathways.

9.4.1 Deployment of Renewable Energy

Renewable energy resources are the backbone of any energy transition to a clean energy future (IEA 2021a). As the world rapidly moves away from carbon emitting fossil fuels, renewables need to be used in multiple sectors to ensure a smooth pathway to net zero emissions. Some countries have already proven that renewable energy can account for a major share of the electricity mix. For example, wind power alone accounts for 58.6% of the total generation in Denmark (REN21 2021).

The deployment of renewable energy in recent years is quite substantial. Renewable energy reached its highest recorded share (29%) in the global electricity mix in 2020 due to low operating costs and preferential access to electricity networks during periods of low demand for electricity (REN21 2021). Figure 9.2 shows that renewables are expected to account for around 50% of the world total energy supply by 2050 (USEIA 2021a). Solar and wind could likely become the most dominant renewable resources in future energy scenarios. Increased demand and procurement required more of these technologies to be manufactured and developed, which should not only reduce costs due to economies of scale but should also increase the incentive for additional procurement and create additional investment for expanding manufacturing capacity. Electricity from new solar photovoltaic (PV) plants and onshore wind farms is now cheaper than electricity from new coal-fired power plants. The cost of electricity from solar PV plants has decreased by 90% since 2009 (Bhutada 2021).

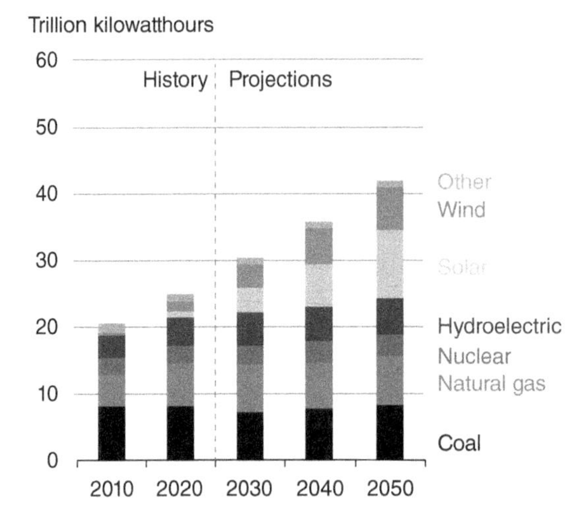

Figure 9.2 World energy mix, 2010–2050 (USEIA 2021a).

Apart from economies of scale, there has been a sharp decline in costs over the last decade, mainly due to the efficiency improvement of photovoltaic modules (now 25.2%, up from 19.2% in 2019) (MIT 2021) and the reduction in hardware and inverter costs. Since 2010, there has been a 64%, 69%, and 82% reduction in the cost of residential, commercial-rooftop, and utility-scale PV systems, respectively (NREL 2021). The cost of onshore wind fell 39% between 2010 and 2019 and it provides a substantial share of electricity in a growing number of countries (IRENA 2020). In 2020, globally, it is estimated that wind power capacity in operation accounted for more than 6% of total electricity generation.

9.4.2 Improvements to Fossil Fuel Based Power Generation

Combined cycle power plants as discussed in Chapter 7 are the improved version of conventional power plants, where the exhaust gas that leaves the turbine after generating electricity is utilised instead of being released to the atmosphere. The thermal efficiency (i.e. the ratio of heat equivalent mechanical energy transmitted to this turbine shaft to rotate it to produce electricity to heat for fuel combustion) of combined cycle gas turbine (CCGT) power plants (i.e. 50%) is more than other fossil fuel power plants (i.e. 33% for coal steam turbine plants, 35% for fuel oil steam turbine plants, 33% for diesel generator sets and 32% for natural gas single cycle plants, and 48% for pulverised coal advanced steam power plants) (IPCC 2007). In the case of coal and fuel oil steam power plants, these fuels are directly burnt to produce heat to convert water to steam to run a turbine to generate electricity; they follow the principle of the steam Rankine cycle as discussed in Chapter 7. Unlike combined cycle plants, the heat in the exhaust gas that leaves the turbine is not recovered and is released to the atmosphere. In a diesel generator set, diesel is combusted with air in the cylinder of an internal combustion engine (ICE), where the combusted gas creates pressure on the piston which is reciprocated inside the cylinder and this reciprocating motion is converted to rotary motion via a shaft to run the generator to generate electricity. The exhaust gas either discharges into the atmosphere or can be used to run the compressor to compress the incoming air to further improve the thermal efficiency by 10–20% (IPCC 2007). Pulverized coal firing ensures complete combustion of the coal, thus leading to higher efficiency for steam generators. This is predominantly adopted in large coal-fired utility boilers. The finer the grinding of coal, the more efficient the combustion. The total time required from entry of a coal particle into a furnace to combustion of the particle is very short. This timeframe, however, is dependent on various factors. The pulveriser, which is also known as the mill, is the heart of a pulverized coal-fired boiler. Depending on the speed, pulverisers are classified as low-speed, medium-speed, or high-speed mills.

As the plant becomes more efficient, it offers long-term savings in energy costs, which can be used to offset the initial extra costs incurred in these plants. Apart from this long-term financial benefit, resources are conserved for future generations and the atmospheric emissions per unit of electricity produced are lowered. There are sustainability implications for these efficient conventional plants as the increase in thermal efficiency saves fuels for future generations. This offers long-term benefits as there are operational savings to pay back the incremental costs in efficiency improvement.

The electricity generation technologies with the lowest capital costs are CCGT plants as more power is produced per unit of power generation capacity (i.e. US$ 1000/kW for CCGT as opposed to US$ 2100/kW and US$ 5000/kW for coal and nuclear, respectively) (IEA 2021b). Gas-fired CCGT power stations offer lower cost generation compared with coal and produce around only half the CO_2 emissions (Lund and Biswas 2008). Gradual increases in the share of renewable energy in the electricity mix will slow down the exploration rate of natural gas and conserve these nonrenewable resources for future generations.

Fuels that are considered for CCGT plants are usually natural gas, diesel, and heavy bunker fuels. Of these fuels, natural gas will play an important role in the next few decades in decarbonisation due to its abundant supply and lower carbon footprint. The CO_2 emission factor of natural gas is around 500 kg/MWh, in comparison to 850 kg/MWh for diesel. Natural gas alone currently accounts for 25% of the global energy mix, and its consumption is expected to increase in the future (IEA 2016). The large increase in US natural gas production has contributed to its competitiveness in international markets and lowered natural gas prices globally. Energy sector transformation without the gas industry would therefore be more expensive and perhaps more unreliable. Natural gas is the most suitable backup for renewable energy generation in terms of lower emissions it is the cleanest fossil fuel, it has a reliable supply, and is cheaper than oil. Gas is the fastest-growing fossil fuel and is the only fossil fuel expected to grow beyond 2035 (McKinsey and Company 2021). Consequently, many gas companies are in the process revising their business models to help accommodate the energy transition.

Although the emission factor of natural gas is about two-thirds of that coal, it is currently the fastest growing source of CO_2 emissions, and the largest projected source of CO_2 growth over the next decade due to its significant share in the energy mix.

CCGT has lower investment cost, shorter construction time, and new plants can more easily be sited. It can also be fitted with carbon capture equipment either during construction or as a retrofit to act as a potential bridge to a low-CO_2 future, which will increasingly rely on technologies such as wind, solar, advanced nuclear, and carbon capture as those technologies mature (Babaee and Loughlin 2016). A logical approach may be to substitute coal with new CCGT in the near-term and then carbon capture and storage (CCS) can be fitted to capture CO_2 (CCGT-CCS) when low carbon regulatory or economic drivers are in place (IEA 2007).

There are, however, technical and environmental challenges to widespread deployment of CCGT-CCS (Babaee and Loughlin 2018). Firstly, fugitive methane is emitted from the production, transmission, and distribution processes of natural gas. Secondly, cost and energy penalties associated with the attachment of carbon capture retrofit technologies to CCGT affect its cost competitiveness. Thirdly, the lower carbon content of natural gas may yield technical difficulties in capturing CO_2 economically. Fourthly, stringent GHG reduction targets may make natural gas plants less attractive in the long-term, even with carbon capture since these plants are not carbon neutral.

Third party access (TPA) matters when gas is supplied by underground pipes from one region to another which sometimes depends on the geopolitics of energy for a country. Factors that are considered in TPA include geographical location and the role of supply, transit, or demand for energy. For example, Country A imports gas from Country C through pipeline which goes via Country B. Countries A and C maintain geopolitical relationships with Country B to ensure continual transmission of gas.

Until recently, most natural gas trade has been limited to regional scale due to the cost challenges of transporting gas over long distances. Liquefied natural gas (LNG) – an attractive option that reduces the volume of gas by about 600 times allowing for transportation by ship – has created an opportunity for expansion of the international market for natural gas (Clear Seas 2022). Pipeline options show cost competitiveness for short to medium distances and for inland destinations, but suffer from geological and political constraints. For example, Russian gas transit to the EU became a hostage of the conflict between Russia and Ukraine since 2014 due to the latter's change of policy orientation towards European integration. Russia then made a strategic decision to close its gas transit through Ukrainian pipelines by designing and commissioning new gas export pipelines that bypassed Ukraine (Naumenko 2018).

Diversification is highly valued as it enhances supply security and offers competitive prices for natural gas. It includes introducing carbon-neutral or carbon-negative gas types such as biomethane, hydrogen, and synthetic natural gas (SNG) into the gas network (Devaraj et al. 2021). As the production of these alternative gas sources is still in its infancy, natural gas holds place as the ideal backup fuel source. Diversifying the gas supply can assist in supply security whilst accelerating low-carbon energy.

9.4.3 Plug in Electric Vehicles

Given that road vehicles account for around one fifth of global greenhouse gas (GHG) emissions arising from fuel combustion, the adoption of plug-in electric vehicles (PEVs) is likely to be an important pathway for mitigating climate change. The electric vehicle initiative by the International Energy Agency (IEA) estimates that PEVs need to meet at least 30% of new light-duty vehicle sales by 2030 to achieve the climate targets in the Paris Agreement (IEA 2017). In addition to climate change benefits, there could be savings in operational costs due to the lower cost of electricity relative to the higher efficiency and lower maintenance cost of electric drivetrains (motor, power electronics, and batteries). Additional benefits include air quality improvement for populated areas due to zero tail pipe emissions. The pollutants VOC, CO, NO_x, particulate matter, SO_x, CH_4, N_2O, and CO_2 emitted from the combustion of petrol or diesel in the ICE are responsible for respiratory disease problems. The near-silent operation of the drivetrain also reduces traffic noise. The replacement of the imported fuels with domestically produced electricity is also expected to create more local employment.

An electric vehicle (EV) mainly uses electricity as energy for propulsion is a means for creating force for movement. When compared to an ICE vehicle, EVs have an electric motor instead of an ICE, which stores electricity in a battery or in a charging station rather than in a fuel tank and receives energy via a plug and cable rather than a petrol station.

PEVs are of two types: battery electric vehicles (BEVs) and plug-in hybrid electric vehicles (PHEVs). The former incorporates a large on-board battery, which can be charged during parking via a cord to a power grid or in a charging station. The battery provides energy for an electric motor to propel the vehicle. The latter is also equipped with an ICE generator that runs the motor to produce electricity once the initial battery charge is exhausted.

A large number of PEVs connecting to a grid might cause deterioration in power quality, voltage imbalance and security of supply for essential purposes since the power distribution networks are not usually designed for considerable mobile loads. Typically, recharging lithium-ion batteries takes considerably longer than refuelling of the ICE car. Consumer acceptance of electric vehicles (EVs) could be facilitated by a recharge ("refuelling") experience similar to that of an ICE powered car, roughly 8–10 minutes.

There are three categories of charging equipment based on how quickly each can recharge a car's battery (USEPA 2020). Levels 1 and 2 use alternating current which is slower, while Level 3 uses DC direct current which is faster.

Level 1 EV has a 120 V battery which takes 1 hour to charge to run the car for 3–8 km. It can be suitable in a small country town to perform basic day to day transport activities. The car can be charged directly from any power point in the garage, apartment, or workplace, as long as there is a reasonable length extension charge cord to charge the battery. It will be more environmentally friendly if the battery is charged by a rooftop solar panel installed in the home/office building.

Level 2 (240 V) is faster than level 1, and 15–30 km can be driven on charging the battery for an hour. Instead of drawing electricity directly from the power point, a charger is needed. There is a quicker home option, known as a wall-box charger, which increases the charging power to 7.2 kW (Voelcker 2019).

The inclusion of park, charging, and riding (PCR) systems in every train station could allow opportunities for this charging. The car is charged while it is parked and it gets charged by the time the driver has returned from work by train. The car is charged by the electricity produced from photovoltaic panels installed on the car shade at the carpark. This PCR approach not only reduces fuel consumption and fuel costs but it also enables people to maximise the use of public transport. The rider may have to pay a nominal amount to enable the transport authority to recover the capital and maintenance costs of this carpark charging station.

Level 3 is considered as direct current fast/rapid charging and about 100 km can be driven after 20 minutes of charging. Charging time may be shorter depending on the station power capacity. These are the public DC chargers (480-volt/direct current) including Tesla Superchargers that are crucial in making EVs viable for driving long distances with little downtime for charging.

Whilst electric cars are more environmentally friendly than petrol cars, there are socioeconomic implications. For example, Australians spend on average AU$ 40,729 on new cars, while an electrical vehicle after the government's current subsidy costs around AU$ 68,000 (Solar Quotes 2021; Electrek 2021; Canstar Blue 2021). The average mileage is 13,301 km/year in Australia. The average cost of traveling per km by petrol, EV (powered by 100% grid electricity), and EV (powered by 50% grid and 50% solar electricity) are 16.7, 4.5, and 3 cents per km, respectively (Electrek 2021). Based on this information, the annual saving (AU$ 16,00–1800) per year could be sufficient enough to recover the additional EV costs (AU$ 28,000) over a reasonable period of time.

The predictions for PEV sales are sensitive to energy and battery prices. The main barriers include long charging times, shorter driving ranges, and higher initial investment. Even the running costs are lower for EVs when compared to ICE vehicles. However, the lack of the availability of charging stations (CS) for electric vehicles due to high investment cost poses an additional barrier to EV users, especially those who cannot charge their vehicles at home. Secondly, there is a low level of consumer awareness of PEVs as still people cannot identify specific PEV makes and models and do not understand how they are fuelled. Thirdly, PEVs require the development of infrastructure for charging batteries. A broad network of easy-to-access public chargers may encourage consumers to purchase an electric vehicle. Fourthly, PEVs are more expensive to purchase than similarly performing ICE vehicles. Upfront prices for many PEV designs may remain above those of conventional ICE vehicles even with prices are continuing to drop as a result of increased production and technological improvements. Options for financial incentives include point-of-sale incentives, rebate programs, tax exemptions, and tax credits. Fifthly, the used PEV market is immature in most regions, leading to uncertainty about resale values, battery performance, and battery recycling that may further deter potential consumers. Lastly, lithium-ion batteries are used in applications that need high energy or power densities. They are ideal for electric vehicles. The EV battery and electric motor may indirectly cause harmful impacts on land water and air quality for using raw material inputs from locations that have weak or poorly enforced environmental regulations. Rare earth battery materials could run out soon with the rapid growth of EVs, EV batteries, and renewable energy technologies and so recovery of rare-earth material is needed to improve the overall sustainability performance of EVs.

Policy instruments are required to make PEVs viable. For example, Zero Emissions Vehicles sales mandates require that these vehicles account for a certain percentage of an automaker's overall production in a given region. The policy allows flexibility enabling manufacturers to meet the requirements in different ways (such as earning more credits from selling long-range battery-electric vehicles and less credits from short-range plug-in hybrid electric vehicles) and by purchasing credits from automakers that exceed the emissions standard.

9.4.4 Green Hydrogen Economy

In a "green hydrogen economy", emissions-free hydrogen is produced using an electrolysis process from two simple ingredients: electricity and water. In the case of electricity, it is generated from renewable resources to power the electrolyser. In addition, there is a gradual drop in the price of electricity from renewable energy resources. These technological characteristics can make hydrogen production cleaner or greener. It is also important to take into account that there is enough water to run the electrolysers. The use of water for electrolysers should not result in water stress for other essential purposes. Accessible freshwater makes up just less than 1% of the planet's water and it is best to avoid creating any additional burden on freshwater usage, especially in areas where drinking and irrigation water is difficult to obtain.

Since the ocean contains 97% of the Earth's water, this water can be desalinated for use in electrolysers without creating pressure on inland water resources (e.g. rivers, creeks, groundwater). However, this desalination process is an energy intensive process. Therefore, either onshore or offshore wind turbines can potentially be used to generate electricity to power the energy-intensive desalination process. This offers two environmental benefits. Firstly, the replacement of fossil fuel power with wind turbines can reduce greenhouse gas emissions and other associated environmental impacts associated with fossil energy. Secondly, renewable energy power plants do not use water for cooling like fossil fuel plants. Therefore, in order to developattain a green hydrogen economy, renewable energy use is required to desalinate seawater for the electrolysis process as well as to operate the energy-intensive electrolysers. A country with access to seawater and plenty of wind and solar could develop a green hydrogen economy.

A country with an abundant natural gas resource can generate hydrogen conventionally via the steam methane reforming (SMR) process. For example, countries like Australia and Middle East nations are blessed with both renewable and nonrenewable sources and have access to seawater to produce hydrogen using either electrolysers or by SMR processes to meet local as well as international demand for hydrogen fuels.

Hydrogen offers more viable alternatives to BEV and advanced biofuels when it comes to long haul heavy-duty transport, aviation, and maritime transport. Many nations have already considered hydrogen as an integral part of their future energy system in transitioning to a low carbon economy.

Hydrogen can be combusted in cars/ICE to produce power or used in fuel cells to produce electricity via an electrochemical process to run the car. The combustion of pure hydrogen is difficult due to challenges related to its storage, transportation, and use, but this is not an issue with the fuel cell electric vehicles (FCEV). Hydrogen gas along with oxygen in fuel cells are used to produce electricity to power the vehicle. A hydrogen-oxygen fuel cell is an electrochemical device that utilises the energy released from the chemical reaction of hydrogen and oxygen to generate electricity with water and heat as the by-product.

Replacing conventional fuels with H_2 fuels in an ICE in a ship requires larger engine dimensions as well as special hydrogen storage requirements. Also the high pressure storage system is a safety concerns. The use of pure hydrogen as the fuel in existing aircraft still remains infeasible due to this. However, hydrogen fuel cell trains are already operational in Europe with Germany, Austria, the

United Kingdom, and France having at least one train or having plans to expand their number of hydrogen trains (Okonkwo et al. 2021).

Hydrogen has the potential to contribute to the sector, either by blending it into the existing natural gas network or by 100% use of hydrogen to indirectly cool or heat district networks (Okonkwo et al. 2021). Hydrogen can easily be used in existing buildings either in a fuel cell or in hydrogen boilers. Alternatively, hybrid co-generation units can use fuel cells as storage or as part of a district network that provides cooling, heating, or electricity along with renewables at low prices.

Although pure hydrogen is not commonly seen as a fuel for power plants, some exceptions exist as hydrogen-fired combined cycle power plants are operational in Italy and Japan.

9.4.5 Smart Grid

The conventional power system allows electricity to flow from the generator to customer in one direction. This limited one-way interaction makes it difficult for a grid to respond to the ever changing and rising energy demands of the twenty-first century. The Smart Grid (SG) introduces a two-way dialogue assisting the power utility and its customers to make the grid more efficient, cheaper, reliable, and greener by avoiding the wastage of surplus energy, maximising the use of renewable energy and also by ensuring uninterruptable electricity supply. A network of communications, controls, computer automation, and new technologies and tools work together in the SG to achieve these objectives. Smart Homes communicate with the grid and enable consumers to manage their electricity usage by measuring the home's electricity consumption more frequently through a smart meter. Utilities can also provide their customers with much better information to manage their electricity bill and maximise their renewable electricity production.

Inside the smart home, there is a home area network (HAN) connecting smart appliances thermostats and other electrical devices to an energy management system. Smart appliances and devices will adjust their run schedule to reduce electricity demand on the grid at critical times or peak hours when the demand and price of electricity is high, which lower customers' electricity bills. These smart devices can be controlled and scheduled online or even via a television.

The SG enables the integration of newer technologies such as solar and wind energy production and plug-in EV (PEV) charging into the grid. Renewable resources such as wind and solar energy are clean sources for electric power, but they are intermittent by nature (i.e. not always available to generate the required level of power) and add complexity for normal grid operation. SG provides the data and automation needed to enable solar panels and wind farms to feed electricity into the grid and optimise its use to adapt to constantly changing energy demands. Utility management turns power on and off depending on the amount of power needed at a certain period of the day. The cost of the power depends on the time of the day it is used. When the demand is high, known as the peak period, additional amounts of electricity need to be generated from expensive sources like diesel power plants as these power plants are dispatchable or can generate electricity instantly. In addition, electricity is most costly to deliver at peak times because additional often less efficient large power plants (e.g. coal) must be run to meet the highest demand of the day. In each instance, SM makes sure that the amount of electricity generated by the power system is equal to the consumption across the entire grid within the network by providing detailed information that enables grid operations to see and manage electricity consumption in real time. This greater control reduces outages and lowers the need for peak power loads. In control rooms across the grid, engineers are able to perform and manage electricity production more efficiently by avoiding or reducing the need to fire up costly secondary power plants. The SG distribution systems counter these energy fluctuations and outages by automatically identifying problems and rerouting and restoring power delivery.

The charging of PEV can be managed over a HAN. The HAN can balance the demand for electricity across the household and prioritise between PEV and other appliances to manage electricity usage and reduce costs. With SG technology and consumer participation, utilities can more easily handle the increasing demand for power to run the PEVs and ensure charging needs are met cheaply and consistently by adding more PEVs to the grid. This is how the potential to reduce fuel costs, lower GHG emissions and reduce the dependency on foreign oil will be achieved.

There could be some negative impacts from SGs. Firstly, the reliance on real-time sharing of household data raises concerns regarding privacy violations (Cuijpers and Koops 2013). Secondly, the increased automation in digital systems might result in reduced user autonomy by giving energy companies more control over households' electricity use (Michalec et al. 2019). Thirdly, there are risks associated with the changing of actor roles, e.g. the greater importance of software providers and more active roles of households, because these raise uncertainties regarding the distribution of responsibilities and risks (Connor and Fitch-Roy 2019). The blockchain concept that is discussed later in this chapter might help overcome most of these risks.

9.4.6 Development of Efficient Energy Storage Technologies

Renewable energy systems have been intensively developed in recent years. However, considerable frequency and voltage fluctuations exist when integrating these systems, as a consequence of the intermittent nature of many renewable energy sources. Therefore, an energy storage system is often needed in a renewable system to buffer the electricity generation and consumption demands.

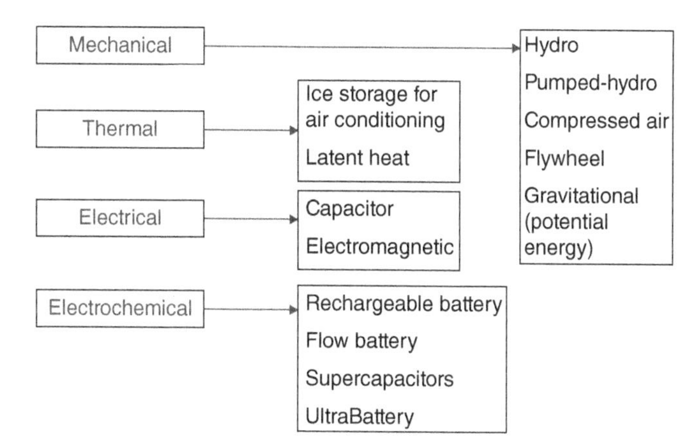

Figure 9.3 Different types of storage systems

Energy storage plays a crucial role in load levelling, energy arbitrage (storing energy to use when the price is at a premium), primary frequency regulation, and shaving the power peak.

With most renewables (such as wind and solar), the electricity demand and availability (production) may not take place at the same time and hence storage of excess energy is mandatory as renewable energy sources are intermittent. Photovoltaic panels can be used during the day when both direct and diffuse radiation is available, wind and wave vary with wind speed and wave height, respectively, and tidal energy is available only during the lunar cycle of the tides. Therefore, storage is important for excess electricity so that the electricity stored can be used when renewable energy sources are not available. Depending on their availability, it could be possible to generate 100% of energy from renewable resources. For example, excess solar electricity during the day can be stored for use at night to reduce the peak load. Figure 9.3 shows the types of storage systems used for storing electricity.

Flywheel systems store kinetic energy by constantly spinning a compact rotor in a low-friction environment (Figure 9.4). At the most basic level, a flywheel contains a spinning mass at its centre that is driven by a motor – and when energy is needed, the spinning force drives a device (motor/generator) similar to a turbine to produce electricity, slowing the rate of rotation. A flywheel is recharged by using the motor to increase its rotational speed once again. Flywheels have been successfully applied in Marble Bar, Western Australia, in a hybrid micro-grid, installed by a local utility company, Horizon Power. The system includes four 320 kW diesel gensets alongside a 300 kW single-axis tracking solar PV array and a 500 kW PowerStore flywheel storage system. The solar array supplies 60% of the grid's daytime electricity demand, saving 405,000 l of diesel fuel a year.

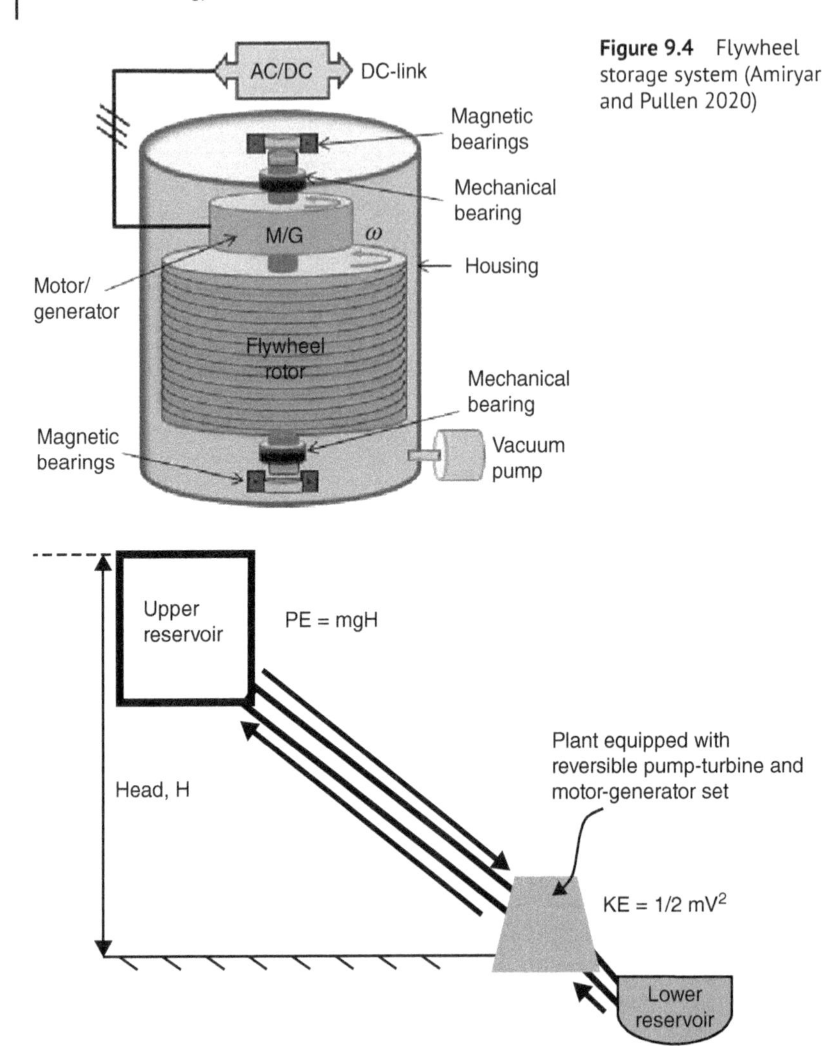

Figure 9.4 Flywheel storage system (Amiryar and Pullen 2020)

Figure 9.5 Pumped storage system

One of the keys to the Marble Bar system's success is its flywheel-based microgrid stabilisation technology, called PowerStore, which was developed by ABB engineers in their research facility in Darwin. Energy storage devices absorb surplus electricity generated by renewable sources during low-demand periods and inject this electricity back into the grid when the demand rises. The technology can hold 18 MWh of energy and shift from full absorption to full injection in 1 ms.

Pumped storage hydroelectricity works on a very simple principle. Two reservoirs at different altitudes are required (Figure 9.5). When the water is released from the upper reservoir, energy is generated by the down flow, which is directed through high-pressure shaft, linked to turbines. In turn, the turbines power the generators to produce electricity. Water is pumped back to the upper reservoir by linking a pump shaft to the turbine shaft, using a motor to drive the pump.

This kind of plant generates energy for peak load, and at off-peak periods water is pumped back for future use. During off-peak periods, excess power available from other plants in the system (often a run-of-river, thermal, or tidal plant) are used for pumping the water from the lower reservoir. A typical layout of a pumped storage plant is shown in Figure 9.5.

Superconducting Magnetic Energy Storage (SMES) systems have pure electrical energy conversion, whilst other energy storage devices involve either electrical-chemical or electrical-mechanical energy conversion, which is much slower. SMES has deep discharge (i.e. about 80% of the energy can be drained without affecting the performance and life of the battery) and recharge ability – unlike batteries, as it can be discharged and recharged fully an unlimited number of times. SMES have a good balance between power density and energy density – although SMES does not have the highest power density or the highest energy density, which is very important for aircraft systems.

It should be noted that the battery with a higher energy density will be able to store a larger amount of energy while battery with a higher power density will release higher amounts of electricity more quickly. Supercapacitors (SCs) have high power density and exceptional durability. They have additional advantages, such as low internal resistance, a wide operating temperature window, and high efficiency, despite having relatively low energy density. These advantageous characteristics are particularly suitable for applications requiring many rapid charge/discharge cycles, rather than long-term compact energy storage (i.e. automobiles, buses, trains, cranes, and elevators,) where they are used for regenerative braking, short-term energy storage, or burst-mode power delivery. Both SMES and SCs are suitable for use in electric vehicles (Table 9.2).

For applications where providing power in short bursts is the priority, flywheel, SMES, and supercapacitors appear to be the most attractive, as a result of their high power density, high efficiency, and long lifespan.

Tesla has built the world's largest lithium-ion battery in South Australia, which is 60% larger than any other large-scale battery energy storage system in the world and is now connected to the Hornsdale wind farm in South Australia.

This huge battery now stores 195 megawatt hours (MWh) of energy to avoid load-shedding blackouts in South Australian summers if electricity demand is forecast to outstrip supply. AEMO predicts that the growing uptake of rooftop solar by homes and businesses will reduce grid demand in South Australia in certain periods to zero by 2023. This creates a rapid change in the nature of energy markets and a growing shift from centralised baseload generation. Figure 9.6

Table 9.2 Performance characteristics of energy storage technologies.

Energy storage technologies	Energy density (J/kg)	Power density (W/kg)	Efficiency (%)
Batteries[a), b)]	10^4–10^6	10–800	70–85
SMES, Fly wheel, supercapacitor[a), b)]	10^2–2×10^4	30–2×10^4	85–100
Pumped hydro-storage[b), c)]	60–120	—	70–85
Compressed air energy storage[b), d)]	2000	—	45–70
Tesla Megapack rechargeable lithium-ion battery[e), f), g)]	9.4×10^5	65	80–90

a) Christen and Carlen (2000).
b) USDOE (2011).
c) Ahshan (2022).
d) KTH (2022).
e) Reuters (2022b).
f) Sun (2010).
g) Normile (2017).

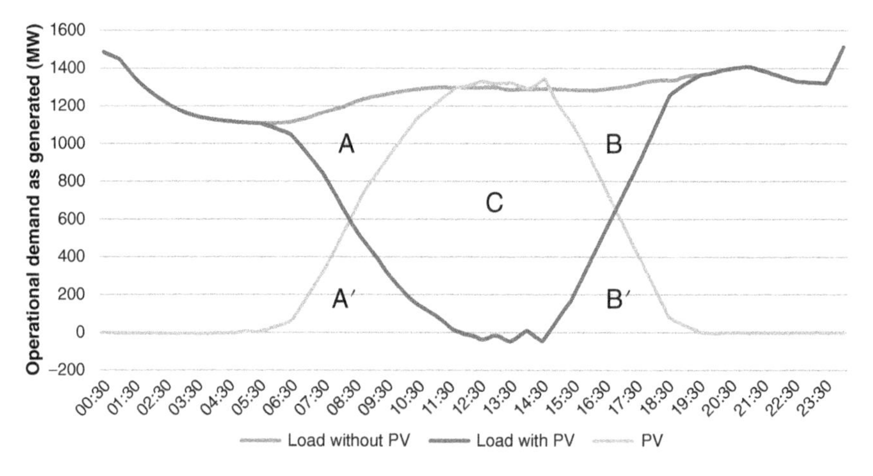

Figure 9.6 Typical daily solar generation curve and load curve (South Australia) (Renew Economy 2015)

shows the amount of solar electricity that can directly meet the load and the excess amount that can be stored for use during the peak period in the South Australian September.

$$\text{Amount of grid demand reduced} = \text{Amount of PV electricity}$$

$$A + B + C = A' + B' + C$$

9.4.7 Energy Storage and the Californian "Duck Curve"

The California Independent System Operator (CASIO) duck chart presents the challenges associated with the accommodation of solar energy and the potential for over generation (Burnett 2016). In the chart, each line represents the net load, which is equal to the normal load minus the electricity generated by wind turbines and solar PV. The "belly" of the duck represents the period of lowest net load, where PV generates the highest possible amount of electricity for a given solar radiation potential in California. The belly has grown as PV installations increased between 2012 and 2020.

In 2020, the net load drops goes 1.3 GW below the demand load during the hours of peak sunlight (see Figure 9.7) resulting an over-generation risk, in which the system's power demand is lower than the minimum power production possible. During over-generation conditions, the excess supply of power could potentially damage generators and motors connected to the grid, so CASIO is forced to implement curtailment, in which variable generation sources like wind and solar are scaled back to keep the net load above the minimum generation value. However, this curtailment reduced the share of renewable electricity in the total energy mix and so it nullifies the zero-emission benefits of solar and wind power.

Another problem arises when sunset approaches. The neck of the duck represents the evening ramp and net peak. During this time, solar suddenly goes offline

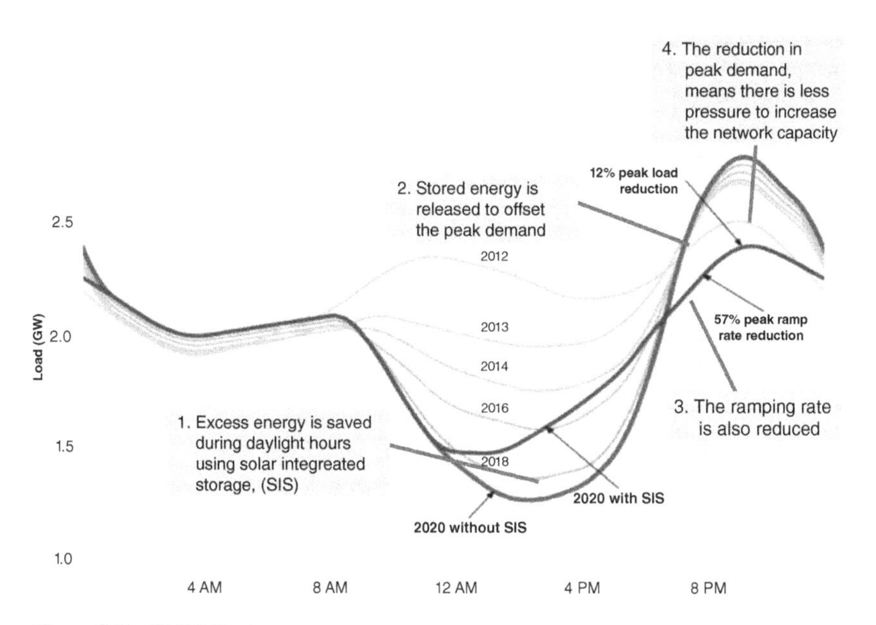

Figure 9.7 CASIO Duck curve

while customer demand rapidly increases to create peak load as people return from work. As a result, energy service providers try to support this ramp with hydroelectric and other greenhouse gas-free resources and natural gas peakload power plants resulting in carbon emissions. Many of these natural gas facilities operate at around 15% capacity 24/7 to meet these short periods of ramping need. When the evening ramp occurs, these often outdated and inefficient facilities are pushed to go from 15% to 100% generation for just a few short hours. This not only increases the GHG emissions from peak power generation but also increases the electricity cost.

The evening ramp is one major concern when considering the reliability of California's energy grid. The duck curve makes this harder for providers as the gap between generation and demand widens.

The aforementioned two issues, including over-generation and evening ramping can be addressed by deploying energy storage technologies to smooth out renewable energy production across the entire day and creating behavioural change in consumers to lessen their demand during peak hours.

Deployment of energy storage technologies or solar integrated storage (SIS) has been found to be one of the most effective ways to reduce the 'duck belly'. To meet stated climate goals, solar overproduction has to be taken advantage of with distributed and utility-scale energy storage in the form of lithium-ion batteries, green hydrogen, and other technologies that can harness and store midday solar production (annual net load = supply load − demand load = 0) or can be used during the evening peak, reducing the need for carbon-intensive natural gas plants. Pairing of solar and storage technologies also reduces cost for consumers and utility concerns, while increasing grid reliability. For residential and commercial customers, batteries can reduce monthly bills while also keeping the lights on during an outage.

9.4.8 Sustainability in Small-Scale Power Generation

Complex geography, long transmission distances, and diffuse populations limit grid extension in many rural and remote areas because of the high marginal cost of connection compared with expected usage (Bhattacharyya 2013). The electricity price in remote areas is cross-subsidised in many developed countries. It means that the customers in remote area pay the same electricity tariff as those connected to a centralized electricity network. The amount of money spent on cross subsidising can potentially be used in developing decentralised electricity generation systems. Recent estimates show that about 70% of rural people need to gain access through decentralized solutions: 65% via mini-grids, 35% via SHS and intra-household or 'pico-solar' products (Alstone et al. 2015).

In the case of developing countries, the rural poor have low energy consumption, struggle to pay connection fees, and face challenges in procuring household wiring and appliances. This requires government subsidy and rebate to incentivise the decentralized power generation systems to provide support to rural areas. In fact, many households and businesses in 'electrified' areas lack access, even if they are directly beneath power lines. On the other hand, centralized grid extension often requires a degree of political power that is a barrier for disadvantaged rural and urban populations. In the case of developed countries like Australia, a remote area power provider like Horizon Power in Western Australia sells electricity at a price which is lower than its generation cost. However, Western Power covers these losses through a Tariff Equalization Fund (TEF) (Boswell and Biswas 2018).

9.4.8.1 Types of Decentralised Electricity Generation System

A 'Mini Grid' is a system having a RE based electricity generator (with capacity of 10 kW and above), and supplying electricity to a target set of consumers (residents for household usage, commercial, productive, industrial and institutional setups, etc.) through a Public Distribution Network (PDN) (GREDA 2019). Since SHS are widely used technologies across regions, the mini-grid is often known as a SHS with a capacity to provide 10 W–10 kW for lighting, radio communication reception, two-way mobile communication, television, fans, additional lighting and communication, with limited motive and heat power.

A 'Micro Grid' system is similar to a mini grid and they generally operate in isolation to the electricity networks of the distribution grid (standalone), but can interconnect with the grid to exchange power. If connected to the grid they are termed as grid connected mini/micro grids (Alstone et al. 2015). The capacity ranges between 10 kW–1 MW and they are capable of providing electricity for all the applications mentioned above with opportunities for community-based service with higher power requirements: for example, water pumping or grain milling. They require financing or investment aggregation for the large capital outlay, but offer relatively low marginal cost electricity to users. This means that the unit cost (i.e. $/kWh) or the price of electricity gets cheaper with increased production of electricity. This requires a critical density of population in order to obtain the required level of revenue to offset the capital expenditure on incremental assets such as transmission and distribution. Therefore, community and local political support are very much required.

Regional grids are 1 MW–1 GW depending on the quality of connection, they fulfil the electricity demand of the applications as mentioned above up to the full range of electric power appliances, commercial and industrial applications. In this case as well, there are often high initial connection costs, but low-cost power after connection. However, the cost of power lines can add significantly to the connection cost in sparsely populated areas.

High-performance, low-cost photovoltaic generation, paired with advanced batteries and controllers, provide scalable systems across much larger power ranges than central generation, from megawatts down to fractions of a watt. The rapid and continuing improvements in end-use efficiency for LED lighting, DC televisions, refrigeration fans, and ICT (a 'super-efficiency trend') enables decentralized power and appliance systems to compete with legacy equipment, on a basis of cost for energy service, for basic household needs. This is because the energy intensity of renewable energy technologies is not as high as conventional generators so that the size of the former increases to meet the same demand as the latter. If end use energy consumption is reduced due to energy efficiency measures, this helps reduce the size of the renewable energy power generator and maximises the use of locally available renewable resources (e.g. sun, wind, biomass).

Pay-as-you-go (PAYG) is a good example of how ICT enables new strategies for financing and managing energy systems off the grid. PAYG is a combination of electronic hardware and software systems that typically rely on mobile phones as a platform for making payments (or verifying the transfer of money), and most include a remotely activated cutoff switch in the system hardware that prevents use when fees or loan payments have not been completed. This reduces the transaction costs for providing and enforcing small loans, and essentially passes retail working capital finance on to the consumer.

Levels of decentralization: There are three broad phases in transitioning from the traditional, fossil fuel-based energy system to one that is decentralised, renewable based, and energy efficient (Burger et al. 2020).

Phase I. Grid-based and centralised system with decentralised renewables as a niche phenomenon (contribution to total power generation less than 10%)

Phase II. Decentralised renewable power network is growing (contribution up to 40% of total power generation)

Phase III. Decentralised renewable network is the dominant player with fully autonomous solutions not requiring connection to a central grid

The economic instruments that are often required to convert centralised generation to decentralised generation are as follows:

1. Subsidy
2. Quota
3. Renewable portfolio standards
4. Capacity payment
5. Capacity auctions
6. FiT (feed-in tariff)
7. NEM (national energy market)
8. DSM (demand side management)

Governance in Phase I is initially driven by incentivising the deployment of renewables at both the supply and demand sides. In the case of supply side, incentives are provided to manufacturing industries of renewable energy technologies to create lead markets for solar cell manufacturers or wind turbine producers. For example, the Danish wind turbine industry and the German photovoltaic manufacturers greatly benefitted from direct and indirect subsidies, as it reduced the capital cost and the unit cost of electricity to compete in the market with conventional sources.

Using a feed-in tariff (FiT), governments can provide their support towards individual renewable sources, which would not be competitive in market-based approaches, for examples auctions. Historically, solar power has benefitted from FiTs. Wind generators were usually found more suitable in remote coastal areas than urban areas. A FiT is a credit customers receive for any unused electricity that their solar power system sends back to the power grid. The price is same as the price for the grid electricity and is paid as a credit on electricity bills. Net energy metering (NEM) is similar to FiT as it allows residential owners of photovoltaic panels to sell their excess electricity to the grid at retail rates – the meter in effect turns back for every kWh that is fed by the PV owner into the grid. In return, households benefit from a net reduction of their utility expenses.

Renewable Portfolio Standards require utilities to produce a certain proportion of electricity from renewable energy technologies within a given time frame. Depending on their design – for example, through tendering or auctions – they are often intended to encourage competition between external utility providers as well as between renewable energy technologies to achieve the maximum reduction of GHG at a competitive price.

FiTs, net metering, or demand-side response (DSP) tend to attract decentralised infrastructure enabling a faster transformation towards the next phase. During this time, the cost comes down, the project becomes more feasible, and investor's confidence increases.

Phase II: policymakers start to revise their incentive systems, as the costs of renewable technologies decrease and their deployment increases. The countries with higher penetrations of renewable energy reach start to abandon the initial subsidy schemes and establish more market-based mechanisms or more competition in the energy market, such as auctioning, in parallel to FiTs for smaller scale technologies. This not only encourages higher capacity extension of renewable energy technologies but also reduces the cost of the electricity generation from renewable resources.

In Phase II, the government lays the foundations for flexibility and energy efficiency, integrating consumers, developing mechanisms to deliver meaningful consent from people and society for the transformation, and evaluating network costs versus decentralised solutions. DSM or efficiency improvement is important for reducing the end use energy consumption (e.g. lighting, heating) given the fact that it will reduce the required size of renewable energy technologies.

Phase III. Governance – meaning market design, network rules and incentives, tariff policy, its coordination and so forth – has been changed by the time a country becomes a Phase III country.

In Phase III, private platforms, autonomous residents, and decentralised networks will co-exist with the central grid, leading to an increased diversification of business models that range from convenient standard packages for energy services with flat rates, similar to insurance, to fully customised solutions for self-producing individuals or communities. Multinational companies, such as Google, Amazon, or Apple, will co-exist with start-ups and local initiatives.

Digitalisation will help cope with the complexities of the energy grid using sensors, smart meters, drones, and augmented/virtual reality in fields as diverse as predictive maintenance, customer care, and weather forecasts. Companies that embrace Artificial Intelligence and Neural Networks to detect patterns in their data may have a blockchain-based application platform to facilitate peer-to-peer trading, monitoring of flows in the transmission grid, and plans to implement many other transaction-related functionalities.

9.4.9 Blockchain for Sustainable Energy Solutions

Blockchain technology is a digital database that enables households and buildings to trade excess renewable energy peer to peer to make power more distributable and sustainable for consumers. All these transactions are recorded or stored in 'blocks' and these blocks are interconnected to form a blockchain. The blockchain tracks generation and consumption by all trading participants (i.e. consumers, prosumers, producers) and settles energy trades on predetermined terms and conditions in near real time using the internet of things (IoT), including smart meters or remote sensing technologies, and machine learning to improve data verification, the identification of data errors, and fraudulent behaviour. These devices collect and assess the real-time energy demand and generation of the prosumer. When the prosumer generates surplus energy, this surplus is transmitted to the local grid and sold on the blockchain-based marketplace to another local prosumer in demand for energy (Figure 9.8). When the prosumer has energy demand, these devices automatically purchase the needed energy on the blockchain market or reduce the energy demand by powering down and turning off devices. These devices react to the dynamic energy price by using more energy during low-cost periods and using less energy during high-cost periods. In this way, these devices allow adjusting energy use over the day while delivering better end-user services (e.g. cold or warm air, hot water, electric vehicle charging).

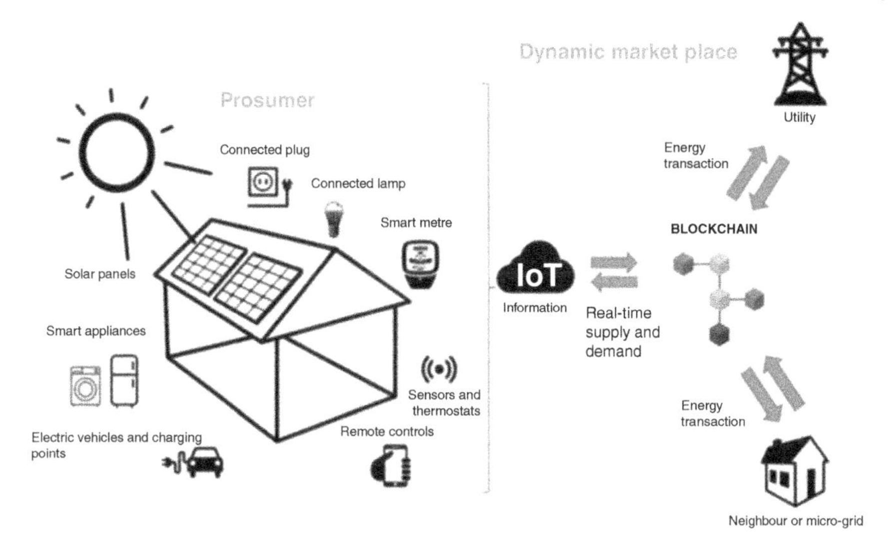

Figure 9.8 Use of blockchain use in a decentralised electricity system. Source: Schletz et al. (2020).

This automation further reduces costs and increases the flow of information by removing intermediaries such as verifiers and auditors. Secondly, it also automates the dissemination and synchronisation of trusted data across a network of participants and provides a tamper-resistant and immutable log. These features enable parties to transact directly, eliminating the need for a trusted, authoritative third party thereby become time efficient and reduce operational costs. Thirdly, it reduces unnecessary production and consumption and maximises the use of renewable energy technologies using real-time energy supply and demand data.

9.4.10 Waste Heat Recovery

Energy cascading and pinch analysis also apply to the energy industry. These include 'cascading' of energy uses, where waste heat from a high temperature process is used to provide energy for a lower temperature process. It keeps going like this until the temperature is not sufficiently high for heat recovery for useful applications (Figure 9.9).

This cascade can happen until the pinch point is reached. This pinch occurs at the smallest gap in temperature across which heat transfer will occur in the system. While technically any value greater than zero can allow for heat transfer, very small values are often not feasible. Low temperature differences decrease the need

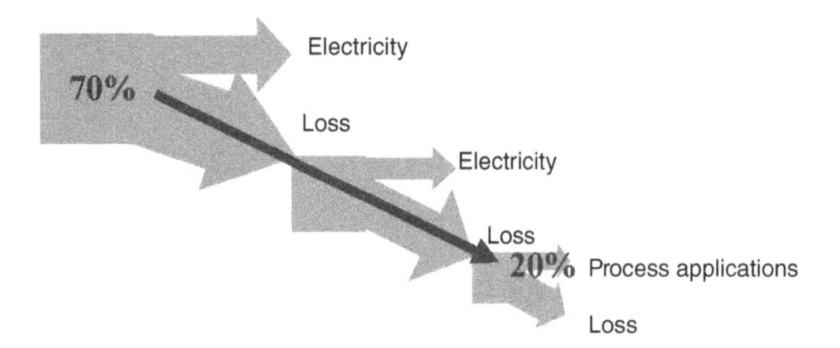

Figure 9.9 Cascade energy recovery

for additional utilities, but require increasingly large heat transfer areas, meaning larger heat exchangers. The trade-off between operating costs and capital costs should be used to choose the minimum approach temperature for the network. Typical choices for minimum temperatures are between 5 and 30 °C.

9.4.11 Carbon Capture Technologies

Over the past decades, coal has been a popular energy choice due to its wide spread geographic availability, tradability, stability of supply, and cost competitiveness. However, coal emits more CO_2 than other fuels, and so the cost of mitigation of CO_2 emissions from coal power plants is higher that the mitigation measures for other fossil fuel plants. In some regions, it is not an easy task to just decommission a coal power plant as it will affect the economy and jobs. The best way in this situation would be to fit CCS to these plants so that the CO_2 emissions can be reduced during the transitional period. Scientist and engineers have developed the carbon capture and utilization (CCU) system, where CO_2 is used for producing products instead of being sequestered underground. Through this upcycling, CO_2 can become a renewable resource and meet energy demands in a less polluting manner.

Technologies that have been developed to capture and sequester CO_2 from combustion and gasification technologies are classified as post combustion capture (PCC) and pre-combustion capture.

9.4.11.1 Post Combustion Capture

CO_2 is captured from flue gas using a scrubber after the combustion of pulverized coal in a conventional power plant. This is known as CO_2 removal PCC.

Ultrasupercritical (USC) pulverized coal combustion is a recent coal technology which is fitted with post-combustion CCS technology. USC is the upgraded version of conventional steam plants, being characterized by extreme operating conditions (steam temperature up to 600–620 °C and cycle maximum pressure higher than 30 MPa). USC plants can attain overall efficiencies up to 45–46%, which is higher than conventional plants with an efficiency of around 40%. USC power plants are integrated with a flue gas treatment section, which includes a selective catalytic reduction (SCR) denitrification system, an electrostatic precipitator (ESP) for particulate removal and low temperature flue gas desulfurization (FGD) system. The flue gas treatment section requires electrical power of 10.2 MW (more than 2% of the overall USC power), penalizing the USC net efficiency by about 1% in separating CO_2 from the flue gas. Therefore, CCS is not considered as a carbon neutral mitigation option. Here CO_2 is captured in a chemical absorption process with an aqueous solution of mono-ethanolamine (MEA).

The flue gas containing CO_2 rises through the absorption column, counter current where the solvent absorbs CO_2 from the flue gas. Purified flue gas is discharged from the top of the column and sent to the stack, while the CO_2-rich solvent is withdrawn from the bottom, heated, and sent to the regeneration column. A reboiler provides thermal energy to the solvent allowing the CO_2 release. Separated CO_2, together with water vapour, rises along the column; in the upper section, the main fraction of steam condensates, whereas the CO_2 rich flow is sent to the compression section. The CO_2 lean solvent is extracted from the bottom, cooled, and recirculated to the absorption column.

9.4.11.2 Pre-combustion Carbon Capture

The integrated gasification combined cycle (IGCC) with a shift reactor to convert CO to CO_2 (as discussed in Chapter 7), followed by CCS, which is often called pre-combustion capture, is also known as IGCC-CCS.

Oxy-fuel (Oxyf) combustion is where pulverised coal is combusted with oxygen rather than air. The oxygen is diluted with an external recycle flue gas (RFG) to reduce its combustion temperature, and to add gas to carry the combustion energy through the heat transfer operations. This is the current first generation technology.

All three technologies are associated with higher generation costs with energy penalties for CO_2 compression, for O_2 production for IGCC and Oxyf, and for CO_2 capture for PCC and IGCC. Compression of CO_2 is challenging because it represents a potentially large auxiliary power load on the overall power plant system.

These figures show that the price of electricity increased by 60%, 31%, and 47% for USC, Oxyf, and IGCC power plants due to the incorporation of CCS. On the other hand, CO_2 was found to be reduced by 0.67, 0.87, and 0.66 tonne per MWh for USC, Oxyf, and IGCC power plants due to the incorporation of CCS. Currently, the carbon price in Europe (2022) is around €100/tonne (Reuters 2022a), which means in the current carbon climate, the cost of CO_2 reduction resulting from the use of CCS would be €67, €87, and €66 for USC, Oxyf, and IGCC power plants, respectively. Interestingly, these carbon reduction costs are higher than the

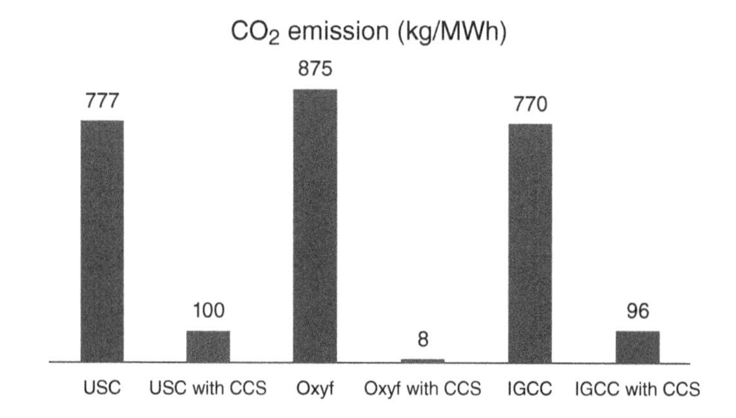

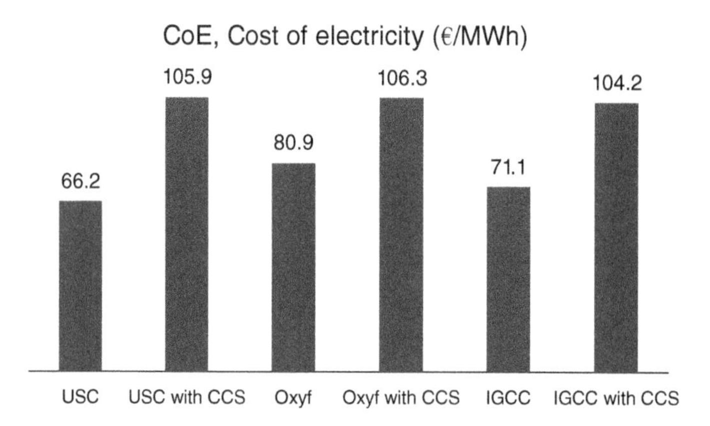

Figure 9.10 Economic and environmental implications of CCS

additional costs of these plants due to the incorporation of CCS (Figure 9.10). However, this data trend is expected to change in the near future with the changes in government policies, socioeconomic pressures and technological change. This example shows how the use of an economic instrument can make these carbon abatement technologies cost-competitive compared to conventional ones.

9.4.12 Demand-side Management

DSM mainly deals with the reduction of energy consumption from end-use appliances. Once energy is saved at home or in offices or by industry, it reduces the supply of electricity/generation of electricity, which means that less fossil fuels are being burnt and less environmental impacts created.

Figure 9.11 shows the flow-chart for assessing the technical potential of efficient appliances for avoiding electricity generation and GHG mitigation. The technical potential of avoiding electricity generation (G_{ipt}) from the use of efficient appliance

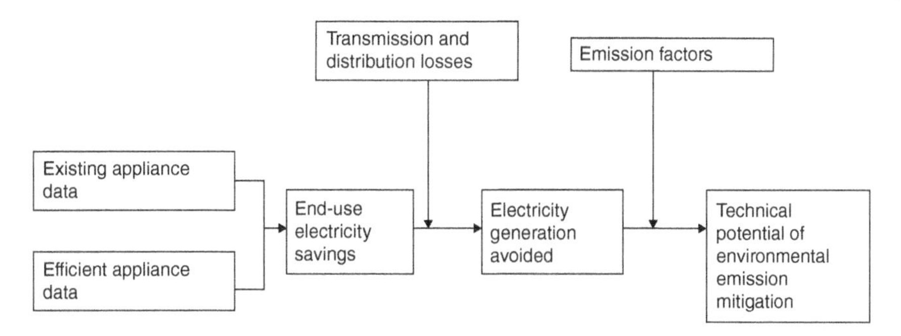

Figure 9.11 Demand side management for power sector emission reduction

types, instead of inefficient ones, in a period (p = peak, mid-peak, and off-peak) of the year t is calculated as:

$$G_{ipt} = \frac{\Delta P_i \times H_{ip} \times N_{ipt}}{1 - \mathrm{TDL}_t} \tag{9.1}$$

where ΔP_i is the power demand avoided with the use of the appliances of type 'i' instead of an inefficient one. H_{ip} is the annual operating hours of the efficient appliance type 'i' in period p, N_{ipt} = number of appliance type i used in period p of year t; and TDL_t = Transmission and Distribution Losses as a fraction of the total generation in year t.

The total generation avoided by the selected efficient appliances in period p of year t (G_{pt}) can be expressed as:

$$G_{pt} = \sum G_{ipt} \tag{9.2}$$

The peak load avoided at the point of generation by the ith efficient appliance in year t (MW_{it}) is,

$$G_{ipt} = \frac{\Delta P_i \times N_{ipt} \times \mathrm{PCF}_{it}}{1 - TDL_t}, \tag{9.3}$$

where PCF_{it} is the peak coincidence factor (= ratio of the total power demand of the appliance during the peak period to the system peak load) for the appliance i in year t

The economic potential is determined by calculating the electricity generation that could be avoided with the use of the cost-effective efficient technologies. The criteria used for assessing the cost-effectiveness of appliances is as follows (Shrestha et al. 1998).

9.4.12.1 National Perspective

Cost per unit of conserved energy (CCE) is used as an indicator to assess the cost-effectiveness of efficient utilization technologies. Cost per unit of conserved energy due to an efficient appliance is defined as the incremental cost (IC) of an efficient appliance over the annual electricity consumption avoided due to use of an efficient appliance. Thus, CCE due to use of an efficient appliance i (CCE_i) can be expressed as:

$$CCE_i = \frac{IC}{e_j} \tag{9.4}$$

$$IC = [AAC_i - AAC_o] + [AOMC_i - AOMC_o] \tag{9.5}$$

where AAC_i and AAC_o = annualized capital costs of the efficient and inefficient appliances, respectively; $AOMC_i$ and $AOMC_o$ = annualized operation and maintenance costs of the efficient and inefficient appliances, respectively; and e_i = annual electricity consumption avoided per unit of efficient appliance of type i. Appliances with CCE below the long-run marginal cost (LRMC) of electricity supply are known as cost-effective options.

9.4.12.2 User Perspective

An efficient appliance is considered to be cost-effective from the user perspective if the net annual benefit (NAB) is positive. The NAB of efficient appliance i (NAB_i) here is expressed as

$$NAB_i = SEC_i - [(AAC_i - AAC_o) + (AOMC_i - AOMC_o)] \tag{9.6}$$

where SEC_i = annual savings in electricity cost with the use of efficient appliance i = electricity price × annual saving of electricity.

9.4.12.3 CO_2 Mitigation per Unit of Incremental Cost

The level of CO_2 mitigation (ε) from the use of an efficient appliance 'i' is,

$$\varepsilon_i = G_{ipt} \times \bar{e} \tag{9.7}$$

$\bar{e}$ = weighted average emission factor.

For example, $\bar{e}$ of electricity a country X with the following energy mix can be calculated as follows:

Coal = 10,000MWh
NG = 15,000MWh

Suppose, emission factor of coal is 1000 kg of CO_2 e-/MWh and 500 kg of CO_2 e-/MWh for gas (it may be different in your case).

The weighted average emission factor of coal is calculated as $\bar{e} = (1000 \times 10{,}000 + 15{,}000 \times 500)/(10{,}000 + 15{,}000) = 700$ kg og CO_2 e-/MWh.

Carbon dioxide mitigation per unit of incremental cost associated with an efficient appliance including its incremental operating cost (COMPIC) is calculated using the relation

$$\text{COMPIC} = \frac{CO_2 \text{ avoided by an efficient appliance}}{\text{Generation avoided by the efficient appliance} \times CCE_g} \tag{9.8}$$

where CCE_g = cost of conserved energy at the point of power generation.

9.5 Practice Example

9.5.1 Step 1

Choose an industry or a commercial centre in such a way that it will give you sufficient information to complete this exercise.

1. Collect the following information:
   End-use appliances
   Operating hours for different end-use appliances
   Energy types and sources
      Technical characteristics of technologies
      Energy consumption (kWh)
2. Create a pie chart which will show the energy consumption of different end-uses. Refer to the example below.

End-use appliances	kWh
CFL	5
Incandescent lamp	8
Boiler	30
Washing machine	15
TV	20
Motor	10

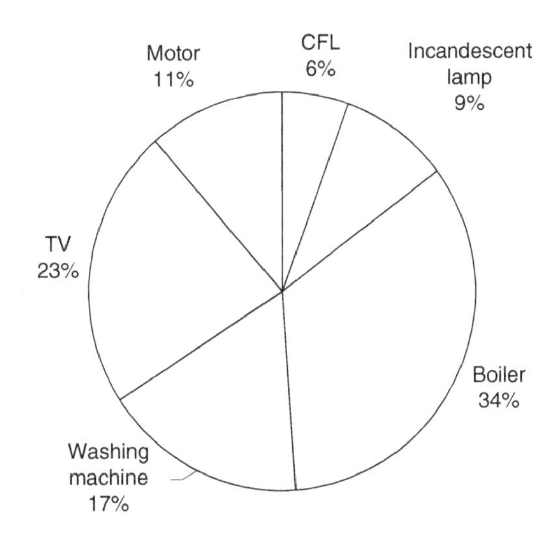

Create an Energy Flow Chart

- Firstly, determine what amount of energy has actually been supplied. You also need to know energy sources and technologies that convert primary energy to secondary energy.
- Secondly, determine what amount of energy has actually been consumed by the appliances and what amount of energy is lost. You need to know the energy efficiency of the appliances to determine the losses.
- Draw a flow chart. A sample flow chart is given below. In your case, it will be a micro level flow chart rather than a macro-level flow chart.

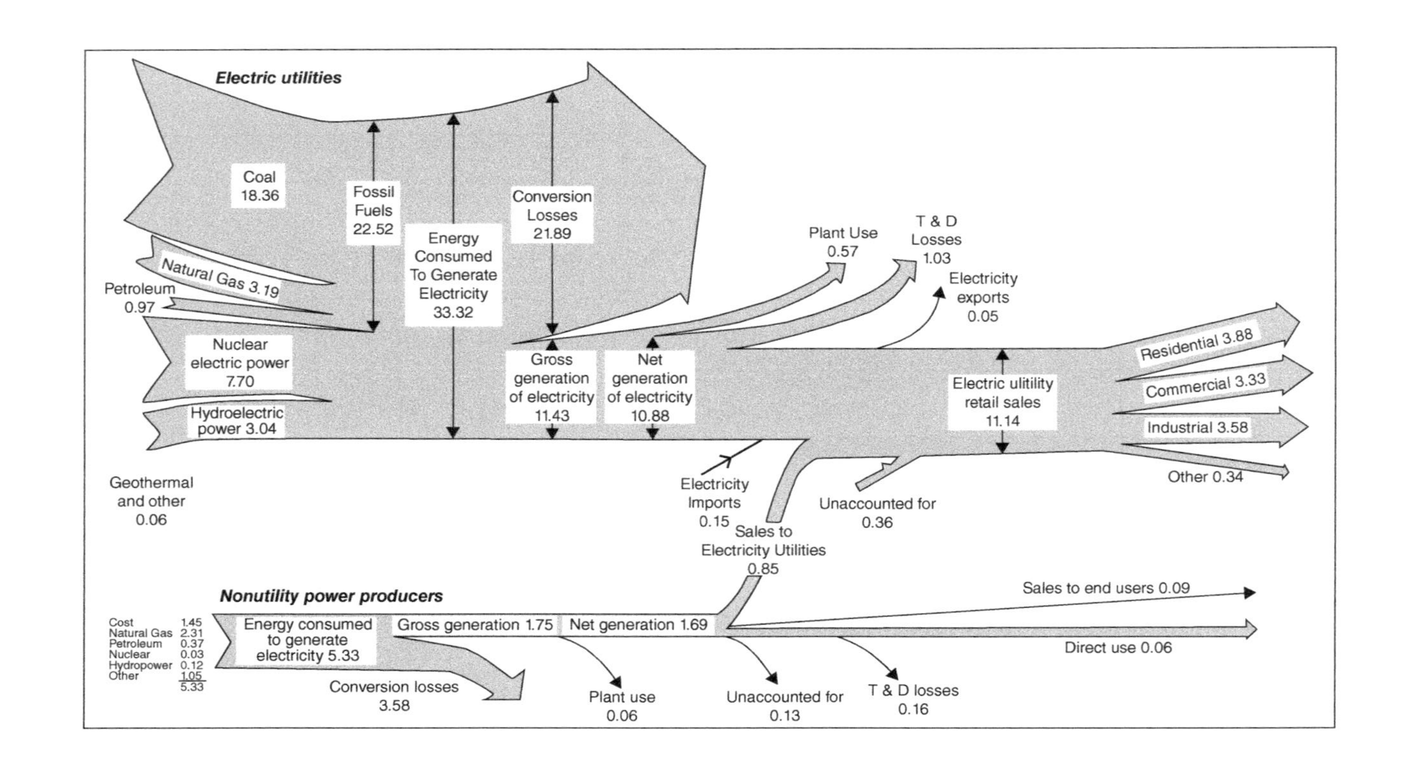

Electric utilities
Coal 18.36
Natural Gas 3.19
Petroleum 0.97
Nuclear electric power 7.70
Hydroelectric power 3.04
Geothermal and other 0.06
Fossil Fuels 22.52
Energy Consumed To Generate Electricity 33.32
Conversion Losses 21.89
Gross generation of electricity 11.43
Net generation of electricity 10.88
Plant Use 0.57
T & D Losses 1.03
Electricity exports 0.05
Electric ulitility retail sales 11.14
Residential 3.88
Commercial 3.33
Industrial 3.58
Other 0.34
Unaccounted for 0.36
Electricity Imports 0.15
Sales to Electricity Utilities 0.85
Nonutility power producers
Energy consumed to generate electricity 5.33
Gross generation 1.75
Net generation 1.69
Conversion losses 3.58
Plant use 0.06
Unaccounted for 0.13
T & D losses 0.16
Sales to end users 0.09
Direct use 0.06
Cost 1.45
Natural Gas 2.31
Petroleum 0.37
Nuclear 0.03
Hydropower 0.12
Other 1.05
5.33

Create a Reference Energy System (RAS)

Identify

I. energy sources
II. type of energy
III. sectors
IV. demand-side technologies
V. end-uses

Refer to the example below for a supermarket energy reference system. Just noting that this exercise uses the RES concept that is introduced later in this chapter, in section 9.7. Discuss the chosen industry and its energy consumption patterns, energy sources, and technologies.

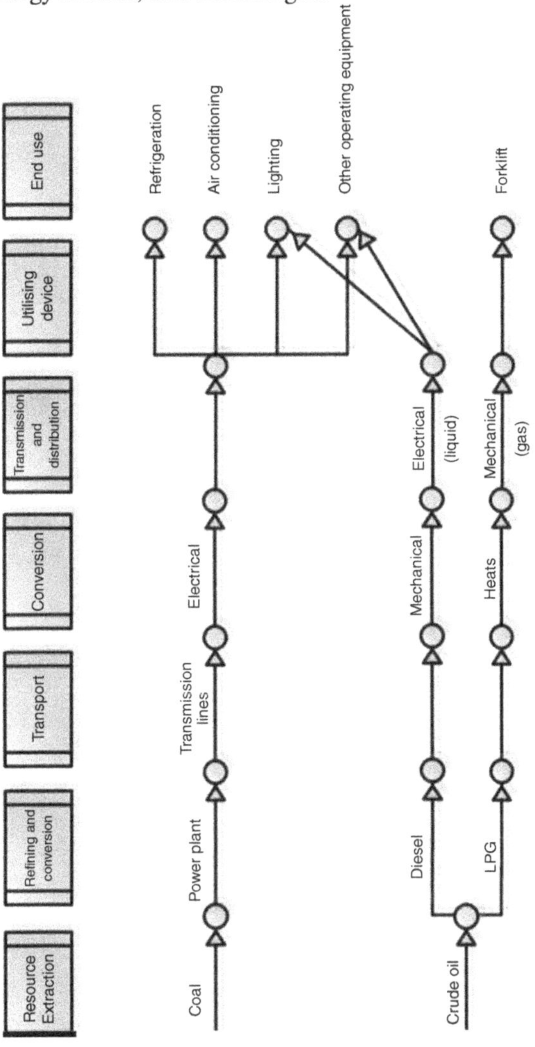

9.5.2 Step 2

- Select alternative technologies for energy saving and energy recovery opportunities.
- List all possible alternatives or new technological options.
- Choose the best one from a number of options for a particular application and discuss why you have chosen them.
- You need to know the efficiencies of both the existing and new technologies to determine the energy savings due to replacements of the former with the latter.

	Alternative or new technologies			Energy recovered	
Existing technologies	Thermodynamic principles/theory	Technological characteristics	Energy consumption	Unit	Amount

Provide sources or references for the information in the above table.

- Give a sample calculation for one case.
- Describe the table.
- Draw an updated energy flow chart (like the one shown in Step 1) where efficient technological options have been used.

9.5.3 Step 3

- Create a bar chart that compares previous and present energy consumption patterns. See the example below.
- Describe this chart.

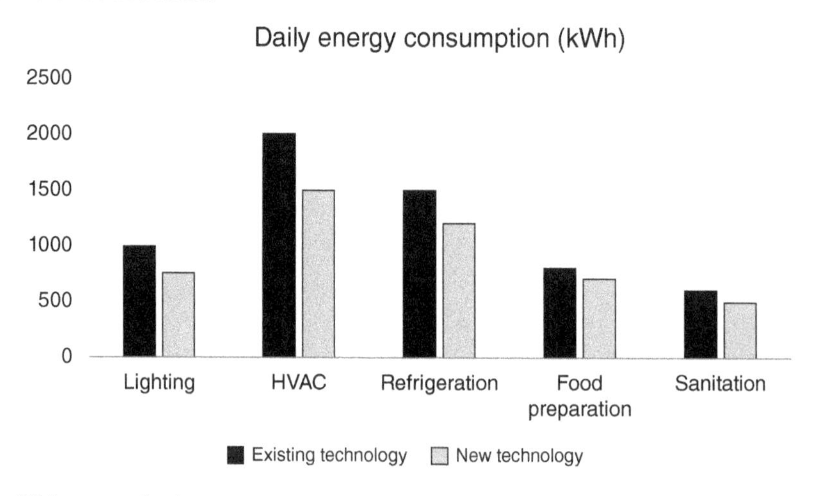

- Write a conclusion

9.5.4 Step 4

You are assessing the economic and environmental implications of the energy saving options. You will need to use an excel spread sheet for this task.

Identify economically viable energy saving options from the following perspectives.

National perspective. The cost per unit of conserved energy (CCE) will be used to assess the economic viability of energy saving options from the national perspective. CCE is defined as the incremental cost of an efficient appliance, as well as operation and maintenance costs, over the cost of inefficient appliance replaced per unit of annual electricity consumption avoided.

If CCE of an energy saving option is greater than the electricity or gas price, then the option is economically viable from the national perspective. Please obtain the electricity price from your local supplier or provider.

Create a graph as shown below.

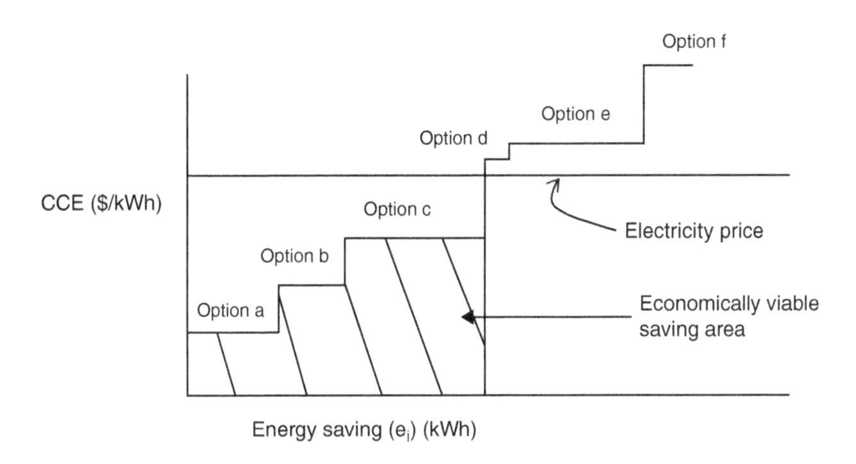

Then, calculate the total amount of energy which can be saved by economically viable options from the national perspective.

Consumer perspective: An efficient appliance is considered to be cost-effective from a user perspective if the NAB due to the use of the efficient appliance is positive.

Get the value of 'IC' from previous analysis. Draw up a table as shown in below.

If the value of NAB is positive, the option is economically viable from a consumer's perspective.

Options	NAB
Option a	−VE
Option b	−VE
Option c	+VE
....	

Finally calculate the total amount of energy which can be saved by the economically viable options from the consumer perspective.

9.5.5 Step 5

Determine the environmental implications, in this case greenhouse gas (GHG) issue, of economically viable energy saving options from national and consumer perspectives.

Get your local energy mix value to calculate the emission factor.

Draw up a table that shows the CO_2 mitigation potential of economically viable options from both national and consumer perspectives.

	GHG mitigation (kg or tonne of CO_2 e-)	
Options	**National**	**Consumer**
Option a		
Option b		
....		
Total		

Calculate the total CO_2 mitigation by the economically feasible options from a consumer's perspectives, and total CO_2 mitigation by the economically feasible options from a national perspective.

9.5.6 Step 6

Calculate the CO_2 mitigation per unit of incremental cost (COMPIC) for economically feasible options from national and consumer perspectives.

Like before, draw a table that shows the COMPICs for different options for both economic perspectives.

Discuss all the results of this section in two paragraphs.

9.5.7 Step 7

Analyse the social benefits, policies, and institutional framework.

Social Benefits

- List a few direct and indirect social benefits due to the implementation of the energy saving options.
- Include inter- and intra-generational equity aspects that are relevant to your results and analysis.
- Include a brief discussion on the affordability and accessibility issues in relation to the energy saving options.

Policies

- From your analysis, what policies including pricing, economic incentives, carbon tax, and market mechanisms are appropriate to make viable the energy saving options, which were not viable from your analysis.
- What policies and strategies would you recommend for propagating the economically viable options that you found outfrom your analysis?
- Suggest some goals and targets which are based on the state of art of the technologies, market, research & development etc
- Discuss some successful programs (which canbe obtained from the web & literature) that are relevant to your analysis and will help you to propose useful policies (Max 400 words) Institutional framework
- Stakeholders involved/relevant to your casestudy (direct and indirect eg gov, research, NGO, business, people...)
- What should be their roles and responsibilities in this particular case?
- How can they be integrated to meet the policies, goals and targets as stated above

9.6 Life Cycle Energy Assessment

Using a LCA methodology, environmental performance indicators, including life cycle embodied energy (LCEE), energy intensity, energy payback time, and environmental impact, can be determined for energy technologies.

The embodied energy consumption for all stages is added to determine the LCEE consumption using Eq. (9.9).

$$\text{LCEE} = \text{EE}_{\text{material}} + \text{EE}_{\text{transport}} + \text{EE}_{\text{manuf}} + \text{EE}_{\text{use}} + \text{EE}_{\text{eolife}} \tag{9.9}$$

Using the methodology discussed in Chapter 3, LCEE can be calculated. In this case, the environmental emission factors are replaced with the energy factors of inputs used during the life cycle of the product.

The energy intensity is defined as the ratio of the energy requirement for construction, operation and decommissioning of a power plant and the electricity output of the plant over its lifetime (Lenzen and Munksgaard 2002). This is mathematically expressed as

$$\eta = \frac{E}{E_1} \tag{9.10}$$

where E = the energy requirement for construction, operation, and decommissioning of a power plant.

$$\text{Electricity output, } E_1 = \text{P} \times 8760\,\text{h}/y \times \lambda \times \text{T} \tag{9.11}$$

where, P, λ and T are power rating, capacity utilization factor and life time, respectively.

The inverse of energy intensity (i.e. $1/\eta$ or E_1/E) is known as the energy payback ratio (EPR). Because the energy payback ratio is less affected by upstream choices of energy supply, it is considered as one of the most reliable indicators of environmental performance (IEA 2000). A high energy payback ratio indicates good environmental performance. If a system has an EPR between 1 and 1.5, it consumes nearly as much energy as it generates, so it should never be developed.

The energy payback time (EPBT) is an useful indicator for renewable energy power systems. It is the time that is taken by the renewable energy technology to generate the primary energy requirement to construct, operate, and decommission it (Schleisner 2000). This is mathematically expressed as,

$$\text{EPBT} = \frac{E}{E_g} \tag{9.12}$$

where Eg is the annual electricity generation by the renewable energy technology

As the value of E in Eq. (9.12) increases, EPBT as well as the environmental emissions generated by a renewable energy technology over its life cycle will increase. The concept of energy intensity and energy payback period, based on the life cycle paradigm, are operational and easy to interpret from the sustainability point of view. Both concepts focus on how much conventional energy we use today to obtain energy tomorrow. The value of E is affected by many factors including duration, power ratings, load factor, type and maturity of technology, and country of manufacture.

9.7 Reference Energy System

The reference energy system (RES) is a network representation of all of the technical activities required to supply various forms of energy to end-use activities (Beller 1976). RES defines the total set of available energy conversion technologies, refer to

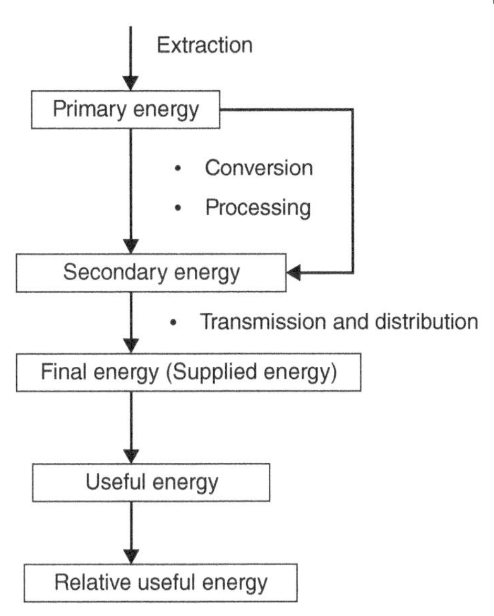

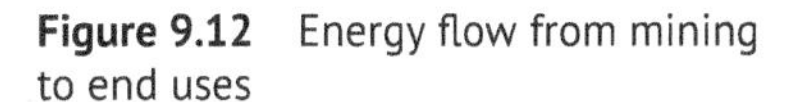
Figure 9.12 Energy flow from mining to end uses

all types of energy technologies from resource extraction to transformation, transport, distribution of energy carriers, and end-use technologies.

Figure 9.12 shows that the fuels that are directly obtained from nature (i.e. crude oil/coal/gas) or primary energy sources that are converted to secondary energy (e.g. refined oil/diesel/petrol). These secondary energies are a more suitable form of energy and are used directly for energising the economy and producing electricity and heat. Electricity and heat are the final forms of energy, but there are losses associated with their transmission and distribution of the final energy to customers in industry, commercial sites, shopping centres, and residential areas for heating, cooling, and machinery operation purposes. In some instances, electricity may not be the final form of energy as it is converted to heat to operate end use appliances (e.g. electrical oven).

A RES is a way of representing the activities and relationships of an energy system. Also, it shows the flow of energy (amount) from source to end-use applications. Diagrammatically, it consists of the following symbols:

- Directed paths
- Nodes
- Links

In the following RES, the directed path starts from the extraction of resources and then follows all the way through to the end use appliances. Each node represents a change in the form of energy. In this example, the nodes are refining and conversion, transportation, conversion, transmission and distribution, utility

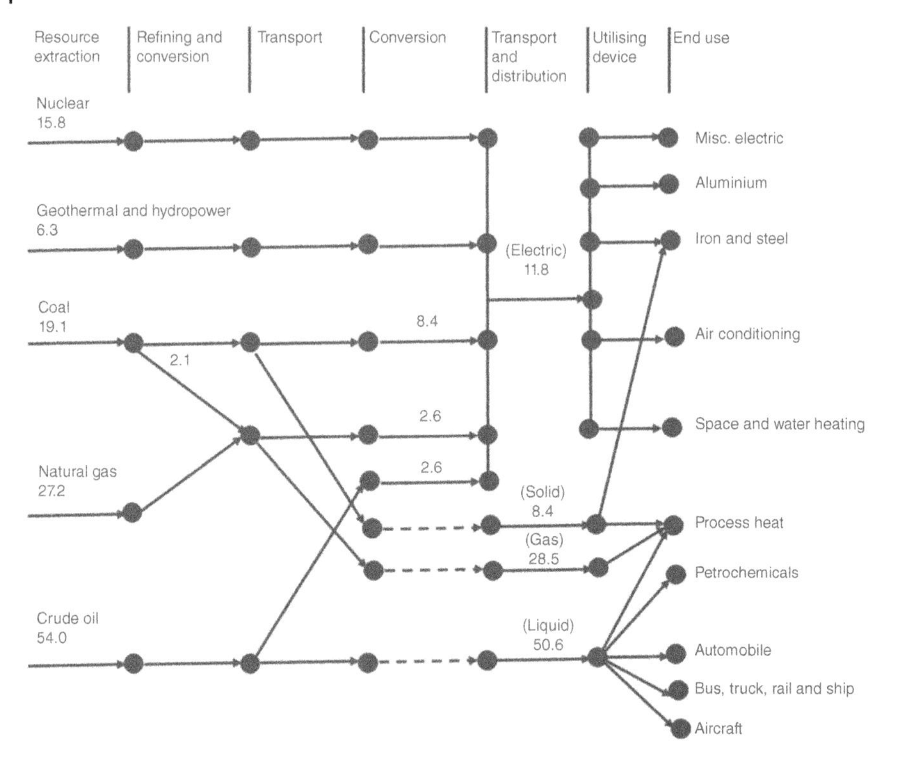

Figure 9.13 Reference energy system

services and end uses. The arrows between the nodes are known as links. A RES describes what is actually happening in the energy system. As we can see in Figure 9.13, nuclear, geothermal and hydroelectric plants produce electricity directly and 100% of these plant capacities are used for electricity generation (Figure 9.13). Dotted arrows in the RES represent some hidden flows meaning that some portion of energy is coming from the external sources.

Out of 19.1 units of coal energy, 10.7 units are used in metal processing industries and a gasification process (2.1 units of coal energy are used for purifying the natural gas and the remaining 8.6 units of energy is used for converting coal to coking coal for metal processing (iron and steel) industries). In order to produce 8.4 units of coking coal for both energy and reduction of iron oxides (Yang et al. 2014), 12.6 units of coal are needed as the conversion rate is 66.6% (Corsacoal 2022). Therefore, an additional 4 units (12.6–8.6 units) of coal is coming from other energy sources for producing coking coal.

Out of 54 units of crude oil, 2.6 units are directly used for generating electricity and the remaining 51.4 units are refined to produce refined oil. Around 50%

of the crude oil is converted to gasoline in petroleum refineries (USEIA 2022). Therefore, 25 units of gasoline (i.e. 50.6 units -0.5×51.2 units) is coming from or is imported from other regions or countries. 28.5 units of liquid oil is used as transportation fuels (road and aviation), air conditioners, and petrochemical applications.

About 95% of crude natural gas has been refined to produce 25.5 units of refined gas (i.e. $95\% \times 27.2$). Of these 25.5 units, 2.6 units are used for electricity generation and 23.54 units are used for industrial process heating and residential and commercial space and water heating purposes. An additional (28.5–23.54) units or 5.26 units of gas energy was imported from elsewhere.

Suppose the overall efficiency of converting all energy sources (coal, geothermal, hydro, nuclear, coal, and gas) to electricity is 32.5%. The electricity produced is about 11.8 units (i.e. $32.5\% \times (15.8 + 6.3 + 8.4 + 2.6 + 2.6)$). This electricity is mainly used in household electrical appliances (lamps, television, ovens, computers, chargers, etc.), residential and commercial heating and cooling and industrial applications such as aluminium smelters and steel and iron mills.

9.8 Conclusions

The chapter discusses the causes and consequences of current energy consumption patterns and their climate change implications. Various technologies, tools, and management approaches have been discussed that help to achieve resource efficiency, cost competitiveness of energy technologies and for decision-making purposes. Sustainable energy is a critical part of twenty-first century engineering. Without a focus on renewable energy knowledge and development, the world will struggle to achieve a low carbon future. Sustainable energy is the key to sustainable development.

References

Ahshan, R. (2022). Pumped hydro storage for microgrid applications. In: *Recent Advances in Renewable Energy Technologies* (ed. M. Jequirim), 323–354. UK. Chapter 8: Academic Press.

Alstone, P., Gershenson, D., and Kammen, D.M. (2015). Decentralized energy systems for clean electricity access. *Nature Climate Change* 5: www.nature.com/natureclimatechange.

Amiryar, M.E. and Pullen, K.R. (2020). Analysis of standby losses and charging cycles in flywheel energy storage systems. *Energies* 13 (17): 4441. https://doi.org/10.3390/en13174441.

Atilgan, B. and Azapagic, A. (2015). Life cycle environmental impacts of electricity from fossil fuels in Turkey. *Journal of Cleaner Production* 106: 555–564. https://doi.org/10.1016/j.jclepro.2014.07.046.

Babaee, S. and Dan Loughlin.(2016) The potential role of natural gas power plants with carbon capture and storage as a bridge to a low-carbon future. Presented at 34th USAEE/IAEE North America Conference, Tulsa, OK.

Babaee, S. and Loughlin, D.H. (2018). Exploring the role of natural gas power plants with carbon capture and storage as a bridge to a low-carbon future. *Clean Technologies and Environmental Policy* 20: 379–391. https://doi.org/10.1007/s10098-017-1479-x.

BBC (2016). Air pollution which is linked to 40,000 premature deaths a year and deemed the biggest environmental risk to public health in the UK. https://www.bbc.com/news/health-35629034 (accessed 20 January 2022).

Beller, M. (1976). Reference energy system methodology. *Presented at 81st Natl. Meeting of the Am. Inst. of Chem. Engr.*, Kansas City, USA.

Bhattacharyya, S. (2013). *Rural Electrification through Decentralised Off-grid Systems in Developing Countries*. Springer.

Bhutada, G (2021). Electricity from renewable energy sources is now cheaper than ever. https://www.visualcapitalist.com/electricity-from-renewable-energy-sources-is-now-cheaper-than-ever/ (accessed 20 March 2022).

Boswell, B. and Biswas, W.K. (2018). Techno-economic and environmental implications of electricity generation from solar updraft chimney power plant in Meekatharra in Western Australia. In: *Transition towards 10% Renewable Energy* (ed. A. Sayigh), 31–47, UK:. Cham: Springer.

Brown, J.H., Allen, C., Burger, J.R. et al. (2014). Macroecology meets macroeconomics: resource scarcity and global sustainability. *Ecological Engineering* 65: 24. https://doi.org/10.1016/j.ecoleng.2013.07.071.

Brundtland, G.H. (1987). Our common future: report of the world commission on environment and development. Geneva, UN-Dokument A/42/427.

Burger, C., Froggatt, A., Mitchell, C., and Weinmann, J. (2020). *Decentralised Energy — A Global Game Changer*. London: Ubiquity Press https://doi.org/10.5334/bcf. License: CC-BY 4.0.

Burnett, M. (2016). Energy Storage and the California "Duck Curve". Submitted as coursework for PH240, Stanford University, Fall 2015.

Canstar Blue (2021). How much do Australians spend on new cars? https://www.canstarblue.com.au/vehicles/average-car-price/ (accessed 25 February 2022).

Centre for Public Impact (2017). The Solar Home Systems initiative in Bangladesh. https://www.centreforpublicimpact.org/case-study/solar-home-systems-bangladesh (accessed 10 February 2022).

Christen, T. and Carlen, M.W. (2000). Theory of Ragone plots. *Journal of Power Sources* 91: 210–216.

Clear Seas (2022). LNG & Marine shipping. https://clearseas.org/en/lng/ (accessed 2 March 2022).

Connor, P. and Fitch-Roy, O. (2019). The socio-economic challenges of smart grids, Chapter 14. In: *Pathways to a Smarter Power System* (ed. A. Taşcıkaraoğlu and O. Erdinç), 397–413. Academic Press.

Corsacoal (2022). Coal in steel making. https://www.corsacoal.com/about-corsa/coal-in-steelmaking/#:~:text=Approximately%201.5%20metric%20tons%20 (accessed 12 of February 2022).

Cuijpers, C. and Koops, B.J. (2013). *Smart Metering and Privacy in Europe: Lessons from the Dutch Case European Data Protection: Coming of Age*, 269–293. Dordrecht: Springer.

Devaraj, D., Syron, E., and Donnellan, P. (2021). Diversification of gas sources to improve security of supply using an integrated Multiple Criteria Decision Making approach. *Cleaner and Responsible Consumption* 3: 100042.

EIA (Energy Information Administration) (2020). Global electricity consumption continues to rise faster than population. US Energy Information Administration. https://www.eia.gov/todayinenergy/detail.php?id=44095 (accessed 10 January 2022).

Electrek (2021). EV vs. ICE: How far can you travel in each state for $100? https://electrek.co/2021/07/27/ev-vs-ice-how-far-can-you-travel-in-each-state-for-100/ (accessed 25 February 2022).

Huang, H., Du, Z., and He, Y. (2018). The effect of population expansion on energy consumption in Canton of China: a simulation from computable general equilibrium approach. *International Journal of Engineering Sciences & Management Research* 5 (1): 19–26.

IEA (2007). Renewables in Global Energy Supply, IEA, Paris. https://www.iea.org/reports/renewables-in-global-energy-supply (accessed 8 February 2022).

IEA (2016). Medium-Term Gas Market Report 2016, IEA, Paris. https://www.iea.org/reports/medium-term-gas-market-report-2016 (accessed 20 February 2022).

IEA (2017). New CEM campaign aims for goal of 30% new electric vehicle sales by 2030. https://www.iea.org/news/new-cem-campaign-aims-for-goal-of-30-new-electric-vehicle-sales-by-2030 (accessed 14 January 2022).

IEA (2021a). Renewables 2021. https://www.iea.org/reports/renewables-2021 (accessed 5 February 2022).

IEA (2021b). *World Energy Model*. Paris: IEA https://www.iea.org/reports/world-energy-model (accessed 5 February 2022).

International Energy Agency. (2000). Hydropower and the environment: Present context and guidelines for future action (Subtask 5 report). Paris.

IPCC (2007). Climate Change 2007: Working Group III: Mitigation of Climate Change. https://archive.ipcc.ch/publications_and_data/ar4/wg3/en/ch4s4-4-3-1.html (accessed 25 February 2022).

IRENA (2020). *Renewable Power Generation Costs in 2019*. Abu Dhabi: International Renewable Energy Agency.

KTH (2022). Renewable Energy Technology Energy Storage modes are determined by the particular end-use applications. file:///C:/Users/230077I/Downloads/Solar%20Energy%20Part%201%20section%203.1-3.4%20Energy%20Storage.pdf (accessed 10 February 2022).

Lenzen, M. and Munksgaard, J. (2002). Energy and CO_2 life cy analysis of wind turbines—Review and applications. *Renewable Energy* 26: 339–362.

Lund, C. and Biswas, W. (2008). A review of the application of lifecycle analysis to renewable energy systems. *Bulletin of Science, Technology & Society* 28: 200–209.

McKinsey & Company (2021). Global gas outlook to 2050 – Summary report. https://www.mckinsey.com/~/media/mckinsey/industries/oil%20and%20gas/our%20insights/global%20gas%20outlook%20to%202050/global%20gas%20outlook%202050_final.pdf (accessed 18 February 2022).

Michalec, A., Hayes, E., Longhurst, J., and Tudgey, D. (2019). Enhancing the communication potential of smart metering for energy and water. *Utilities Policy* 56: 33–40.

MIT (Massachusetts Institute of Technology) (2021). Researchers improve efficiency of next-generation solar cell material. MIT News. February 24, https://news.mit.edu/2021/photovoltaic-efficiency-solar-0224.

Naumenko, D (2018). Russian gas transit through Ukraine after NORD stream 2: scenario analysis. Konrad-Adenauer-Stiftung. Kyiv, Ukraine.

Normile, D (2017). Tesla to build titanic battery facility. Science.

NREL (2021). Documenting a decade of cost declines for PV systems. https://www.nrel.gov/news/program/2021/documenting-a-decade-of-cost-declines-for-pv-systems.html (accessed 22 March 2022).

Okonkwo, E.C., Al-Breiki, M., Bicer, Y., and Al-Ansari, T. (2021). Sustainable hydrogen roadmap: a holistic review and decision-making methodology for production, utilisation and exportation using Qatar as a case study. *International Journal of Hydrogen Energy* 46 (72): 35525–35549.

Outhred, H. (2000). Impacts of electricity restructuring on rural and regional Australia. *Proceedings of the new competitive energy market: how co-operatives and regional Australia can benefit*, Institute for Sustainable Futures, University of Technology, Sydney.

REN21 (2021). Renewables 2020 – global policy report. https://www.ren21.net/wp-content/uploads/2019/05/GSR2021_Full_Report.pdf (accessed 25 March 2022).

Renew Economy (2015). Rooftop solar to cut total grid demand to zero in South Australia. http://reneweconomy.com.au/rooftop-solar-to-cut-total-grid-demand-to-zero-in-south-australia-32943/.

Reuters (2022a). Europe's carbon price nears the 100 euro milestone. https://www.reuters.com/business/energy/europes-carbon-price-nears-100-euro-milestone-2022-02-04/ (accessed 10 February 2022).

Reuters (2022b). Tesla's Musk hints of battery capacity jump ahead of industry event. https://www.reuters.com/article/us-tesla-batteries-idUSKBN25L0MC (accessed 10 March 2022).

Schleisner, L. (2000). Life cycle assessment of a wind farm and related externalities. *Renewable Energy* 20: 279–288.

Schletz, M., Cardoso, A., Dias, G.P., and Salomo, S. (2020). How can blockchain technology accelerate energy efficiency interventions? A Use Case Comparison. *Energies* 13 (22): 5869. https://doi.org/10.3390/en13225869.

Shrestha, R.M., Biswas, W.K., and Shrestha, R. (1998). The implications of efficient electrical appliances for CO_2 mitigation and power generation: The case of Nepal. *International Journal of Environment and Pollution* 9 (2-3): 237–252.

Solar Quotes (2021). Electric cars vs petrol cars: how much can you save? https://www.solarquotes.com.au/blog/electric-vs-petrol-car-savings/ (accessed 25 February 2022).

Sun, J. (2010). *Car Battery Efficiencies. Introduction to the Physics of Energy*. USA: Stanford University.

Tester, J.W. et al. (2005). *Sustainable Energy Choosing Among Options*. The MIT Press.

The World Bank (2022a). Energy use. https://data.worldbank.org/indicator/EG.USE.PCAP.KG.OE (accessed 10 January 2022).

The World Bank (2022b). GDP per capita. https://data.worldbank.org/indicator/NY.GDP.PCAP.CD?locations=NO (accessed 10 January 2022).

Thompson, B. (2008). Decentralised Energy – in the Victorian Context. Brian Robinson Fellowship 2007-08 Report. Moreland Energy Foundation, Australia.

USDOE (2011). *Energy Storage Program Planning Document*. US Department of Energy: Office of the Electricity Delivery and Energy Reliability.

USEIA (2021b). How many gallons of gasoline and diesel fuel are made from one barrel of oil?. https://www.eia.gov/tools/faqs/faq.php?id=327&t=9 (assessed 5 March 2022).

USEIA (2022) Oil and petroleum products explained. https://www.eia.gov/energyexplained/oil-and-petroleum-products/ (accessed 6 February 2022)

USEIA (Energy Information Administration) (2021a) International Energy Outlook 2021 Narrative. US. Department of Energy Washington, DC 20585.

USEPA (2020). Plug-in electric vehicle charging. https://epa.gov/greenvehicles/forms/contact-us-about-green-vehicle-guide (accessed 31 December 2022).

Voelcker, J. (2019). Porsche's fast-charge power play: the new, all-electric Taycan will come with a mighty thirst. This charging technology will slake it. *IEEE Spectrum* 56 (09): 30–37. https://doi.org/10.1109/MSPEC.2019.8818589.

Walker, G. (2008). Decentralised systems and fuel poverty: are there any links or risks? *Energy Policy* 36 (12): 4514–4517.

y Leòn, S.B. and Lindberg, J.C.H. (2022). The humanitarian atom: the role of nuclear power in addressing the United Nations sustainable development goals. In: *Nuclear Law*. The Hague: T.M.C. Asser Press https://doi.org/10.1007/978-94-6265-495-2_13.

Yang, Y., Raipala, K., and Holappa, L. (2014). Ironmaking. In: *Treatise on Process Metallurgy. Volume 3: Industrial Processes* (ed. S. Seetharaman), 2–88. Science Direct. Chapter 1.1.

Part IV

Outcomes

10

Engineering for Sustainable Development

10.1 Introduction

This chapter presents the role of engineers in addressing key sustainability challenges utilising the concepts principles, mechanisms, and technologies discussed in this book. Life cycle thinking, cleaner production and eco-efficiency strategies, industrial symbiosis, and design for environment could potentially change the mindset of engineers and enable them to produce sustainable engineering solutions. Their solutions can help achieve sustainable production and consumption, circular economy, dematerialisation, address water and energy scarcity, and climate change challenges and achieve the sustainable development goals (SDGs). The contents of this book are also aligned with the sustainability codes of engineering ethics discussed at the end of this chapter.

10.2 Sustainable Production and Consumption

Engineers play a pivotal role in achieving sustainable consumption and production (SPC) as it is about promoting resource and energy efficiency, sustainable infrastructure, and providing access to basic services, green and secured jobs, and a better quality of life for current and future generations. The implementation of SPC helps create sustainable development pathways, utilising innovative engineering solutions to reduce future economic, environmental, and social costs, strengthen economic competitiveness, eco-efficiency, and eco-sufficiency. Whilst eco-efficiency strategies are focused on doing more with less, they are not as sustainable as eco-sufficiency strategies, as the former assumes substitutability between different forms of capitals (i.e. gas plant as a replacement of coal power plant), as well as focus on relative efficiency gains (e.g. moving away from 50% efficient fossil fuel plant to 60% plant), largely ignoring the rebound effect (for example, if householders install improved insulation, they should reduce their

Engineering for Sustainable Development: Theory and Practice, First Edition.
Wahidul K. Biswas and Michele John.

heating bills, but they instead maintain their homes at a higher temperature than before, or heat them for longer periods, so the savings may be wholly or partly negated). By contrast, eco-sufficiency is a strategy that takes into account limits to growth, and overall consumption levels and patterns are addressed (Heikkurinen et al. 2019).

Eco-efficiency aims to get the same (or more) product from less material and eco-sufficiency seeks to get the same welfare out of fewer goods and services. Modern engineers should not only assist in decarbonisation by designing electric cars and renewable energy systems but they also need to consider the minimisation of nonrenewable resources (e.g. R strategies) used for producing decarbonisation initiatives. Ride sharing which is based on information technology in recent times is one of the great examples of eco-sufficiency as it provides the same service as an efficient transport or a hydrogen powered taxi with less cost, less environmental impact and less time, and reduces the number of vehicles on the road.

Sustainable consumption and production therefore is the increase of net welfare gains from economic activities by reducing resource use, degradation, and pollution throughout the whole lifecycle, while increasing the quality of life. SPC is the production and use of products and services in a manner that is socially beneficial, economically viable, and environmentally benign over their whole life cycle. Life cycle assessment can therefore make engineers critical thinkers as it enables them to find ways to reduce energy, material consumption, and waste generation during the overall life of the products or services. Hydrogen is a clean fuel and more kilometres can be driven with far less emissions compared to the cars driven by petroleum products, but at the same time, it needs to be remembered that the production of hydrogen is not causing the scarcity of water and the emission of pollutants from powering the electrolysers. LCA is a decision-making tool enabling engineers to choose the right pathway or design to achieve resource efficiency by taking both upstream (i.e. until production) and downstream processes into account. The concept of life cycle thinking facilitates engineers' awareness and knowledge on sustainable consumption by taking into actors operating in the supply chain, from producer to final consumer and also the outputs of life cycle analysis enable engineers to provide justifications for environmental standards and labelling and to engage in increasing sustainable public procurement processes. The success of life cycle sustainability depends on the involvement of different stakeholders, including business, consumers, policy makers, researchers, scientists, retailers, media, and government development cooperation agencies, among others. If policy makers do not develop climate changes policies, engineering innovation on carbon footprint technology will not fully develop. The earlier the policies are developed and implemented, the better the long-term outcome.

SPC has a close link to many sectors in the economy including food, transport, tourism, construction, and health for reducing air, water, and sanitation, energy, and waste. The economic growth of a nation or a community should not result in the disruption of eco-systemic interrelations due to extensive levels of extraction, production, and consumption affecting the inter- and intra-generational sustainability. Responsible consumption and production is thus required by the people who are directly or indirectly involved and effected during the product or service life cycle stages. An engineer or an engineering organisation needs to be a role model to prompt others and other organisations to change the way they behave and work so that we can achieve good economic growth without an ecological crisis. In an increasingly technologically dependent society, engineers should eliminate the negative sustainability implications of engineering decisions on people's lifestyles and provide technical solutions to help achieve SPC outcomes.

SPC looks uniquely at the interactions between technology, consumption, policy, and the environment to help identify more-sustainable solutions for both production and consumption systems. Behavioural and technological changes are equally important to influence our production and consumption, but engineers have more control over technology development, which is in fact one of the key factors contributing significantly to the social changes required.

10.3 Factor X

'Factor X' is a technology factor that stands for an intelligent, efficient, and environmentally friendly use of natural resources by increasing resource efficiency through a Technological Factor of X. This is an important determinant of sustainable production and consumption. The term Factor X goes back to the 'Factor 10' concept developed by Friedrich Schmidt-Bleek in the early 1990s (Umweltbundesamt 2021). If sustainable development is to be achieved globally by continuing the flow of resources to humankind, industrialised countries must reduce their consumption of resources by a factor of 10 (= roman X), or 90%, within 50 years. As discussed in Chapter 1, a factor of 4 can double wealth and halve resource consumption. The idea behind Factor X is that the use of natural resources must become X times more intelligent and efficient to generate X times more wealth from the same ton of raw material (GRDC 2022).

The growing intensive use of natural resources is increasing our ecological footprint, exceeding earths biocapacity and exacerbating global environmental problems and resource scarcity. In the last 40 years, the global extraction of raw materials has more than tripled to around 85 billion tonnes per year (Umweltbundesamt 2021). Today, this already significantly exceeds the earth's regenerative capacity and endangers the development opportunities for future generations. The quality of ore has been decreasing and now more investment is needed than

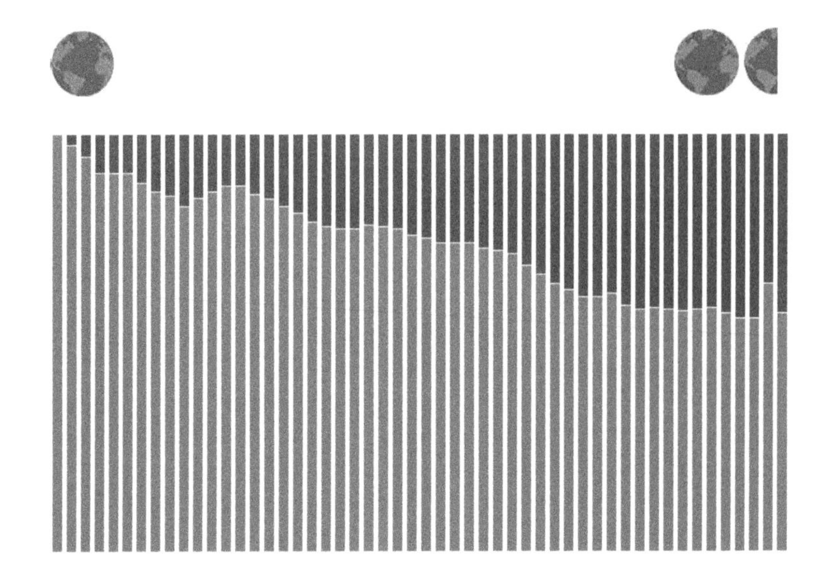

Figure 10.1 Overshoot day (Global Footprint Network 2021).

in the past to obtain the same amount of minerals, requiring increasingly large areas of land causing the loss of biodiversity.

Ecological Footprint tracks the use of productive surface areas and is a potential tool to jointly measure planetary boundaries and the extent to which humanity is exceeding them. On the supply side, a city, state, or nation's biocapacity represents the productivity of its ecological assets (including cropland, grazing land, forest land, fishing grounds, and built-up land). If a population's Ecological Footprint exceeds the region's biocapacity, that region runs a biocapacity deficit. Recent data shows that our current level of consumption, resource demand or ecological footprint exceeded the world's bio-capacity. We are currently consuming 1.7 earth equivalent amounts of resources, and yet we were consuming resources within the earth's carrying capacity in 1970 (Global Footprint Network 2021). Technological factor X is needed to be increased through engineering innovation to help us reach an equilibrium where our consumption or ecological footprint can at least meet the world bio-capacity. However, Engineers cannot solve the problem if policy makers do not implement policies (e.g. tax, subsidies) to change people's behaviour and industry's consumption patterns.

Both Ecological Footprint (EF) and Bio-capacity (BC) are expressed in global hectares and are globally comparable. A formula for calculating them are provided in Box 10.1 based on Borucke et al. (2013) (Figure 10.1).

Box 10.1 Calculation of Bio-capacity and Ecological Footprints

$$\mathrm{BC}_j = A_J \times \left(\frac{Y_L}{Y_W}\right)_J \times \mathrm{EQF}_J$$

$$\mathrm{EF}_j = \frac{Q_j}{Y_W} \times \mathrm{EQF}_j$$

$$\mathrm{EF}_C = \frac{P_C \times (1 - S_{\mathrm{Oacen}})}{Y_C} \times \mathrm{EQF}_\mathrm{j}$$

$\mathrm{BC} < \mathrm{EF}_{\mathrm{Tot}}$ Deficit or overshoot

$\mathrm{BC} > \mathrm{EF}_{\mathrm{Tot}}$ within carrying capacity

j (= 1, …, 5) indicates the 5 land-use types for bio-capacity calculation (e.g. cropland, grazing land, fishing grounds, forest, and built-up land).
A_J = Area (Ha) of type j; Y_L and Y_W = Local and global yields (t/Ha).
EQF_J = Conversion of land use type j into global average bio-productivity.
j (= 1, …, 6) indicates the 6 land-use types for ecological footprint calculation (e.g. cropland, grazing land, fishing grounds, forest, built-up land, and carbon footprint).
Q_j = annual production in tonne.

The following data is provided to calculate the direct ecological footprint and bio-capacity of a country X.

	A	Q_c	Y_c	Y_w	EQF	BC	EP
Cropland	0.20	0.43	2.31	1.1	2.21	0.93	0.86
Forest	0.15	0.82	8.2	2	1.34	0.82	0.55
Grazing	0.10	0.16	2.64	1.2	0.49	0.11	0.07
Fishing	0.07	0.68	6	2	0.36	0.07	0.12
Build	0.08	0.08	1	1	2.21	0.166	0.17
C-footprint[a]							2.1
						2.1	3.87
						Deficit	−1.77

a) This was calculated using the equation of EF_C (see above) and the information below.

Pc (CO_2 emissions)	15 tonne/Ha
Sequestration by the Ocean	82%
Y_{seq} by the vegetation	1.8 tonne/Ha
EQF for sequestration	1.4

There still exists significant resource efficiency potential in many areas of production and consumption which so far has largely remained untapped and is in some cases not yet sufficiently known such as the recovery of some rare earth materials, further efficiency increases in renewable energy technologies, complete conversion and or capture of emissions and wastes to useful products (zero waste) and reverse logistics. In order to achieve resource efficiency goals or Factor X, scientists, economists, politicians, product designers, consumers, and the service sector must work hand in hand with creativity and stewardship dedication. If policy makers and other stakeholders like manufacturers and consumers do not take action to promote, make, and use innovative engineering solutions, global sustainability will continue to decline. Ultimately, the tasks for engineers will increase or may even go beyond their capacity to address the resource scarcity problems. If a collapse of carrying capacity or irreversible damage of an ecosystem take place, neither hi-tech solutions nor trillion-dollar investments will work to bring back nature or establish homeostasis in an ecosystem. The economy must become more efficient in its use of resources, and the supply of and demand for resource-efficient products and services must increase for the benefit of future generations.

Countries around the globe are consuming resources at an untenable rate, with developed nations consuming more than their fair share. As such, developed nations are promoting an unsustainable model of consumption not only by consuming more than their share of world resources but also by creating additional ecological footprints in other nations in order to obtain raw materials to run their businesses. Consequently, if nations want to ensure they do not exceed the planet's carrying capacity and want to provide adequate resources for future generations, changes in resource use, behaviour or consumption patterns, government policies, and technological advancement are required. Engineering input to this process is essential.

10.4 Climate Change Challenges

The greenhouse effect is when heat is trapped close to the Earth's surface by 'greenhouse gases'. These heat-trapping gases can be considered as a blanket wrapped around the globe, keeping the planet warmer than it would be without them (NASA 2022). Greenhouse gases include carbon dioxide mainly from fossil fuel combustion, methane mainly from pasture and rice production, nitrous oxides mainly from fertiliser application, and water vapour, predominantly resulting from the human activities. They are known as anthropogenic emissions.

As can be seen in Figure 10.2, global temperature (coloured bars) and atmospheric carbon dioxide (grey line) increased more slowly during the first half of the late nineteenth and early twentieth centuries. Atmospheric carbon dioxide levels

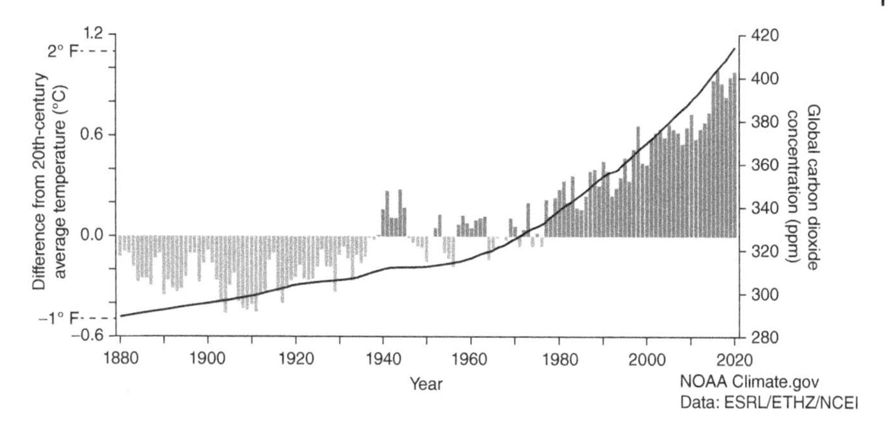

Figure 10.2 Relationship between global atmospheric CO_2 and surface temperature. Source: Lindsey 2021/U.S. Department of Commerce/Public domain.

began to rise by around 20 parts per million over the 7 decades from 1880–1950, while the temperature increased by an average of 0.04 °C per decade. Over the next seven decades (i.e. from 1950 onward), however, carbon dioxide climbed nearly 100 ppm – five times as fast as the rate of warming averaged 0.14 °C per decade. This mirrors quite well the population growth and associated energy consumption experienced during that time (Figure 10.3). A massive technological development began after the first half of the twentieth century when people began to have access to energy, better healthcare facilities, hygiene, transport, and home comforts. Consumption and production have significantly increased with population growth over last 7 decades. With the increase in production, the combustion of fossil fuels increased to power the modern economy and even still in early twenty-first century a signification share of energy is still coming from fossil fuels which are the main cause of global warming.

The concentration of atmospheric CO_2 is currently 450 ppm that has resulted in an increase of temperature by 1.4 °C above pre-industrial levels (National Research Council 2011). New findings have strengthened the link between climate change and wildfires. Even if warming stays below 1.5 °Celsius, 40% more land could still experience more wildfires. Temperature, fuel load, dryness, wind speed, and humidity associated with climate change all affect fire risk and are compounded by global warming. Strong winds and scorching heat have prolonged fires which is not normal. Climate change is lengthening the bushfire season causing longer and hotter heatwaves. Bushfire can be actually important to rejuvenate an ecosystem, but anything excess is not good. These fire conditions are now more dangerous than in the past. Instead of burning a certain amount of forest area, it is now burning surrounding areas affecting cropland, pasture,

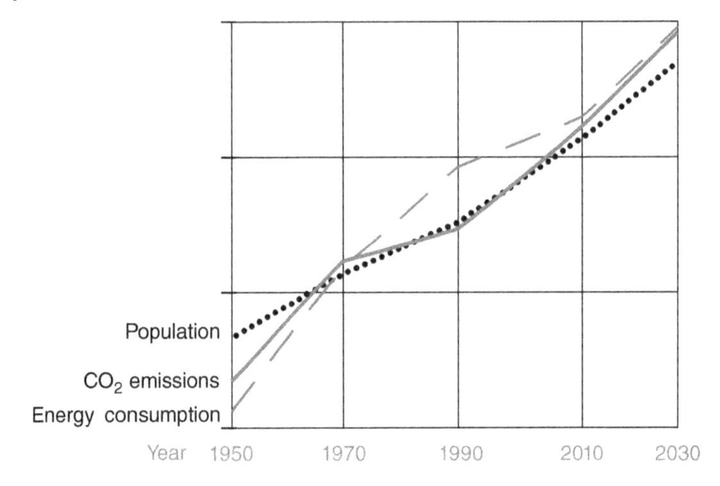

Figure 10.3 Relationship between population, energy, and CO_2 emissions. Source: Markham 2008.

tourism industries, and human settlements. At the same time, the costs for preventing fires are increasing.

For every degree of warming, seven percent more water can be absorbed by the air and therefore, there is an increased risk of flooding. The world may lose half of its sandy beaches by 2100 due to sea level rise that will affect many island and Asian nations and countries like Australia and Bangladesh where many people live on the coast. Average temperatures in Kazakhstan have increased by 0.3 °Celsius every decade and caused significant animal deaths (Reuter 2021). Also some 42% of California's population is now living under a drought emergency (ABC 2021).

There has been a significant increase in the rate of average glacier melt since the 1990s. The Greenland ice sheet is losing more ice than it used to gain by snowfall. Ocean levels are now more than 20 cm higher than in 1870. Melting ice sheets in Greenland and Antarctica contributed to 14 mm of sea level rise between 2003 and 2019 (ABC 2020). By the end of the century, the global mean sea level is likely to rise at least 0.3 m above 2000 levels, even if greenhouse gas emissions follow a relatively low pathway in coming decades (NOAA 2022).

The three most common 'at-risk' definitions associated with SLR are as follows (Hauer et al. 2020).

Populations living in low-elevation coastal zones (LECZ). This zone encompasses any area under 10 m in elevation and sometimes within a 100 km of a coast. This group of people might need to relocate as a result of SLR. In the case of Australia, Australians mainly live on the coast as a large portion of the country is dry and arid. If coastal areas are greatly impacted, the displacement of human settlements and businesses will create very significant impacts.

Populations living in a 100-year floodplain. Flood plains are suitable for agriculture as they possess fertile soil; however, residents might experience various SLR associated hazards that could influence migration decisions, such as severe storm surges, and flooding, saltwater will gradually intrude into surface waters and soils and groundwater wells, and there will be a shift in sediment regions and coastal erosion.

Populations living in areas that would be inundated under selected SLR scenarios. This group of people is 'at risk' of relocation due to SLR, resulting in human migration and permanent inundation. Bangladesh, Netherlands, and the Pacific Island nations are more vulnerable to SLR compared with other countries.

Engineers could potentially respond to the above SLR hazards using the following approaches: protection, accommodation, and migration. Protection refers to armouring design to prevent the hazards. Engineers can build seawalls, groins, and other infrastructure that maintain and expand the current shoreline, and in some instance, this can provide protection against storm surges. Secondly, accommodation which refers to measures designed to facilitate living with the SLR. Engineers for example can help build houses on raised platform while ensuring the structural performance of the building. Thirdly, migration or retreat, refers to the relocation of individuals or communities away from the hazard. The relocation can be expensive for developing nations as new settlement, infrastructure, and electricity networks need to be re-established. Engineers can make prefabricated or modular types of houses without changing the landscape and install decentralised power generators using renewable energy technologies in a relocated area.

Also coral reefs will be affected due to thermal stress, sedimentation due to SLR (i.e. toxic pollutants come with floodwater to the seabed), reduction of pH (i.e. acidic water), and also due to global warming and increased concentration of CO_2 in the atmosphere.

A global mean temperature increase of 2.5 °C would result in the average person losing 1.3% of her/his income. Developing countries are more vulnerable to climate change (CC) for three reasons: First, they are typically exposed to adverse weather like drought, SLR, and flooding. They depend on climate as they are often primary producers. Therefore, the adverse weather resulting from climate change will affect their productivity and economy more. Secondly, these countries tend to be in hotter places, which will increase the required cooling load. Thirdly, poorer countries tend to have a limited adaptive capacity, and the affordability to access new technologies and new businesses is more difficult.

If the temperature increases from 1.4 to 4.5 °C by the end of the century, the sea level will rise and a large portion of low laying countries will go under water and forest fires and intense flooding will increase. Apart from human ecological footprint, climate change consequences will destroy some land resources. Biocapacity will be reduced further and global carrying capacity could collapse. Drastic action is needed to reduce CO_2 concentration to levels that will keep temperature at 1.5 °C above the pre-industrial level. In order to maintain this level of CO_2 concentration in the atmosphere, a net zero emission target needs to be reached by 2050. A significant technological breakthrough is need to achieve this target, which is discussed below. Here no new oil and gas fields can be approved for development. Also there will be no new coal mines or coal mine extensions. An unwavering policy focus on climate change towards a net zero carbon pathway will result in a sharp decrease in fossil fuel demand, resulting in oil and gas producers switching entirely to 2050 zero emissions reductions.

As can be seen in Figure 10.4, the emissions from electricity and heat generations will be reduced by 50% of 2020 levels by 2030 mainly because of the reduction in coal power plant electricity generation from 2020. 60% of total cars produced by 2030 are expected to be electric vehicles, resulting in transportation emission reductions of around 20%. This is because the number of cars produced during this period has increased by 40% and so there will still be quite a large number of ICE cars in 2030 causing CO_2 emissions in the transport sector. In 2030, there will also be some old building infrastructure but any new building built during this period will be zero carbon produced and so the GHG emissions from the building sector slightly decrease. By 2030, most industries will begin to introduce clean technologies at some scale.

By 2035, the emissions from the electricity sector will be almost abated as all unabated coal power plants will be are phased out by 2030. There will not be that many internal combustion engine cars in the market. About 50% of trucks will be powered by electricity. Building energy appliances will be improved significantly especially cooling systems which account for a significant portion of energy during the building life cycle.

By 2040, emissions from the energy and building sector will totally be reduced. Emissions from transportation will be further reduced as 50% of the fuels in the aviation sector will be replaced with green electricity. Although the efficiency of the main end use appliances in the industrial sector (i.e. motor) will be increased, emissions will not be decreased to the required level due to the use of high-grade thermal energy from fossil fuels.

Heat pumps will meet 50% of the energy demand in the residential sector, 85% of buildings will be zero carbon ready by 2050. The hydrogen fuel car along with electric vehicles are replacements for petroleum products, PV and wind account for 70% of electricity. About 7.6 GT of CO_2 will be captured by carbon capture and storage (CCS) and by maintaining a reasonable amount of vegetation coverage.

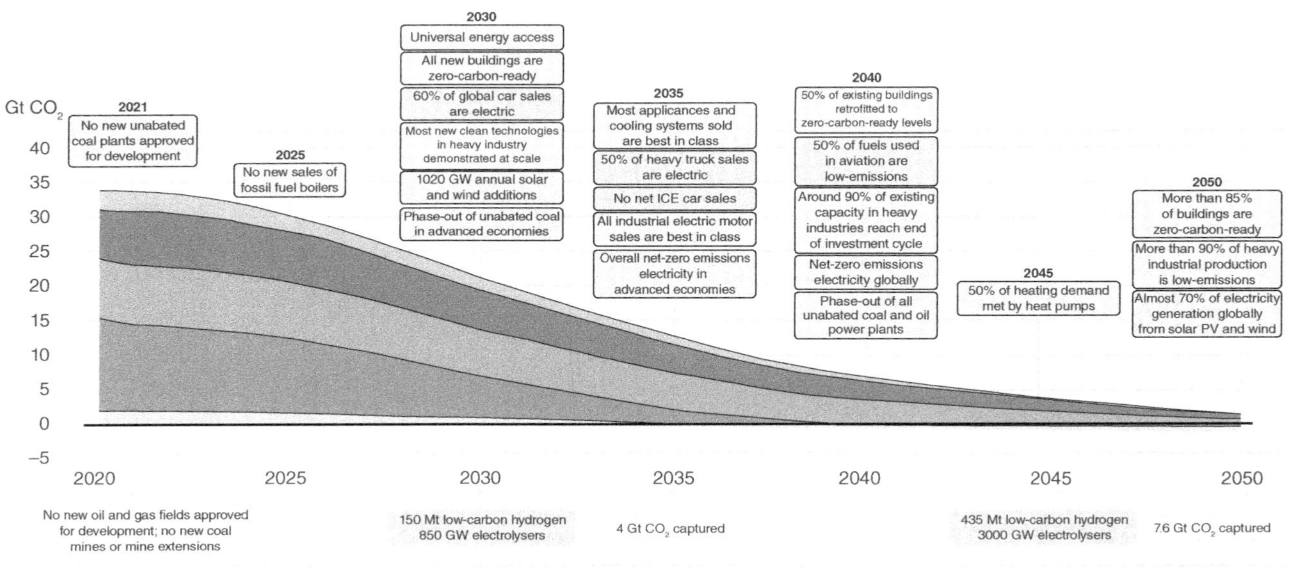

Figure 10.4 Net Zero by 2050, IEA, Paris. Source: IEA 2021.

As discussed above, clean electricity generation, network infrastructure, and end-use sectors are going to be key areas for increased investment. Enabling infrastructure and technologies are vital in transforming the energy system. Annual investment in transmission and distribution grids is expected to be three times more in 2030 (US$ 820 billion) than that we spend now (IEA 2021). The number of public charging points for EVs will be 40 times more in 2030 than the number of charging stations in 2021. Annual battery production for EVs needs to increase from 160 gigawatt-hours (GWh) today to 6600 GWh in 2030. This is in fact equivalent to the addition of almost 20 gigafactories of energy production every year until 2030 (IEA 2021). The required roll-out of hydrogen and carbon capture use and storage units after 2030 will require a large amount of annual investment in CO_2 pipelines, and hydrogen-enabling infrastructure investment is expected to be 40 times higher in 2030 than that in 2021 (IEA 2021).

10.5 Water Challenges

Less than 3% of the world's water is potable, of which 2.5% is frozen in the glaciers of Antarctica and Arctic region (United Nation 2018). Water in rivers and lakes has been polluted faster than nature can recycle and purify due to the discharge of increased levels of industrial effluents containing toxic chemicals (e.g. heavy metals like chromium, nickel, and cadmium) and agricultural effluents (ammonium nitrogen, herbicides, and pesticides) which causes air, land surface, and soil pollution and damage to human health and world ecosystems. Also because of the exponential rise in population and increased levels of industrial operations to meet the growing demand for goods and services, excessive use of water contributes to global water stress as more than one billion people still do not have access to fresh water. The application of Green Engineering principles including atom efficiency, use of renewable feedstock, minimisation of chemical derivatives, process intensification, and bio-degradability, could potentially reduce the toxicity of effluents discharged into the surrounding environment. The concept of industrial symbiosis or wastewater synergy can reduce or avoid the disposal of effluent by enabling one industry to use wastewater or effluent from another neighbouring industry and prevent discharge into natural waterways (rivers).

Water is a free gift of nature, but the infrastructure needed to extract, treat, and deliver it is capital and resource intensive. Water consumption has been increased worldwide by about 1% per year since the 1980s. Over 2 billion people live in countries experiencing high water stress, where groundwater recycling schemes could help reduce water supply challenges by reducing water footprints (Boretti and Rosa 2019). The integration of wastewater treatment with groundwater replenishment (i.e. injection into the groundwater) and groundwater treatment plants (i.e. after the withdraw of water) can help convert wastewater into drinking water. This water extraction takes places after a number of years to overcome the psychological stigma of drinking wastewater. This scheme has a number of benefits in a water scarce region. Firstly, this helps to recharge the underground aquifer, which could potentially prevent land subsidence. Secondly, it avoids the disposal of treated wastewater into the sea. Thirdly, it is less energy intensive than an energy intensive desalination plant. However, further improvement can be made to this groundwater recycling scheme by operating it using renewable energy technologies. Agriculture (including irrigation, livestock, and aquaculture) are the largest water consumer, accounting for about two third of annual water withdrawals globally and the rest is consumed by industrial (including power generation) and residential sectors. Municipal wastewater can be treated to produce treated sewage effluent (TSE) for land scaping and irrigation purposes making sure that heavy metals are non-existent in TSE. Also the same can be used in cooling towers and district cooling systems in a water stress region.

As discussed before, climate change is intensifying rainfall and flooding in some areas. Cities and urban areas produce large volumes of runoff and point and diffuse sources of contamination. The prevalence of impervious surfaces (i.e. pavements, carparks) increases the velocity of water flow and inhibits ponding and the infiltration of water into the soil. During large rainfall events, stormwater flows can rapidly enter urban streams and carry high levels of nutrients, sediment, and heavy metals to nearby water bodies or the ocean. The construction of permeable pavements can reduce surface runoff by more than 50% and have an influence on the reduction of flood peaks and the hysteresis of flood peaks (Zhu et al. 2019). Permeable roads help reduce the runoff coefficient and flood peak flow.

10.6 Energy Challenges

Despite technological advances that have promoted energy efficiency gains, global energy consumption is projected to increase by about 30% and electricity production is expected to double by 2050 (IAEA 2020). Therefore, developing ways to maximise the share of renewable energy in the electricity mix in a cost-competitive manner is an enormous challenge for engineers not only to address global warming issues but also to address energy scarcity. The use of the block chain and smart grid concepts could potentially maximise the use of renewable energy in the grid in a cost-effective manner. Commercial and residential energy use is the second most rapidly growing area of global energy use after transport. In addition to lighting efficiency, another appliance which needs to be targeted is heating, ventilation, and air conditioning (HVAC) units as typically heating and cooling accounts for 80% of the total energy consumption in a building (Biswas 2014). Heat pumps, as discussed in Chapter 7, are an electrical climate-control system which provides heat as well as refrigerated cooling and are also known as reverse-cycle air conditioning. Heat pumps are machines that can pump heat in both directions – from the inside to the outside (cooling) and from the outside to the inside (heating). Reverse-cycle air conditioning is 300–600% efficient, which means that it can take one unit of electrical energy and turn it into three to six times as much heating or cooling energy (DIESR 2022). Under mild conditions, some products can achieve efficiencies of over 1000%. Temperatures will be increasing due to climate change, the cooling load is likely to increase. It is therefore necessary to consider the use of heat pumps to reduce the energy consumption for cooling.

Worldwide car sales will grow to around 66.7 million automobiles in 2021, which is expected to increase to about 3 billion by 2050 (Statista 2021; Fuel Freedom Foundation 2022). With the International Energy Agency predicting an 800% increase in EVs over the next decade to meet net zero emissions targets, electric car battery recycling will soon become a major task for engineers not only to address gradually decreasing availability of lithium resources but also to avoid electric car battery waste and electric car battery disposal pollution problems (Krause et al. 2020). This battery recycling is equally important for renewable energy power systems as they are storage dependent due to the intermittent nature of renewable resources and also these resources have less energy intensity compared to conventional energy sources. Mimicking nature is important as engineers modified an enormous aircraft to mirror the anatomy of the owl to reduce turbulence, noise, emissions and the fuel cost.

The global electrification rate reached 89% in 2017 (from 83% in 2010), still leaving about 840 million people without access to electricity (United Nations 2018). The global population without access to electricity fell from 1.2 billion in 2010 to 840 million in 2017. Decentralised electricity generation using locally available resources has to be introduced through an innovative financing system as discussed in Chapter 9 to make it more affordable, accessible, and acceptable by disadvantaged community as they account for a significant portion of rural populations in developing nations.

10.7 Circular Economy and Dematerialisation

There is only one earth, but the Western world is currently consuming 1.7 times more resources than the earth can replenish (Global Footprint Network 2021). In the twenty-first century, the earth will not have enough natural resources to sustain population growth. Our economy has seen 'linear' growth for a very long time, with the traditional 'Take-Make-Use-Dispose' functions dominating. This means that raw/virgin materials are used to make products to meet societal demands, and then after use they are thrown away with no recovery and reuse of their finite and gradually depleting resources. The size of both source and sink capacities are continually increasing leaving inadequate areas or bio-capacity to meet the demand of resources for future generations. To ensure there are enough resources for future generations, we need to switch from the current linear economy to a circular economy (CE) model through dematerialisation using possible technological functions of the 6Rs (Reduce, Reuse, Recycle, Recover, Redesign, and Remanufacture), digitisation, enhanced conversion efficiency, and maximisation of the use of renewable resources. In a circular economy, products need to be designed and manufactured in a way that the end-of-life (EoL) products can either be converted to usable products or at least the precious, valuable, or rare earth materials and other resources in the products can be recovered from these products to reduce the rate of extraction of virgin materials. CE focuses on substituting products for services in a way that increases utility and longevity of both the products and the materials used. The concept of design for disassembly or remanufacturing enables some businesses like office equipment, vertical transport to operate, maintain, and take them back after the end of life for remanufacturing. Customers pay for the service instead of buying directly from the shops. For example, Uber and other ride-hailing apps do not own any cars, nor do they employ any drivers. The apps connect drivers around the world to potential clients. On the sustainability front, ride-hailing apps are better for the environment as Uber is planning to take 1 million cars off the streets of New York City (Uber 2015). This ride sharing approach will not only reduce fuel and vehicle kilometres travelled but also a significant amount of energy intensive material consumption can be avoided due to the reduction in the number of cars required.

The life cycle approach or supply chain management (SCM) needs to be considered in all engineering design to reduce energy, material consumption, and waste generation and to achieve a circular economy. CE for SCM focuses on closing resource loops which is not an easy task. There are different levels of SCMs to achieve a CE, including slowing resource loops, narrowing resource loops, dematerialising resource loops, and intensifying resource loops which are discussed below (Hazen et al. 2021).

- In detail, closing resource loops as a practice to reuse the materials through recycling, remanufacturing, retrofitting, refurbishing, and recovering (Geissdoerfer et al. 2018), whereas slowing resource loop designs durable goods or increases the product's service life which depends on the type of material chosen and the maintenance performed during the life of the product (Leising et al. 2018). It is impossible to achieve a 0% circular gap or a 100% CE due to the unavoidable system losses following the second law of thermodynamics. For example, it is not physically possible for a heat engine to convert 100% heat to work and so no matter the efficiency of technology, the materials or energy used cannot be 100% converted to recovered product or service.

- The narrowing loop involves resource efficiency via using fewer resources per product and still being able to conserve resources for future generations (Bocken et al. 2016), and the dematerialising resource loop refers to substituting product utility with services and software solutions (e.g. online delivery, virtual conference, multifunctional devices) with the purpose of increasing longevity or the conservation of resources by avoiding fuel, transportation, and the need for infrastructure. Dematerialisation can include digitising processes, for instance by using communications technology to foster digital collaboration, or servitising products, as in the case of eBooks. However, most supply chain processes and products do not lend themselves to complete dematerialisation. The goal for most companies, therefore, is to rethink their supply chain to ensure that they are using the most eco-effective material processing in the most eco-efficient way possible.
- Finally, the intensifying resource loop motivates a more intensive product use phase that creates more efficient value (Geissdoerfer et al. 2018). For example, instead of buying a dry and wet vacuum cleaner or a lawnmower, they can be rented from a shop as these machines are not used on a daily basis.

10.8 Engineering Ethics

According to the World Federation of Engineering Organizations code, engineering professionals should use their knowledge and skills for the benefit of the world, in order to create engineering solutions for a sustainable future and strive to serve the communities ahead of any personal or sectional interests (WFEO 2022). Engineers need to consider the impact of the project on the ecosystem before they sign off on the contract. The sustainability related code of ethics of World Federation of Engineering Organizations is as follows

Create and implement engineering solutions for a sustainable future: this means that engineers in practice need to be aware that the principles of ecosystem (i.e. interaction and interrelationship between living and non-living organisms), diversity maintenance (enriched flora and fauna), resource recovery, and inter-relational harmony (between species as well as with air, soil, and water) form the basis of humankind's continued existence and that each of these poses a threshold of sustainability that should not be exceeded in order to maintain a liveable environment for future generations. The destruction of ecosystems or the creation of ecological imbalance due to engineering activities such as construction, mining, and refining, could affect the critical capital like water tables and air and soil quality, and cause deforestation and the loss of bio-diversity. This is because air, water, and soil have certain capacity to absorb waste and pollutants and the pollutants and wastes generated from the engineering activities should not exceed the critical limit beyond which the environment is neither liveable for human beings nor for other species like birds, fishes, micro-organisms, etc.

Engineers in particular should consider the triple bottom line consequences of proposals for projects and actions, direct or indirect, immediate or long term, upon the health of people, social equity, and the local system of values. As discussed in this book, it is important to make sure during the design stage that wastes and emissions are reduced during the life cycle stages of the products. Further green engineering principles are to be considered to avoid health hazards, costs, energy, and waste.

Engineers must promote a clear understanding of the actions required to restore the environment in their proposal or engineering design report (EDR) and also, if possible, to incorporate improvement plans for the environment if it is disturbed by their activities.

Be mindful of the economic, societal, and environmental consequences: Engineers need to be mindful of the triple bottom line or economic, societal, and environmental consequences of their engineering designs so that precautionary approaches can be taken into account in order avoid future negative consequences.

Engineers need to make sure in practice that their understanding of potential environmental impacts is as accurate as possible. They need to strive to satisfy the beneficial objectives of their work with the lowest possible consumption of materials and energy and the lowest amount of wastes and pollution generation.

Engineers must study the environment that will be affected by their work, assess the impacts that might arise in the structure, dynamics, and aesthetics of the ecosystems involved – urbanised or natural – as well as pertinent socioeconomic systems, and select the best alternatives for development that are both environmentally sound and sustainable.

Engineers should reject any kind of commitment that involves unfair damage to human surroundings and nature and aim for the best possible technical, social, and political solutions. Engineers should know that clients and employers are often vary aware of societal and environmental consequences of actions or projects and endeavour to present engineering issues to the public in an objective and truthful manner. They should not hide behind contractual clauses (or the minimum requirements of codes of practice) when they know the overall process is flawed and does not deliver quality and/or value for money for the end user.

Promote and protect the health, safety, and well-being of the community and the environment: Engineers in practice should have due regard for the health, safety, and well-being of the public and fellow employees in all work for which they are responsible. They should not accept a narrow or inadequately framed design commission within a design and build delivery model when there is no certainty their design will be appropriately integrated with other parts of the project. Also they should not operate in a commercially competitive landscape with pressures to produce 'leaner' designs to save costs by compromising safety and long-term performance of their design.

Engineers should try to the best of their ability to obtain a superior technical achievement which will contribute to and promote a healthy environment for all people, in open spaces as well as indoors.

Engineers should inform their employer or contractor of the possible consequences of their recommendations on issues of safety, health, welfare, or sustainable development if they are overruled or ignored.

10.8.1 Engineers Australia's Sustainability Policy – Practices

Specific sustainability considerations could include (Engineers Australia 2020):

The use of resources should not exceed the limits of regeneration meaning that engineers should take resource availability or bio-physical limits of resources into their design process.

The use of nonrenewable resources should create enduring asset value (everlasting and/or fully recyclable), and be limited to applications where substitution with renewable resources is not practical. For engineers, the design for the environment concept is important as the product can be designed in a way that it can be disassembled after the end of life for reuse or recycling or remanufacturing.

Engineering design, including product design, should be whole system based, with consideration of all impacts from product inception to reuse/repurposing. As discussed in Chapter 3 in this book, the knowledge on life cycle assessment would make engineers system or critical thinkers in their technology or product design process to achieve triple bottom line sustainability benefits.

Product and project design should consider longevity, component re-use, repair, and recyclability. Engineers should consider the use of recycled materials or byproducts in product design while maintaining the structural integrity of the recycled base materials. For example, concrete using a certain percentage of fly ash and a certain percentage of recycled aggregates has been found to achieve structural performance for both high and low compressive strength applications.

Eliminating waste should be a primary design consideration as unavoidable waste from any process could potentially be examined for recycling potential as an input into another productive process. The concept of industrial symbiosis can potentially be applied to fulfil the sustainability objectives of Engineers Australia.

The rate of release of any substances to the environment should do no net harm, and be limited to the capacity of the environment to absorb or assimilate the substances, and maintain continuity of ecosystem services. In all instances, such releases should be lifecycle-costed and attributed. The application of green chemistry and green engineering principles could potentially address these issues and also engineers should help industries to internalise the external cost by introducing preventative measures like pollution control, waste minimisation, and air pollution technologies.

Proactive and integrated solutions are preferable to reactive, linear, 'end of pipe' solutions, such that there is a net sustainability benefit. As discussed in Chapter 5, the application of cleaner production strategies can prevent waste generation and atmospheric emissions in an economically and environmentally friendly way.

In circumstances where scientific information is inconclusive, or incomplete, the precautionary principle and risk management practices should be applied to ensure irreversible negative consequences are avoided.

References

ABC (2020). Satellites show melting ice sheets in Antarctica and Greenland have contributed to 14 mm sea level rise in 16 years. https://www.abc.net.au/news/science/2020-05-01/nasa-satellites-ice-loss-in-antarctica-greenland-sea-level-rise/12200748 (accessed 17 March 2022).

ABC (2021). California's state's emergency drought proclamation now covers 50 of state's 58 counties. https://www.abc10.com/article/weather/california-drought/gov-newsom-californians-voluntary-cut-water-use-drought/103-36f79aa8-0b84-4abb-b9b5-25f3e5e6c812#:~:text=California\stquotes%20state\stquotes%20emergency%20drought%20proclamation,42%25%20of%20the%20total%20population (accessed 20 March 2022).

Biswas, W.K. (2014). Carbon footprint and embodied energy consumption assessment of building construction works in Western Australia. *International Journal of Sustainable Built Environment* 3: 179–186.

Bocken, N.M., De Pauw, I., Bakker, C., and van der Grinten, B. (2016). Product design and business model strategies for a circular economy. *Journal of Industrial and Production Engineering* 33 (5): 308–320.

Boretti, A. and Rosa, L. (2019). Reassessing the projections of the World Water Development Report. *Clean Water* 2: 15. https://doi.org/10.1038/s41545-019-0039-9.

Borucke, M., Moore, D., Cranston, G. et al. (2013). Accounting for demand and supply of the biosphere's regenerative capacity: The National Footprint Accounts' underlying methodology and framework. *Ecological Indicators* 24: 518–533. https://doi.org/10.1016/j.ecolind.2012.08.005.

DIESR (Department of Industry, Energy Science and Resources) (2022). *Heating and Cooling*. Australian Government: Households.

Engineers Australia (2020). Engineers Australia's sustainability policy–practices. https://www.engineersaustralia.org.au/Communities-And-Groups/Colleges/Environmental/Publications-Policy (accessed 22 March 2022).

Fuel Freedom Foundation (2022). What cars will we be driving in 2050?. https://www.fuelfreedom.org/cars-in-2050/ (accessed 15 March 2022).

Geissdoerfer, M., Morioka, S.N., de Carvalho, M.M., and Evans, S. (2018). Business models and supply chains for the circular economy. *Journal of Cleaner Production* 190: 712–721.

Global Footprint Network (2021). Ecological footprints. https://www.footprintnetwork.org/our-work/ecological-footprint/ (accessed 25 March 2022).

GRDC (2022). Factor X. SD features. https://www.gdrc.org/sustdev/concepts/11-f10.html (accessed 20 March 2022).

Hauer, M.E., Fussell, E., Mueller, V. et al. (2020). Sea-level rise and human migration. *Nature Reviews Earth & Environment* 1: 28–39. https://doi.org/10.1038/s43017-019-0002-9.

Hazen, B.T., Russo, I., Confente, I., and Pellathy, D. (2021). Supply chain management for circular economy: conceptual framework and research agenda. *The International Journal of Logistics Management* 32 (2): 510–537. https://doi.org/10.1108/IJLM-12-2019-0332.

Heikkurinen, P., Young, C.W., and Morgan, E. (2019). Business for sustainable change: extending eco-efficiency and eco-sufficiency strategies to consumers. *Journal of Cleaner Production* 218: 656–664.

IAEA (2020). Energy, electricity and nuclear power estimates for the period up to 2050. Austria. https://www-pub.iaea.org/MTCD/Publications./PDF/RDS-1-40_web.pdf (accessed 25 March 2022).

IEA (2021). Net zero by 2050. IEA, Paris. https://www.iea.org/reports/net-zero-by-2050 (accessed 25 March 2022).

Krause, J., Thiel, C., Tsokolis, D. et al. (2020). EU road vehicle energy consumption and CO2 emissions by 2050 – Expert-based scenarios. *Energy Policy* 138: 111224, 0301-4215. https://doi.org/10.1016/j.enpol.2019.111224.

Leising, E., Quist, J., and Bocken, N. (2018). Circular economy in the building sector: three cases and a collaboration tool. *Journal of Cleaner Production* 176: 976–989.

Lindsey, R. (2021). If carbon dioxide hits a new high every year, why isn't every year hotter than the last? NOAA Climate.gov. https://www.climate.gov/news-features/climate-qa/if-carbon-dioxide-hits-new-high-every-year-why-isn%E2%80%99t-every-year-hotter-last (accessed 10 March 2022).

Markham, V.D. (2008). US population, energy and climate change. U.S. Population, Energy and Climate Change by the Centre for Environment and Population (CEP).

NASA (2022). What is the greenhouse effect? Global Climate Change. https://climate.nasa.gov/faq/19/what-is-the-greenhouse-effect/ (accessed 25 March 2022).

National Research Council (2011). *Climate Stabilization Targets: Emissions, Concentrations, and Impacts over Decades to Millennia*. Washington, DC: The National Academies Press.

NOAA (2022). Climate change: global sea level. https://www.climate.gov/news-features/understanding-climate/climate-change-global-sea-level#:~:text=Even%20if%20the%20world%20follows,2100%20cannot%20be%20ruled%20out (accessed 15 March 2022).

Reuter (2021). Heatwave turns already arid Kazakh steppe into mass grave for horses. https://www.reuters.com/world/asia-pacific/heatwave-turns-already-arid-kazakh-steppe-into-mass-grave-horses-2021-08-06/ (accessed 20 March 2022).

Statista (2021). Number of cars sold worldwide between 2010 and 2021 (in million units) https://www.statista.com/statistics/200002/international-car-sales-since-1990/ (accessed 20 March 2022).

The Umweltbundesamt (2021). Factor X. https://www.umweltbundesamt.de/en/topics/waste-resources/resource-conservation-in-the-manufacturing/factor-x (accessed 15 March 2022).

Uber (2015). Taking 1 million cars off the road in New York city. July 10, 2015/ New York City. https://www.uber.com/blog/new-york-city/taking-1-million-cars-off-the-road-in-new-york-city/ (accessed 18 March 2022).

United Nations (2018). The 2030 Agenda and the Sustainable Development Goals: An opportunity for Latin America and the Caribbean (LC/G.2681-P/Rev.3), Santiago.

WFEO (2022). *WFEO Model Code of Ethics*. France: The World Federation of Engineering Organizations.

Zhu, H., Yu, M., Zhu, J. et al. (2019). Simulation study on effect of permeable pavement on reducing flood risk of urban runoff. *International Journal of Transportation Science and Technology* 8 (4): 373–382. https://doi.org/10.1016/j.ijtst.2018.12.001.

Index

Engineering for Sustainable Development: Theory and Practice, First Edition.
Wahidul K. Biswas and Michele John.